Andaman Articles

Arif Mohd. Mustafa

ISBN
Paperback 979-8-89475-311-9
Hardcase 979-8-89544-850-2

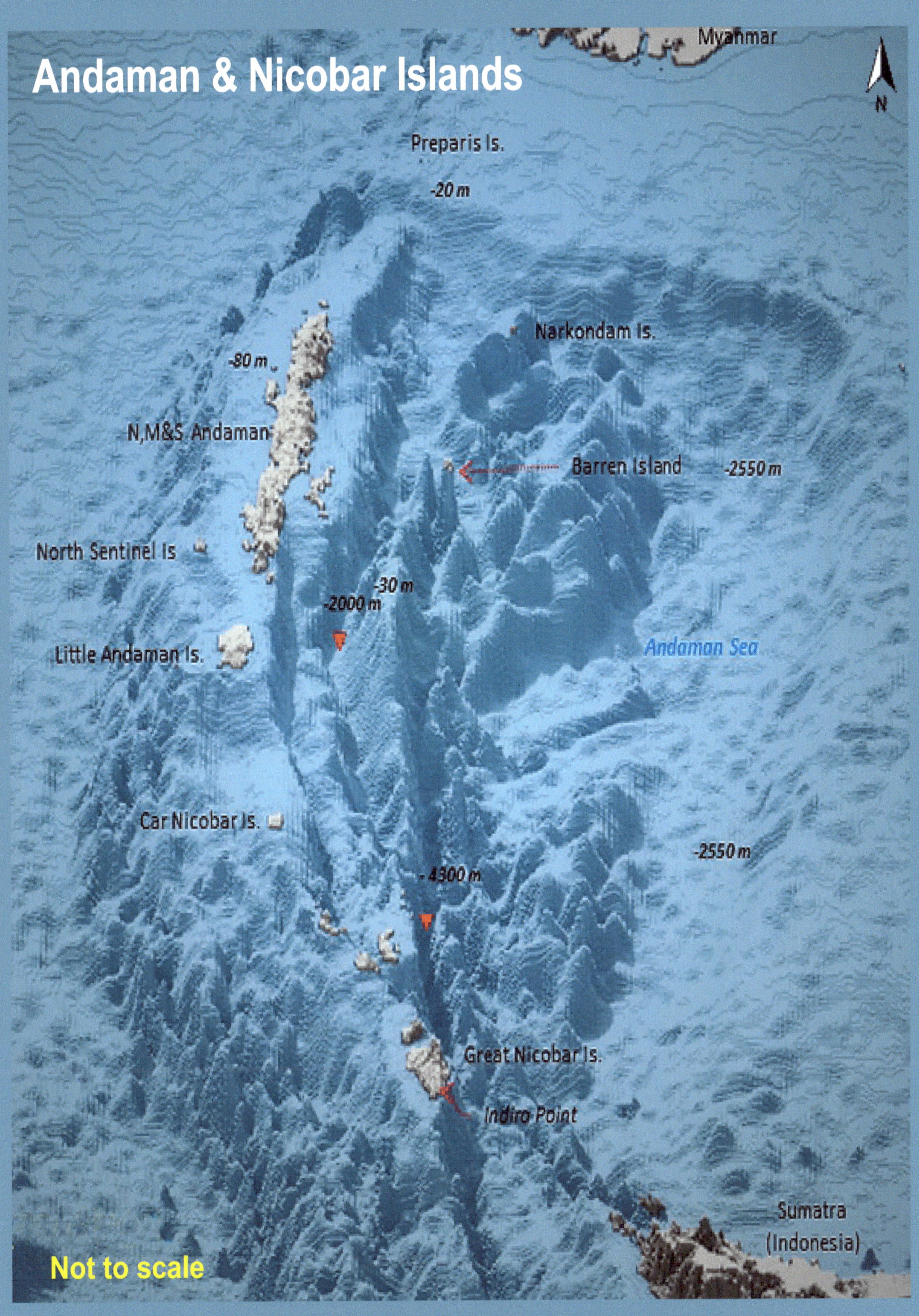
Andaman & Nicobar Islands
Myanmar
N
Preparis Is.
-20 m
Narkondam Is.
-80 m
N,M&S Andaman
Barren Island
-2550 m
North Sentinel Is
-30 m
-2000 m
Little Andaman Is.
Andaman Sea
Car Nicobar Is.
-2550 m
- 4300 m
Great Nicobar Is.
Indira Point
Sumatra
(Indonesia)
Not to scale

Preface

Communication is the fulcrum of scientific & professional development thus this compilation is a step forward in that direction. Significant prime publications spanning over 40 years of selfless work, many self sponsored on ◦ Tourism wonders ◦ Natural History ◦ Volcano ◦ Coral & the Coral reefs and ◦ Fisheries in Andamans are condensed & published. Articles are presented in descending chronological order.

For economical prosperity the national level 'Island Development Authority' (IDA) has identified Tourism & Fisheries as core sectors for development in Andaman and Nicobar Islands. I hope this publication will prove beneficial to stakeholders.

I distinctly remember my father Late M.A.Gafoor and mother Late Hafiz-un-Nisa during my childhood frequently advice *"Beta padho, iss mulk ko aage badhao"* ("Son study, take the Nation ahead"). Perhaps this impregnated advise embedded deep in memory and possibly the inherited patriotic trait is the drive to put in tremendous effort in presenting this publication.

Port Blair,
July,2024

Arif M. Mustafa
Retd. Assistant Director of Fisheries
Formerly Incharge, A&N Centre, Department of Ocean Development
Formerly Consultant Indian Navy, A&N Command, Port Blair
Formerly General Secretary, Andaman Science Association
Formerly Chief Liaison Officer, A&N Administration

24, Shastri Road,
Gafoor Manzil, Near Ganesh Mandir,
Aberdeen Bazaar, Port Blair - 744101,
Andaman & Nicobar Islands.
Mobile - 09434261302, 07063930101
Telephone - 03192-234563
email ID - arif.m.mustafa@gmail.com

- Dedicated to -

My Galaxy

ISHRAT-UN-NISA

AURSHIA

ARFAAN AHMED

AULVEERA

AYMAN NOOR NAWAZ

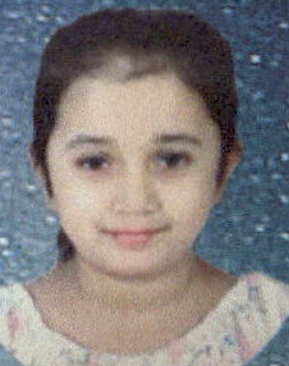

ALTAMASH MUSTAFA

Computer Graphic & Designing
- CH S Srinivas Rao -

Andaman Articles

- Unique blend of write-ups -

Volcano

Natural History

Mud Volcano

Aborigines

Birth of the islands

Lime stone caves

Divine Mountain

Mineral Resources

Oceanic deposits

Penal Settlement

Induced breeding in Andamans

Territorial Waters

Sport Fishing

Life in Forest, Sea and Air

Artificial Reef

Cage Culture

Marine Venom

Perch Fishery

Coral and the Coral Reefs

Sex in corals

Deep Sea Shark and Lobster

GEOGRAPHY GEOLOGY METEOROLOGY OCEANOGRAPHY

INDEX

S.No.	Year	Topic	No. of Pages	Page Sl. No.
1.	2023	Natural History	64	3
2.	2022	Territorial Waters around A&N Islands	04	67
3.	2021	Active Volcano at Barren Island	06	71
4.	2014	Volcano at Barren Island	32	77
5.	2013	Sport Fishing in Andaman Islands	48	109
6.	2011	Perch Fishery in Andaman Islands	30	157
7.	2005	Icthyo-beauties of A&N Islands	04	187
8.	2001	Understanding A&N Is. – A&N Command	48	191
9.	2001	Fisheries of Exportable	15	239
10.	2000	UNDP Expert Report – Coral ecosystem	32	254
11.	1999	Artificial Reef at Port Blair......	10	286
12.	1995	Coastal Sea Cage Culture at Port Blair.......(Abstract)	22	296
13.	1992	Report 'Barren Island Volcano'	23	318
14.	1991	Sex and spawning of corals at Port Blair	03	341
15.	1990	Deep Sea Lobster *Linuparus andamanesis*	05	344
16.	1990	Increasing environmental stress on coral reef	03	349
17.	1990	Coral reef degradation, conservation	04	352
18.	1987	Blue revolution in A&N Islands......	18	356
19.	1987	Coral Reefs, Ornamental fishes of A&N Islands ...	07	374
20.	1987	Human sex hormone in fish breeding	03	381
21.	1986	Deep Sea spiny Dog Fish Shark	02	384
22.	1985	Induced breeding in Andamans	09	386
23.	1983	ICLARM – Fisheries of A&N Islands	04	395
24.	1981	Dissertation – Fish and Fisheries ......(Abstract)	20	399

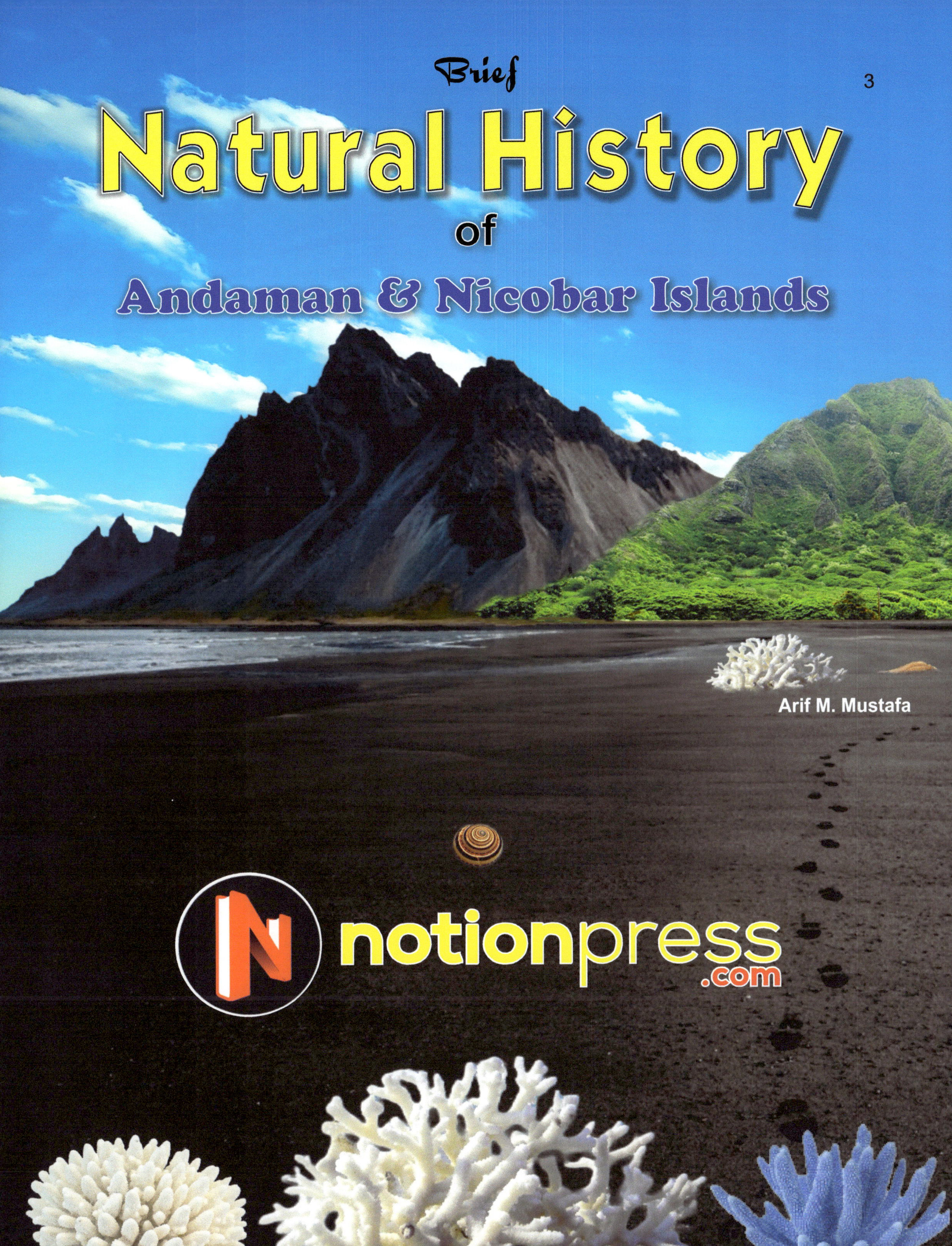

Brief
Natural History
of
Andaman & Nicobar Islands
Arif M. Mustafa
notionpress
.com

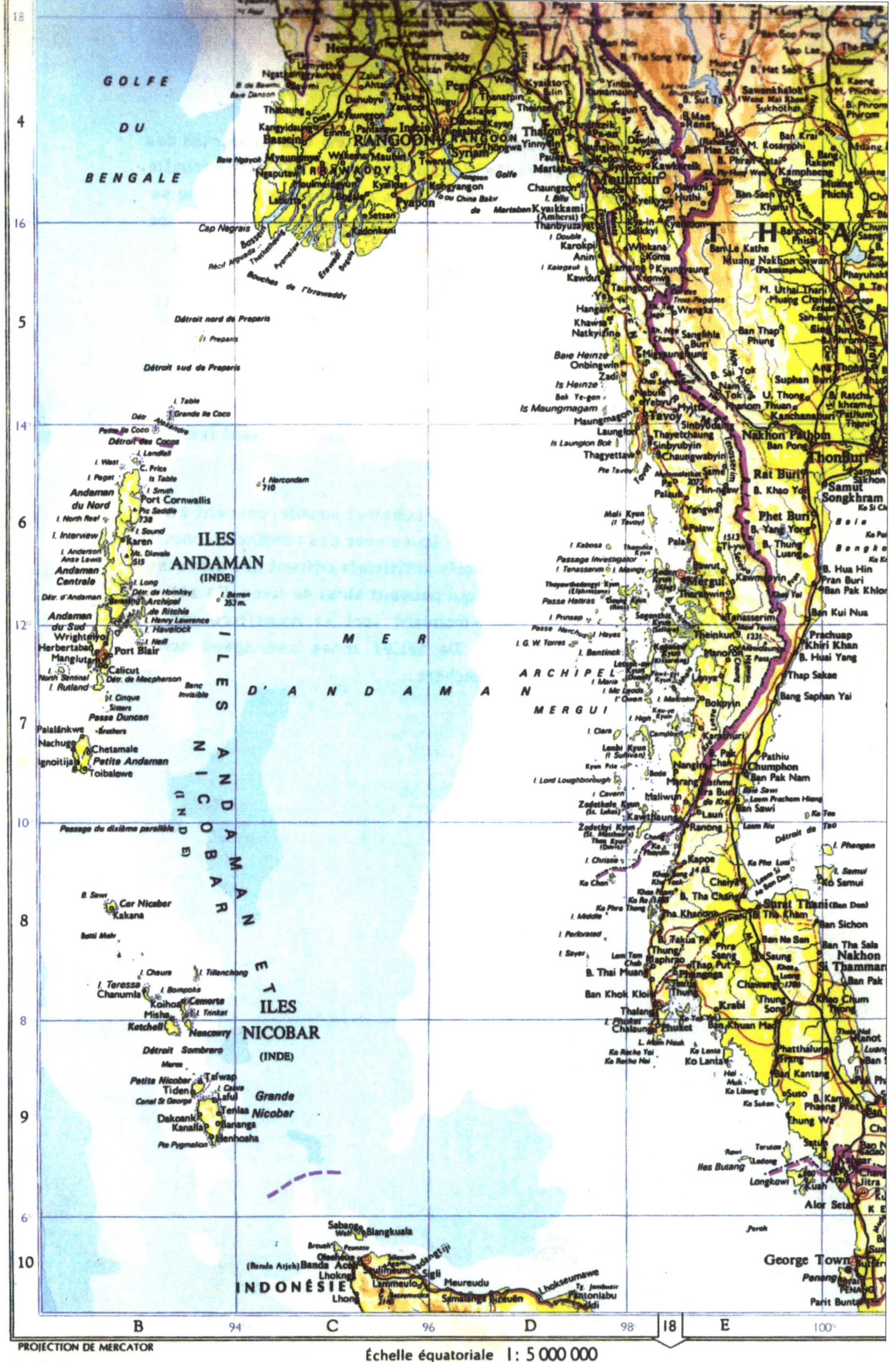

PROJECTION DE MERCATOR

Échelle équatoriale 1: 5 000 000

Brief Natural History of Andaman & Nicobar Islands

by

Arif M. Mustafa

Retd. Assistant Director of Fisheries
Formerly Incharge, A&N Centre, Department of Ocean Development
Formerly Consultant Indian Navy, A&N Command, Port Blair
Formerly General Secretary, Andaman Science Association
Formerly Chief Liaison Officer, A&N Administration

- Published by -

Index

1.	**Introduction**	-	1
2.	**Geology**	-	3
	2.1 Mineral Deposit	-	5
	2.2 Map Non-Living Resources	-	6
	2.3 Lime Stone Caves	-	7
	2.4 Mud Volcano	-	8
	2.5 Barren Island Volcano	-	9
3.	**Earthquake & Tsunami**	-	10
	3.1 Ravages of Tsunami	-	23
4.	**Ancient Inhabitants**	-	24
	4.1 The belief	-	26
	4.2 The Divine Mountain	-	27
	4.3 Battles	-	28
	4.4 Rani of Nancowrie	-	29
5.	**Forest**	-	30
	5.1 Timber	-	31
	5.2 Rudraksha	-	32
	5.3 Mangrooves	-	32
	5.4 Medicinal Plants	-	33
	5.5 Wild Life	-	34
	5.6 Mammals	-	34
	5.7 Marine Mammals	-	36
	5.8 Reptiles	-	37
	5.9 Turtles	-	38
	5.10 Others	-	39
	5.11 Butterflies & Moths	-	40
	5.12 Birds	-	41
6.	**Sea**	-	44
	6.1 Fishes	-	44
	6.2 Shells	-	49
	6.3 Coral and Coral Reefs	-	51
	6.4 National Park - Wandoor	-	55
7.	**Marine Venom**	-	56
8.	**Protected Areas**	-	57
9.	**References**	-	58

(Sections 2.1–2.4) by **Altamash Mustafa,** HOD *(i/c)* B.Tech. (Civil) **DBRAIT**

1

Natural History

Introduction

This brief Natural History treatise of A&N Islands is an attempt to unravel the marvels and mysteries associated with the land of '*Pu-luga*.'[1]

***Puluga* the forgotten, ancient, omnipotent divine power, of that lost Negrito civilisation, which once prevailed and ruled over Andamans since antiquity till mid ninetieth century.**

Throne of the *Lord*, was believed to be atop Saddle peak the highest land point under the occupation of lost regime.

Drops of land like shimmering emerald, circumjacent with dazzling golden beaches interrupted by patchy coral reefs and muddy creeks amid vast blue Bay f Bengal more than a thousand kilometer away from mainland/continental India is the splendid Union territory of Andaman and Nicobar Islands.

Existence of 'still alive' pre historic stone age human civilisation, pristine ecology, dense tropical rain forest & organic plantation, unpolluted bays & creeks, rich & rare endemic bio-diversity, sun bathed silky beaches, amazing dive sites in the shallow sea which glisten in every hue of green and blue, deep mangrove shaded creeks,silent lime stone cave, active volcanoes, abundance of sea food and the traditional hospitable spirit of islanders based on the divine logic of 'Atithi devo bhavah' (Guest is God) have made these islands a favourite eco-tourism destination for the World today.A paradise for naturalists, eco-tourists, explorers, researchers & historians alike.

Since time immemorial highest mountain is believed to be divine, earthquake & Tsunami not an unknown phenomena, sea around notorious, cyclone & thunder storm common, soil rich in minerals, bio diversity hold treasure of bio-active substances. Apart from original inhabitants, modern society is cosmopolitan; free, from social, religious, regional & linguistic barriers representing the India of Mahatma Gandhi's dream.

8 Annually about a hundred thousand tourists visit Andaman & Nicobar islands, with the exception of tribal reserves & wildlife sanctuaries, they are free to visit any site, however restricted areas could also be visited if a permit is obtained from the appropriate authority. The areas thrown open to tourist are well connected either by road, sea or air. One could reach northern most settlement in North Andaman covering about 300 km. by road, sea or air and could also visit southernmost habitation at Great Nicobar which is some 550 km. away from Port Blair by sea or air.

Historically, human greed appears to be the prime motive that made adventurous people venture in the remote wilderness of Andaman & Nicobar Islands for exploring and understanding various myths related to it. Most amazing among all the stories related to these islands was an English sailor's narration related to Nicobars, which spread like a fairy tale abroad some time during 17th century. The story tells "....... a native who was carrying a shell of water accidentally spilt some of the contents on the anchor, and the part so wetted immediately turned into gold ...."[15], indicating presence of a well in Nicobar whose water converts iron into gold. This fantastic myth was kept alive by Gamelli (1605), Keoping (1647), Fernandez (1669) etc[16]. Thus began the race. The first recorded camping was by two French gentlemen Faure and Taillandier for 2 ½ years at Great Nicobar since January 1711, subsequently they stayed at Car Nicobar for 10 months. Later came Father Charles of Pondicherry for a year during 1742, then came Danes at Great Nicobar during 1756, they further moved to Kamorta island. However, illness & unfriendly weather condition uprooted Danish settlement from Nancowry by 1845 along with the completion of their scientific expedition by research vessel Galatea. Nicobar was re-explored by Austrians during 1868, and a year after the territory was annexed by British India.

Contemporary efforts towards then fearsome Andamans began during 1789 when the Governor General of British India commissioned a survey of these islands by Lt. Archibald Blair, who conducted the first ever topo-cum-hydrographical survey and reported suitability for human settlement. Immediately thereafter, the first settlement was established at Port Blair (then Port Cornwallis) in the present day Chatham Island (then Mark Island) by bringing in hard core rebels of British India from undivided India during 1789. However, high mortality due to malaria and frequent attacks by aborigines forced the settlement to be shifted to a new port at North Andaman during 1792 wherein again similar problems cropped up and that settlement was also abandoned by 1796. Later, it was during 1857, after India's First War of Independence - the Sepoy Mutiny, that second successful penal colony was established at Port Blair with the first lot of 200 'rebels/mutineers'. During 1861 the administrative control of A&N islands was transferred to the Chief Commissioner of Burma.

Mutineers of 1857 at Ross Island during 1858.

With the advent of Indian Independence on 15th Aug, 1947 these islands were merged with the Indian mainstream.

Geology

Hypothesis of Genesis.

Geological past, a journey back in time to trace the birth of these islands is amazing. It all occurred some 150 million years ago, much after the dramatic disintegration of the than existing super continent 'Jumbo dweep', called 'Pangaea' into seven smaller continents of today. Spurt of molten rock and oozing of magma, defying enormous overhead water pressure began almost along the zone of subduction, and continued for centuries; the magma violently fought the waves, piled up and majestically surfaced to form today's Andaman & Nicobar islands, they are volcanic born out of volcanism[2], with a few exceptions of subsidiary coral islands. Rising and bending is evident in the rock folds seen today. Soil depth varies from 2 to 5m. Hectic submarine volcanic activity prevails throughout the eastern side, apparent volcanism continues through Barren island volcano 139 km. N-E of Port Blair, submerged volcano Alcock 60 km. East off Barren island and Sewell mount 200 km. East off Car Nicobar. Hostile thermal vents, spitting sulphur & hot water are also present in Andaman sea. A submerged volcanic plateau named Rani Jhansi Is./Invisible bank located 97 km. S-E off Port Blair could be an island of tomorrow.

Present day crescent arch of this largest Indian archipelago, comprising of 572 insular islands islets and rocks sprawls in the northern Indian Ocean at the South-eastern edge of the Bay of Bengal between Longitude 92° to 94° East, stretching from Latidude 6° to 14°North. Northern tip is 280 km from Myanmar and Southern 145 km from Sumatra. These islands have a maximum width of 58 km (average being 20 km) and a total land length over N-S axis is 726 km. Cumulative land area being 8,24,900 ha. (8,249 km^2) out of which 7,09,400 ha (7,094 km^2) i.e. 86% including 1,07,046 ha (1,076 km^2) of tribal reserve is claimed to be under forest cover. Highest point, the Saddle peak in North Andaman is 732 m.. above MSL. Eastern water basin is named Andaman sea where lies the deepest depth of - 4198 m East off Car Nicobar.

Climate is tropical, precipitation heavy, normal rainfall 3180mm, mean relative humidity 79%, mean minimum and mean maximum temperature being 23.5°c & 30.5°c respectively. Island experiences both S-W & N-E monsoon and it rains for about eight months a year. Unpredicted thunder storm and gale is common. Frequently the vast stretch of surrounding sea cradle devastating cyclones, fortunately their ferocity is mostly at ebb while passing over these islands.

Terrain is undulating, main ridges are seen over N-S axis, possibly this being one of the reason that earlier these islands were believed to be the peak of a submerged mountainous range running from Arakan yoma of Myanmar to the Achin head of Indonesia. In between the main ridges, deep inlets and creeks flanked with mangroves is a common site. Flat lands and perennial streams are sparse, similarly ground water reserve and top soil deposit is limited. Soil is mostly acidic, poor in nutrients and clay to loamy sand in nature.

This Union territory is ethnologically and; presently administratively, bifurcated in northern (Negrito) Andaman and southern (Mongoloid) Nicobar, 160 km. apart with 10°channel in between.

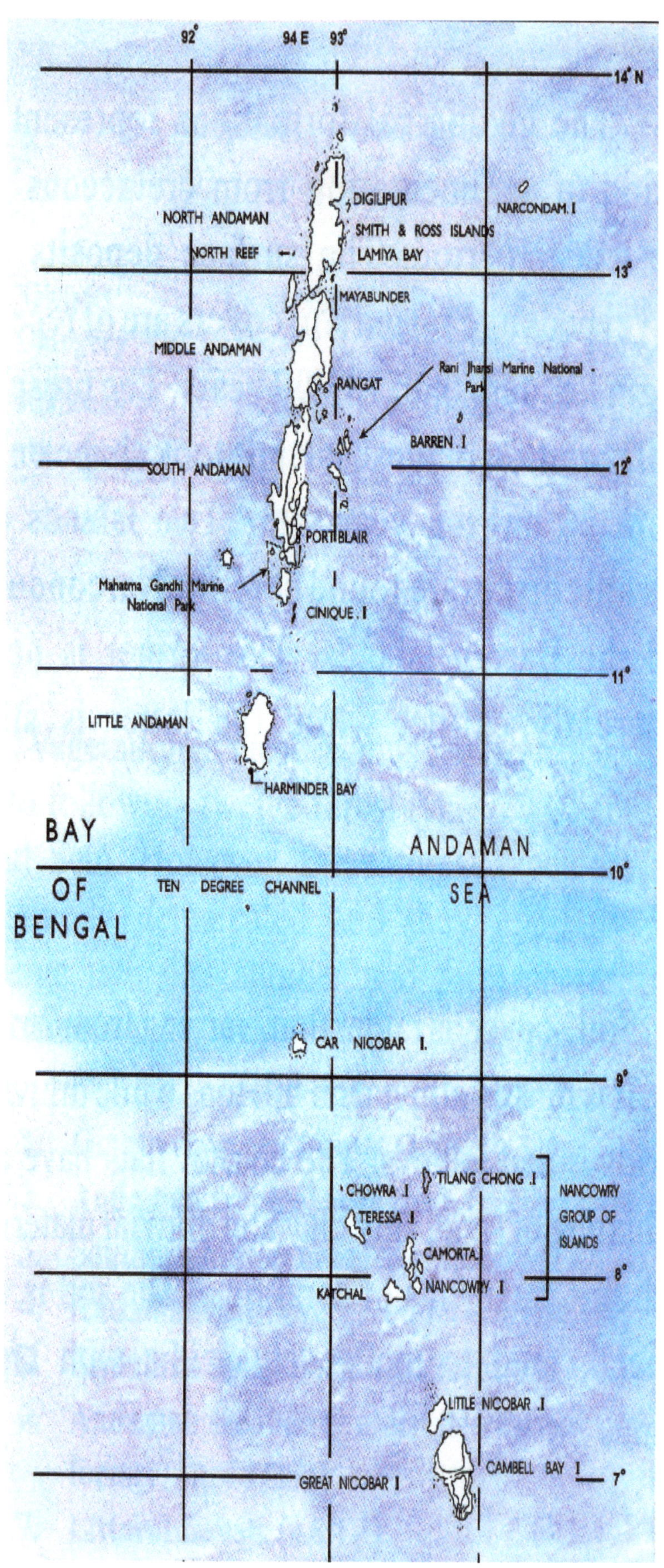

Mineral Deposit

A&N Islands possess mineral rich cretaceous rocks which are 240 million year old. The same rock belt yields valuable Jade in Myanmar & Chromite in Manipur India.Recent studies through particle induced X-ray emission by Bhabha Atomic Research Centre at Hyderabad is indicative of Titanium 228 - 2370 mg/kg. Vanadium 60 - 925 mg/kg., Nickle 1436 - 2000 mg/kg., Chromium 210-3300 mg/kg., and Manganese 1100 -1351 mg/kg. etc. deposit in Andamans.

Perhaps the World's finest limestone crystal from Nancowry Is. complex.

Nancowry island has the finest lime stone deposit known in the World today where Calcium Carbonate ($CaCo_3$)is above 98% followed by 93% $Ca\ Co_3$ deposit throughout Little Andaman.

Based on survey conducted by Late Shri **S.P.Goenka**,
Chairman & Managing Director,
GIMPEX group of Companies, Chennai.

Non-Living Resources

(Indicative)

Legends

 Active volcanism

 U/W Thermal Vent

 Invisible Bank

 Natural Gas / Shale Gas

 Manganese nodules

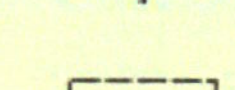 Hydrocarbon deposit

 Methane clathrate / Methane ice

 Jade deposit extension

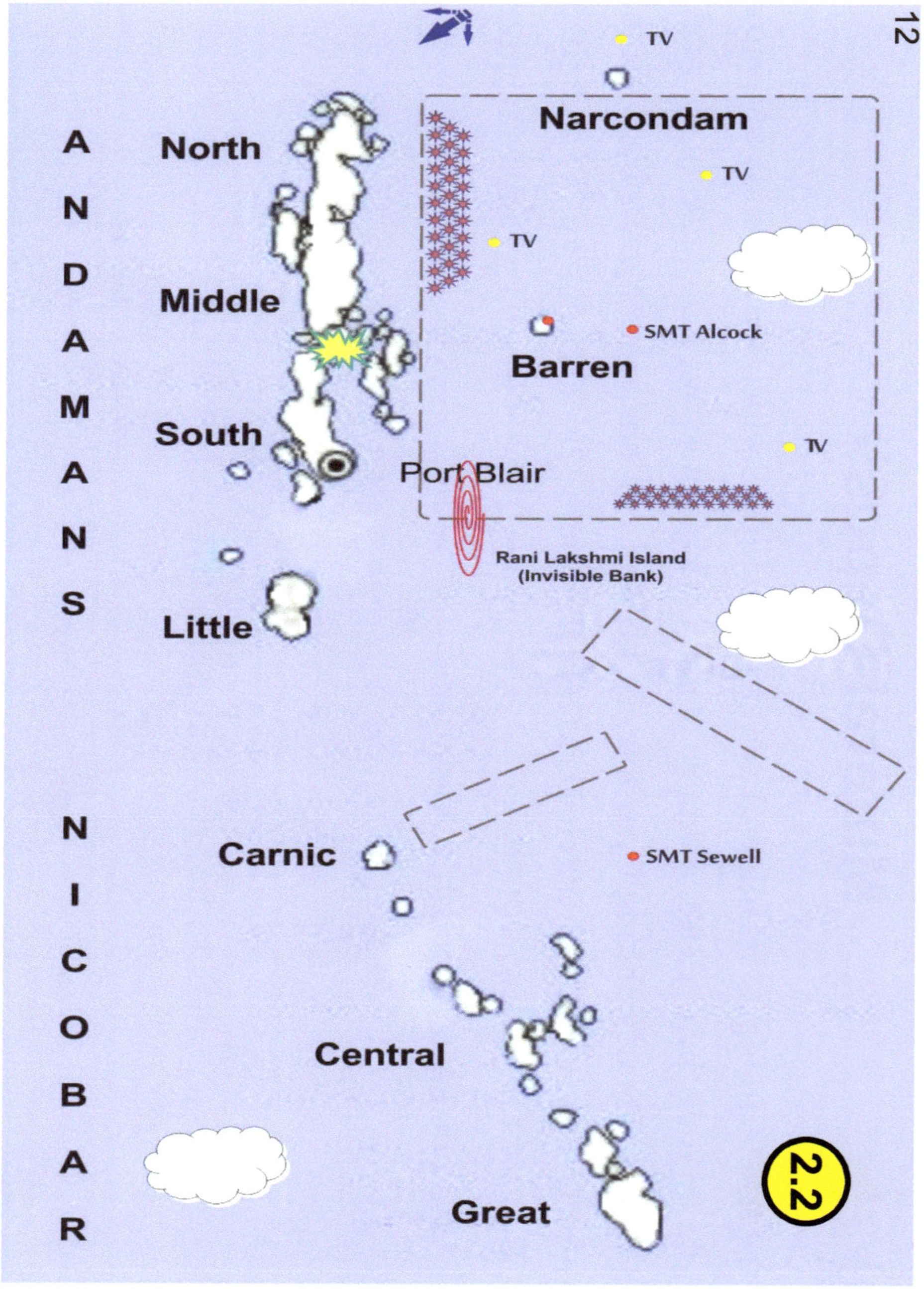

12

A N D A M A N S

North

Middle

South

Little

Narcondam

TV

TV

TV

TV

Barren

SMT Alcock

Port Blair

Rani Lakshmi Island
(Invisible Bank)

N I C O B A R

Carnic

Central

Great

SMT Sewell

2.2

2.3 Lime Stone Caves

Many calcite galleries are found in the hills of Baratang island, Interview island and at Chalis aik located between Cadell Point and Morland rock, south of Saddle peak in Diglipur, North Andaman.

The Baratang caves are famous for magnificent and intricate ceiling hang outs called stalactite.

Photo - Arif M. Mustafa

2.4 14

Mud Volcano

A conical mound of hardened clay and sand etc. spitting fine semi liquid mud with intermittent inflammable gaseous bubbles. It mimics a normal volcano. They are born out of upheaval thrust developed as a consequence of earthquakes forcing suspended mud and gas up the fissures.

Flowing mud ooze - 2003

Photo - Arif M. Mustafa

% Composition 2003 Mud Ooze	
SiO_2	59.67
Al_2O_3	15.49
Fe_2O_3	5.72
TiO_2	0.85
P_2O_5	0.18
CaO	1.23
MgO	2.74
K_2O	1.96
Na_2O	3.21
LOI	8.75

Source : Gimpex Ltd., Chennai

In Andamans there are three such mud volcanoes the most famous being the one at Jarawa Creek under Nilambur panchayat in Baratang Island. It last exploded on 19, Feb,2003, however after the mega techtonic event of 26, December 2004 about nine oozing vents have appeared around it in a radius of about 2 km. The second Mud volcano is at Shyam Nagar in Diglipur, North Andaman. The third, nascent one, appeared on 7, June,2005 in Narcondum island an extinct volcanic island 259 km. NE off Port Blair.

Oozing vent atop Baratang mud volcano.

2.5 Barren Island Volcano

Barren volcano - 2005 eruption.

The only live volcano in the Indian peninsula, Barren lies 139 km north east of Port Blair in the Andaman sea (12° 17' N Lat. & 93° 50' E Long.) This tiny, circular Island covers an area of 8 km^2. It belongs to general Sunda group and is believed to have been born out of an eruption, which occurred during the late post Pleistocene period. Later in the course of geological evolution, the prime giant cone got transformed into the present day Barren Volcano which in fact, is the central part of the blown off cauldron.

3 [16] Earthquake and Tsunami

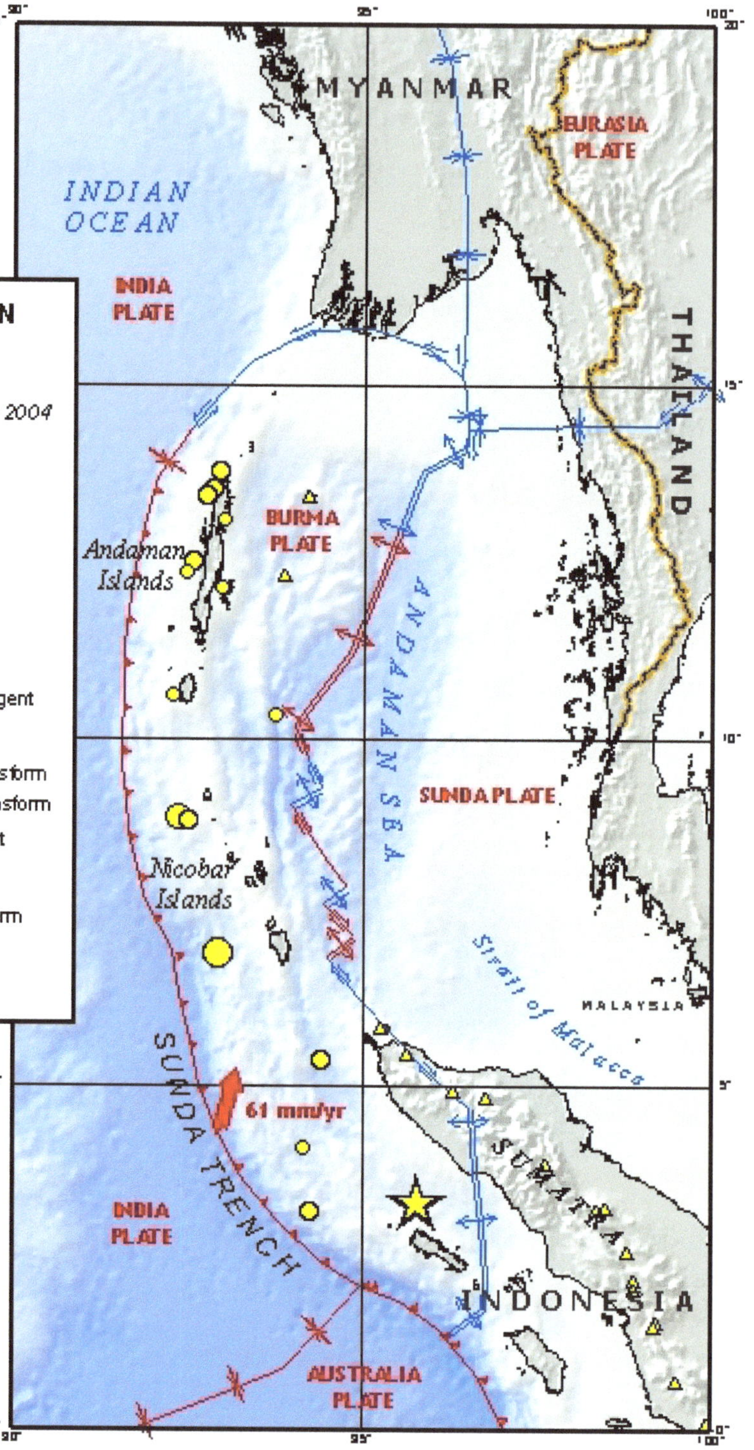

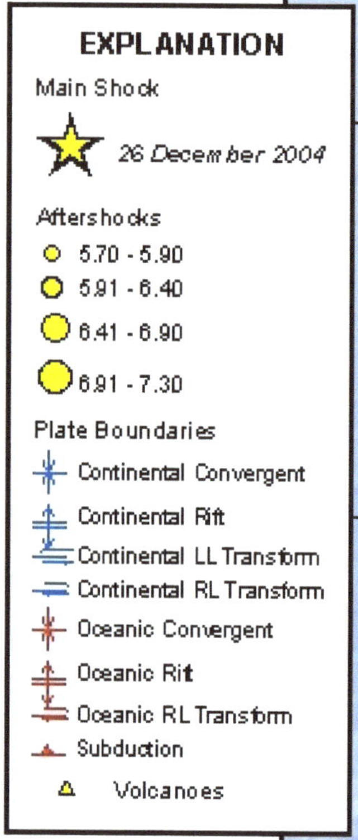

Although the Islands are prone to seismic jolts due to Geological vulnerability inherited by the A&N Islands; most tragic event of recent past occurred on 26th December,2004.

The catastrophy occurred at 6:28 am IST on 26, Dec.,2004 through a 'dip-slip' of ' Indian tectonic plate' underneath the minor Burma plate, upon which A&N islands are located. Focus of this massive tectonic disturbance was in the vicinity of that notorious part of the planet which in the past gave birth to devastating 'Toba explosion' some 74,000 years ago[3] and Krakatoa volcanic explosion of Aug. 1883[4], these two events are believed to have metamorphised the Asian geography and Global human dispersal.

The epicentre was about 15 km down the sea, off tiny island of Simeulue near the West coast of northern Sumatra about 250 km away from the lost city of Banda aceh at 3.3° N & 95.78° E[5] by 06:30 am planet earth shuddered and buzzed over its axis creating regional Deluge or Pralay or Qayamat whatever term one finds appropriate.

Finally arrived magnitude of that earthquake was 9.3 over Richter scale[6]. Estimated energy released was equivalent to 23,000 Hiroshima type Atom bombs or 20 x 10^{17} Joules[8] or 32 billion tones of explosive Trinitroluene (TNT)[9].

Earthquake caused rupture of sea floor for more than 1200 km.; Instant vertical sinking of sea floor made the water line lose its equilibrium for a while. As it moved to regain its equilibrium, the potential energy of displaced water was converted into kinetic energy of horizontal motion. Unusual waves of more than 500 km. long having long periods (10 min to 2 hrs.) traveling with an average speed of 750 km. per hour were generated. These propagated outward as Seismic Tsunami, the initial waves further split into distant tsunami and local tsunami.

Courtesy : USGS

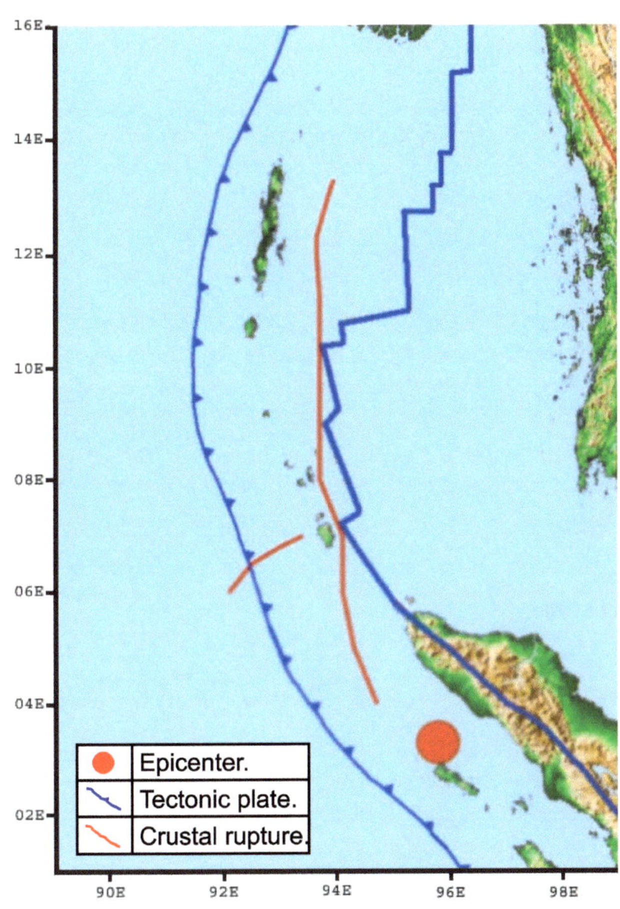

Courtesy - ASC - India.org.

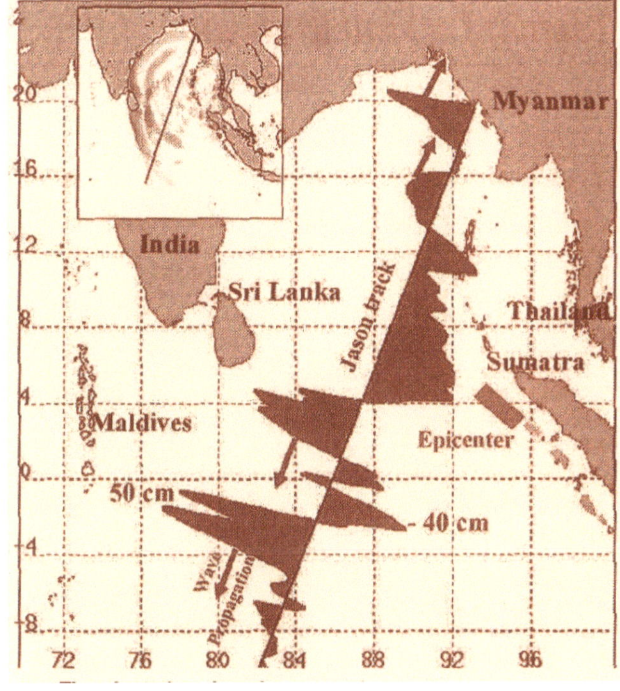

Satellite imagery of Tsunami through Radar altimeter of Jason satellite, USA. taken two hrs. after earthquake.
Courtesy - The Hindu

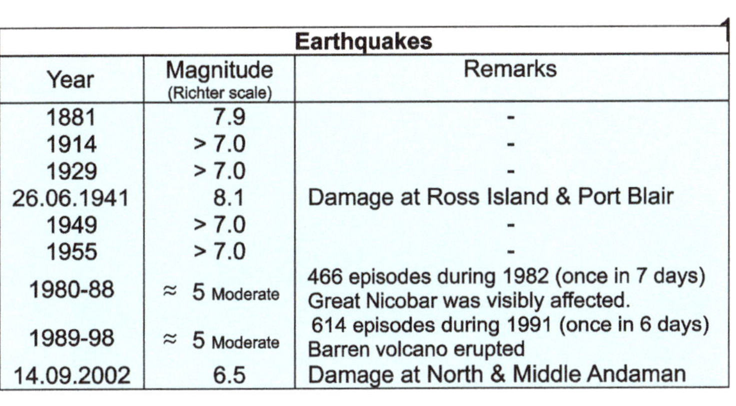

Earthquakes		
Year	Magnitude (Richter scale)	Remarks
1881	7.9	-
1914	> 7.0	-
1929	> 7.0	-
26.06.1941	8.1	Damage at Ross Island & Port Blair
1949	> 7.0	-
1955	> 7.0	-
1980-88	≈ 5 Moderate	466 episodes during 1982 (once in 7 days) Great Nicobar was visibly affected.
1989-98	≈ 5 Moderate	614 episodes during 1991 (once in 6 days) Barren volcano erupted
14.09.2002	6.5	Damage at North & Middle Andaman

Past record of Significant Events

Satellite imagery of devastated Little Andaman.

Earthquake
(26-12-2004)
PORT BLAIR

Passenger Hall Haddo Wharf

Mohanpura Shopping Complex

OH Water Tank Phoenix Bay Power House

OH Water Tank Bambooflat Power House

Earthquake
(26-12-2004)
CAR NICOBAR

Community OH Water Tank, Sawai

ATC, Air Base

Government Quarters - Malacca

Health Centre - Arong

Tsunami

(26-12-2004)

PORT BLAIR

After the first Tsunami Wave

Approach road to Aberdeen Jetty (RGWSC)

Aberdeen Jetty (RGWSC)

Adjacent Cold Storage

Attached Water Sports Centre

Tsunami
(26-12-2004)
PORT BLAIR

Second Tsunami Wave

Aberdeen Jetty (RGWSC)

Mahavir Singh Road

Gymkhana Ground / Netaji Stadium

Tsunami

(26-12-2004)

PORT BLAIR

After Third Tsunami Wave

Inter Island Ferry Harbour, Phoenix Bay

Tsunami

(26-12-2004)

HUT BAY

Survived distal portion of Jetty

Port Control Tower, PMB

Land side portion of Jetty

Tsunami

(26-12-2004)

CAR NICOBAR

Passenger shed, Malacca

Staff Quarters, Malacca

Staff Quarters, Lapathy

Vil. **Arong**

Vil. **Jayanti**

Vil. **Malacca**

Air Force Base

Air Force Base

Air Force Base

Erased village

Stacked coconut stems

Power Plant

Inside Power Plant

EHL Quarter, Chukchukia

EHL Office, Chukchukia

Tsunami
(26-12-2004)
CAR NICOBAR

Teetop Jetty

Teetop Jetty

Fisheries Teeptop

Teetop Road

Teetop Road

Teetop Jetty office

Sr. Sec. School - Mus

Judge Qtr., Lapathy

Near Sr. Sec. School, Lapathy

RG Park, Malacca

203 KL HSD Tank moved 6 km.

5 ton uprooted coral colony

DG Set, Lapathy

Tsunami

(26-12-2004)

NANCOWRY

Vil. Champion

Vil. Champion

Harbour Area Kamorta

Approach to Kamorta Jetty

Power Plant Kamorta

Power Plant Teressa

Signal Tower Teressa

Tsunami

(26-12-2004)

GREAT NICOBAR (CAMPBELL BAY)

Submerged town area

Submerged Fishermen Colony

Ice Plant & Cold Storage

2 Mtr. Sunken Light House at Indira Point

Fishing Boat near Zero Point

Tsunami

(26-12-2004)

GREAT NICOBAR (CAMPBELL BAY)

Airstrip

Adjoining area

Ravages of Tsunami

Epitaph

"On the morning of December 26, 2004, a new word entered the vocabulary of most Indians. It was a word that would suggest that the sea had suddenly gone mad, that the ocean could no longer be trusted, that the water had cannibalised the land.

It was a word that spelt total and unimaginable upheaval - devastation, death, destitution, fear, sorrow and loss. Thousands lost families, homes, livelihoods.

For those who escaped the monstrous wall of water, the real challenge was in rebuilding lives. It was about finding courage in the face of adversity, hope in the face of despair and learning to look ahead while leaving memories behind."

V. Jayanth.
The Hindu
26th Dec. 2005

Impact

Estimated damage (Rs. in Crore)	3,836.56
Population affected (No.)	3,56,000
Villages affected (No.)	192

Death

Poultry	-	98,722
Pigs	-	38,446
Goats	-	16,623
Cattle	-	3,786
Humans	-	3,513

Katchal	-	1551
Car Nicobar	-	854
Great Nicobar	-	549
Nancowry	-	378
Teressa	-	117
Little Andaman	-	57
South Andaman	-	07

Lost / Damaged

Land (Ha.)	- 10,837
House (No.)	- 21,100
Boat (No.)	- 1703
School (No.)	- 85
Jetty (No.)	- 24
Health Centre (No.)	- 34
Power House (No.)	- 21
Ship (No.)	- 09

Relief

- Rs. 821.88 crore granted, out of which about 60 crore disbursed as *ex-gratia.*
- 15,000 ton of material transported from mainland India.

Shelter

- About 10,000 displaced families lived in 8 islands under 58 Intermediate shelter colony.
- Each family occupies 12ft. x 14ft. CGI sheet covered enclosure.
- 9565 such rooms were occupied.

The Tilt

"Since 1830, there have been reports of uplifted coral reefs and submerged forests along the coastline of Andaman and Nicobar Islands. Few geologists believed these were caused by earthquakes. But after the tsunami of December 26, 2004, caused by a quake off the Indonesian coast, scientists reported that the Islands had tilted-the eastern side had gone down, the western side had moved up. There was also an emergence of a western reef off the North Sentinel Island and Western Middle Andaman Island, while parts of southern Nicobar subsided by 2 meters.

To study the phenomenon, Indian and U. S. seismologists have set up a 300 meter long tilt meter, the first in India. a long, hollow tube emptying into sealed water tanks on either end to measure the difference in levels, the tilt meter will provide more exact evidence of the phenomenon after a quake as well as study the impact of the quakes."

Raj Chengappa.
Scientific American, Nov., 2005

4 30 Ancient Inhabitants

The original inhabitants of Andaman & Nicobar Islands today account for only 12% of the total population, whereas, they are believed to have lived here for 20,000 years. The Andaman group of islands presently have six living aborigine groups namely Sentinelese, Jarawas, Onges and Great Andamanese of Negrito origin, Nicobarese and Shompens are offshoot of Mongoloid stock and live in the Nicobar group of islands.

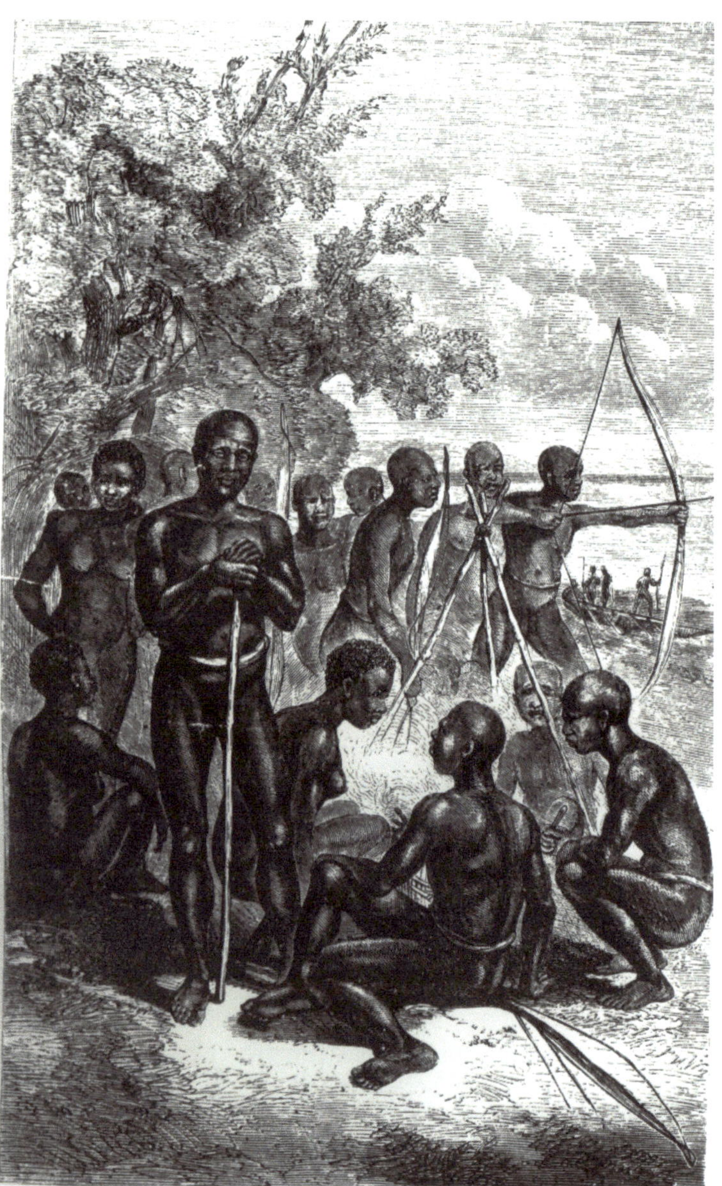

A group of Andamanese.

After - F.J.Mouat. 1863

It is believed that the Andaman's aborigines who were once collectively called Andamanese might have reached these islands very early in time, possibly by boat from mainland India itself[12]. They show strong genetic & phenotype affinity not only to the neighbouring Semangs of Malaysia and Aetas of Phillippines but to distant Africans. Recent DNA match is indicative of having direct links with the Pygmies of Southern Africa[13]. Sentinalese continued to be hostile to outsiders whereas the once ferocious Jarawas are gradually becoming friendly to civilized population. Onges and the Great Andamnese have accepted the presence of outsiders. The Andaman aborigines are primarily hunter-gatherers.

The Nicobarese must have migrated sometime before the Christian era. The origin of Shompens is not known but it appears that the Shompens have a Malayan strain. The Nicobarese have integrated well with outsiders and joined the Indian mainstream whereas Shompens still avoid contact but are non-hostile. These mongoloid people are mainly horticulturists and herders.

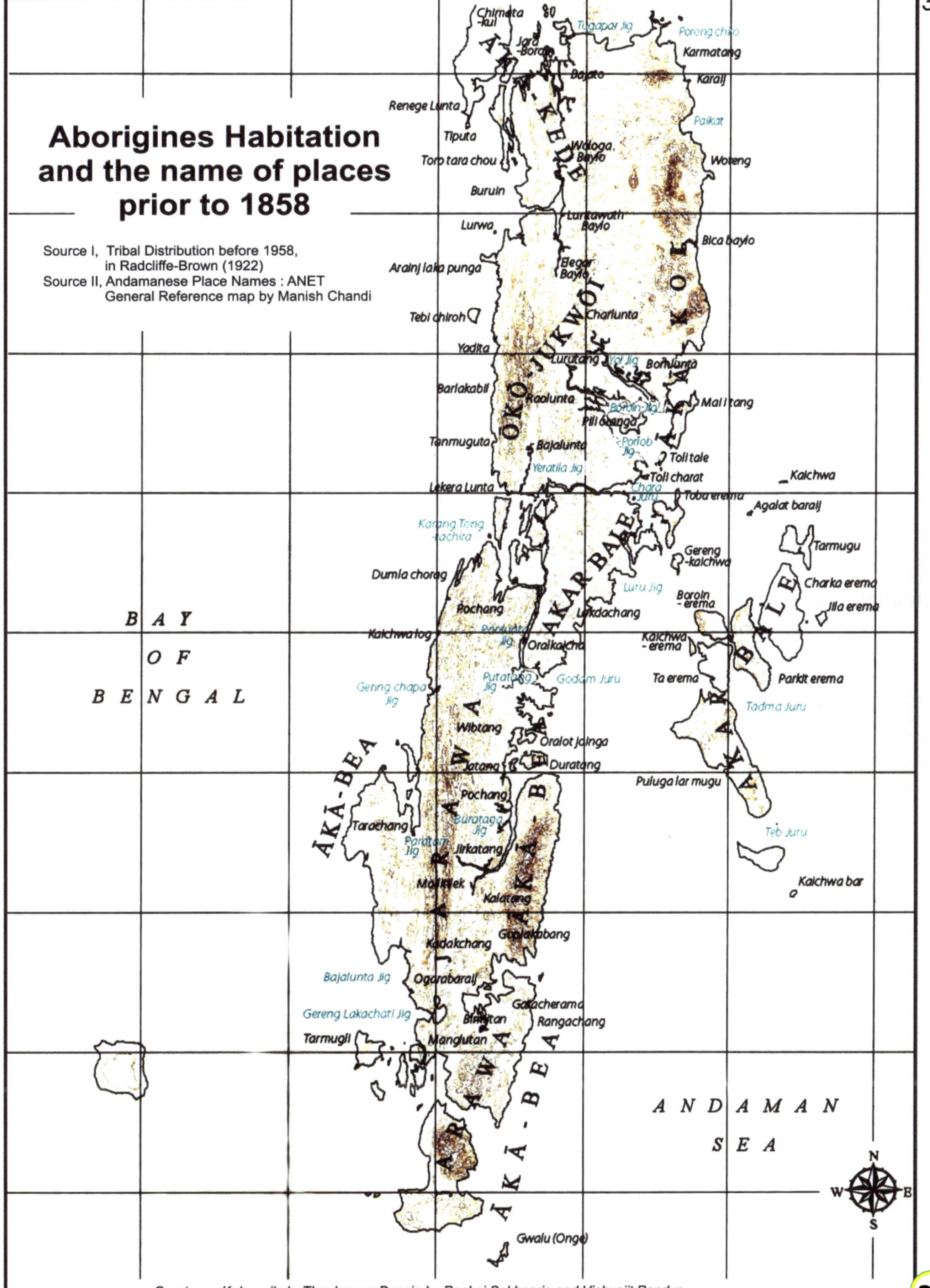

Courtesy : Kalpavriksh, The Jarawa Dossie by Pankaj Sekhsaria and Vishvajit Pandya

4.1 32

The Belief

Andamanese belief on : -

A. The Creator

They believe on the existence of a Supreme being or Creator 'Pu-luga' who created the first man 'to mo' a black, tall, bearded personality, after the creation of this World. 'Pu-luga' is defined as :

I. *He is omniscient.*

II. *He was never born and is immortal.*

III. *By him, the would and all objects were created.*

IV. *He is angered by sins but pitiful to those in pain or distress.*

V. *His appearance is like fire, yet He is (now a days) invisible.*

VI. *He is the Judge from whom each soul receives its sentence after death.*

B. The Doomsday

"This blissful state will be inaugurated by a great earth quake, which, occurring by 'Pu-luga's command will break the Pidga-lar-Chawga (invisible cane bridge between earth and eastern sky) and cause the earth to turn over : all alive at the time will perish......"

C. The Life after death

"The future life will be but a repetition of the present, but all will then remain in the prime of life, sickness and death will be unknown......."

D. The Divine Mountain

Saddle Peak, the highest land point was sacred to Andamanese because as per their tradition 'Pu-luga' once lived here and taught them many arts.

Condensed & quoted from the published work of Edward Horace Man (1846-1929) in the book entitled 'The Aboriginal Inhabitants of the Andaman Islands - 1883'. Reprint-2001 p.89-90, 94-95.pp 224. Mittal Publications, New Delhi.

The Divine Mountain

1

2

3

1

Glimpses of Saddle peak - the divine mountain of Andamanese.
1- Orchid in blossom. **2. Saddle Peak** 3 -Looking down the Peak Centre (R) - An Orchid plant.
Centre (L) - Author being interviewed by National Geography Correspondent Miss Bernice at the top..

Battles

Battle of Aberdeen

Historically a classical example of most uneven battle fought in the valleys & hills of Aberdeen on 17th day of May, 1859.

It was an organised but unsuccessful attempt by thousands of Andamanese to get their traditional home land liberated and further invasion halted by repulsing the colonial foreigners.

Unfortunately modern British weaponry mercilessly crushed the prehistoric, stone age warriors not for the day but for ever.

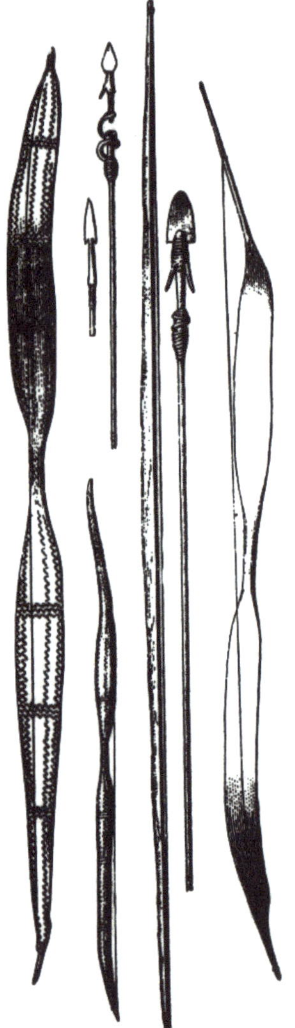

Battle of Little Andaman

A retaliatory battle by British Royal Navy & Army against Onges was fought on the shores of Little Andaman sometime during May, 1867.

As a token of bravery (for revenge) five Victoria Cross was conferred to 24th Regiment of South Walls Borders which attacked through HMS Aracan killing hundreds or thousands of aborigines[20].

Weapons of Andamanese possibly used in battles.

After - E. H. Man.

4.4 Rani of Nancowrie

Rani / Queen of Nancowrie

The tribal population of Nancowrie Islands. was befriended by British navy prior to Ist World War. During the war, a young, brave lady named Islon saved her land by unfurling the British flag and misguiding German armada about the location of Port Blair, thus they never reached the capital of Andamans. When this news reached England, Her Majesty, the Queen of Britain was pleased to honour the lady with the title of 'Rani' and the Nancowrie group of islands were formerly handed over to her.

Nancowrie's village of Malacca - 1880
After - V. Bell

Since then the title Rani is inherited by the progeny. After Rani Islon's death on Dec. 28, 1954 her daughter Rani Lachmi was crowned. Later came Rani Fathima and Ms. Aiyasha Majeed.

The Rani's family enjoys special privilege and protocol status as stipulated by Govt. of India.

Nicobari Hut

Traditional Nicobari Dugout

Shompen Hut

These islands are blessed with a unique luxuriant ever green tropical rain forest canopy covering 92.2% of geographical area, whereas recorded forest area is 7170.69 km^2 which is 86.93% of land area. This forest is a mixed germ plasm bank comprising of, Indian, Myanmarese, Malaysian and endemic floral strain. So far, more than 3500 varieties of plants have been reocrded out of which 223 are endemic and not found anywhere else in world.

Forest vegetation fo these islands are classifeid into different forest type namely: (1) Giant evergreen forest (2) Andaman tropical evergreen forest (3) Southern hilltop evergreen forest. (4) Andaman semi-evergreen forest (5) Andaman moist deciduous forest (6) Andaman secondary moist deciduous forest (7) Littoral forest (8) Tidal Swamp forest (Mangrove forest) (9) Sub-mountane hill valley swamp (10) Cane brakes & 11 Wet bamboo brake. This forest abound in highly commercial species numbering 200 or more.

Forest

Andaman Padauk *(Pterocarpus dalbergioides)* also known as East Indian mahogany is a tall deciduous tree. It is found only in the Andamans. It grows up to a height of 120 feet. The timber of this tree has huge prominence in the making of unique furniture. The wood of Andaman Padauk is red in colour and durable in nature. During early Penal settlement convicts sent to Andamans were made to supply the wood of Andaman padauk to the craftsman worldwide.

5.1 Timber

Types of Wood

Ornamental wood

1. Chooi - Sageraea elliptica
2. Satin Wood - Murraya exotica
3. Silvergrey - Terminalia bialata
4. Padauk - Pterocarpus dalbergiodes
5. Marble wood - Diospyros marmorata

Hard wood

1. Koko - Albizzia lebbek
2. Gangaw - Mesua ferrea
3. Jhingam - Pajanelia rheedii
4. Badam - Terminalia procera
5. Garjan - Dipterocarpus species
6. Chakrisia - Chakrasia tabularis
7. Lakuch - Artocarpus gomeziana
8. Hill Mohwa - Madhuca butyracea
9. Black chuglam - Terminalia manii
10. Jungli aam - Mangifera andamanica

Bamboo

1. Bambusa valgaris
2. Thyrsostachys oliveri
3. Dinochloa andamanica
4. Schizostachyam regersii
5. Denrocalamus calostachys

Cane

1. Calamus palustris
2. Calamus viminalis
3. Calamus longisetus
4. Korthalsia laciniosa
5. Calamus andamanicus.
6. Calamus pseudorivalis.
7. Calamus baratangensis
8. Daemenorops kurzianus

Timber cross section

Andaman forest abound in timbers numbering 200 or more species, out of which about 47 varieties are considered to be commercial. Major commercial timber species are Gurjan (Dipterocarpus spp.) and Padauk (Pterocarpus dalbergioides). Ornamental wood such as Marble wood, (Diospyros marmorata), Padauk (Pterocarpus dalbergioides), Silver Grey (a special formation of wood in white chuglam), Chooi (Sageraea elliptica) and Koko (Albizzia lebbeck) are noted for their pronounced grain formation. Padauk being steadier than teak is widely used for furniture making.

Didu
Bombax insigne

HONEY BEE REPELLENT

Onges spit and smear chewed pulp of *Orphea katshalica*, a jungle shurb known to them as '*tonjoghe*' over honey hives and their body to drive away the Giant rock bees from their hives and prevent stinging.[19]

Andaman Tropical Evergreen Forest

5.2 38 Rudraksha

The super seed Rudraksha *(Rudra - Lord Shiva, Aksha - Eye) Elaeocarps spp.* which is believed to have born out of Lord Shiva's tear also occur naturally at South Andaman and Great Nicobar. It is believed that these sacred insular beads symbolise divine protection for its user and emit beneficial electromagnetic waves to tone up general health.

5.3 Mangrooves

This is a typical group of about 35 varieties of salt tolerant plants, shrubs, climbers and palms, growing along creeks and estuaries. Pre-Tsunami coverage was 966 sq.km (96600 hect.) covering 10.85% of total forest area. However post - Tsunami much of the vegetation lost is now being revived.

5.4 Medicinal Plants

The rich floral bio-diversity in recent past has unveiled 726 varieties of medicinal plants from A & N Islands and the list is still expanding. Valuable, hitherto, hidden plants containing potent medicinal properties are being identified.

Common herbal remedies

Bada daad patti (Cassia alata)	Allergy, Fever, Pain
Bamboo (Bambusa bambos)	Abortifacient
Bail (Aegle marmelos)	Dysentry
Gumchi (Abrus precatorius)	Sorethroat, Cough, Swelling
Jamal Gota (Jatropa gossypifolia)	Tooth decay, Gum ulcer, Constipation
Hawai booti (Chromolaena odorata)	Stop blood from cut & injury
Khari phal (Ardisia solanacea)	Hydrocele, Post delivery
Lajvanti (Mimosa pudica)	Jaundice, Boil, Mumps
Madi pati (Caryota urens)	Indigestion

Bakul/ Khaya
(Mimusops elengi)
Rich in vit. A, C & total soluble solids. A vitamin supplementary.

Khattaphal
(Baccaurea ramiflora)
Rich in vit. C. anti vomiting & antiasthamatic.

Morinda citrifolia / Indian Mulberry / Noni Fruit

Recent pharmaceutical research has unveiled a wonder plant which was known to Onges as 'Pongi' and to Niobarese as 'Lorang' whereas others call it 'Burma Phal' or 'Surangi' traditionally used for Gastrointestinal disorder, Pyorrhoea, Tooth ache and Gum trouble.

Abrus precatorious / Gumchi

40

Wild Life

These islands are endowed with fairly rich terrestrial, aerial and aquatic life forms, the taxonomical indexing keep expanding, thus the territory is also called Zoologist's paradise. Strange endemic wild life thrives in the insular remoteness. Overall faunal endemism is 9%. More than 5100 species of animals found here (2900 marine, 2100 terrestrial, 100 fresh water) which includes 56 mammals, 378 birds, 1576 fishes, 69, reptiles, 16 amphibians and others.

(Source : Forest statistics 2021)

Mammals

*Among mammals, **Rat** constitute the largest group, comprising of 26 species followed by, **Bat,** having 14 species.*

- **Crab Eating Macaque** *(Macaca fascicularis umbrosa)* They are coastal inhabitants of Great Nicobar island, Katchal & Little Nicobar. Omnivorous by nature, marine crabs appear to be their favorite food.

- **Andaman wild Pig** *(Sus serofa andamanensis)* Widely Spread throughout the forest of South, Middle & North Andaman. Hunted by aborigines.

- **Barren Island Feral Goat** *(Capra hircus)* Goat of Indian Origin during summer they do sip sea water and graze vigrously at dawn and dusk so as to avoid sun and minimize sweating.

- **Trinket/Kamorta Island Cow** *(Bos Taurus)* Non-Indian breed left wild after Danes repatriation.

✻ **Spotted Deer** *(Axis axis)*

✻ **Hog Deer** *(Axis porcinus)*

✻ All varieties of Deer were introduced by Britishers prior to Independence. All are protected under Wild Life Protection Act. Widely spread in the forest of South, Middle and North Andaman. Not hunted by the Aborigines.

✻ **Barking Deer** *(Muntiacus muntjak)*

✻ **Indian Elephant** *(Elephas maximus indicus)* - Feral Elephant population at Interview Island, let loose by Forest contractors after Independence.

5.7 42

Marine Mammals

Physeter catodon

Balenoptera musculus

Whale, Dolphin and Dugong abound in Andamans. The common Dolphin *(Delphinus delphis)* and the Dugong or Sea Cow *(Dugong dugon)* are residential, whereas *whales (Balenoptera musculus and Physeter catodon)* are transitional. Whales regularly visit our sea for delivery of their young ones. The whole family enters these waters and stays here till the babies are born. The spent female whale release a sort of fecal matter which is known as ambergris, a fixative used in perfume manufacturing. It is said to have medicinal and aphrodisiac value. The present market value of Ambergris is about ₹ One lakh per kg.

Dolphin

Once upon a time Dugongs were common in these islands. The Dugong creek at Little Andaman has been named after it. Aborigine and tribal population hunt dugongs.

Dugong

5.8 Reptiles

About 80 varieties are know to occur, 70 are terrestrial rest marine. This group exhibit high degree of endemism, 27 of them are found only in A & N Islands

Snakes 46 varieties recorded, 13 endemic and an equal number poisonous.

Regal Python

- **Reticulated Python** *(Python recticulatus)* Largest and heaviest Indian snake found in Nicobar group of Islands, non-venomous, grows to over 10 meter.

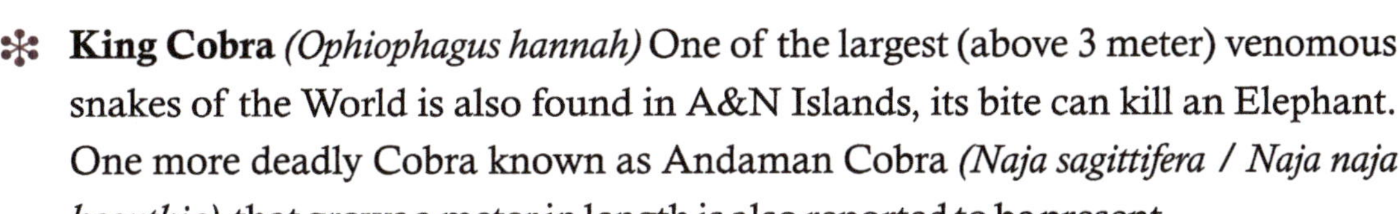

- **King Cobra** *(Ophiophagus hannah)* One of the largest (above 3 meter) venomous snakes of the World is also found in A&N Islands, its bite can kill an Elephant. One more deadly Cobra known as Andaman Cobra *(Naja sagittifera / Naja naja kaouthia)* that grows a meter in length is also reported to be present.

King Cobra

Bamboo pit viper *(Trimeresurus gramineus)* venomous tree snake

Venomous

Fangs & Fang marks

Bite Marks

Non-venomous

Laticuda sp. *is the common sea snake.*
Enhydrina schistosa *highly venomous sea snake is rarely seen.*

Tree Snake Bronzeback *(Dendrelaphis tristis)* Non venomous

Andaman banded krait *(Bungarus andamanensis)* venomous land snake

5.9 Turtles

Four varieties of marine turtles are known to occur in the sea around the islands, whereas only one land Tortoise called Malayan box turtle *(Cura amboinensis)* is found in Great Nicobar. The island beaches are traditional nesting ground for them.

Leatherback *(Dermochelys coriacea)* - The largest turtle of the World.
Photo - Karthik Shankar, Courtesy - Madras Crocodile Bank Trust

Green Sea Turtle
(Chelonia mydas)

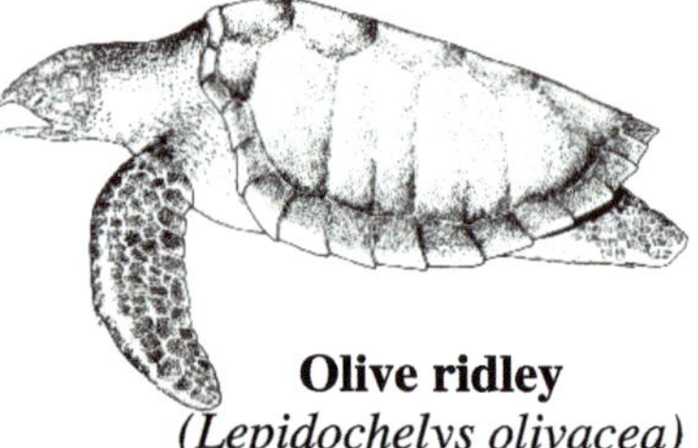

Olive ridley
(Lepidochelys olivacea)

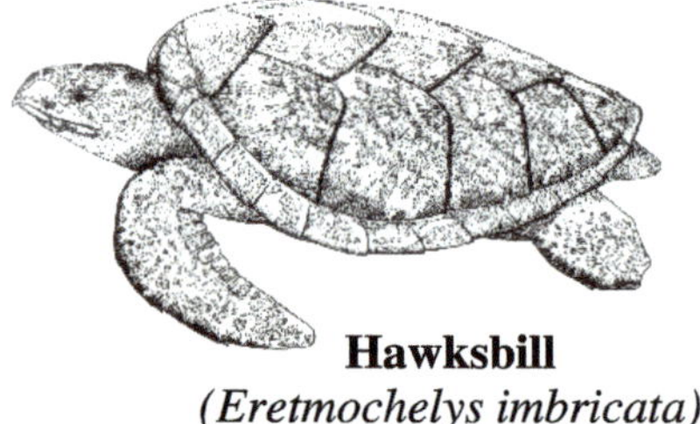

Hawksbill
(Eretmochelys imbricata)

Malayan Box Turtle
(Cuora amboinensis)

Open sea mating
Photo - Bivash Pandav, courtesy - Madras Crocodile Bank Trust

Others

Andaman Day Geeko
(Phelsuma andamanense)

Salt water Crocodile
(Crocodylus porosus)

Giant African Snail *(Achatina fulica)*

Water monitor lizard*(Varanus salvator andamanensis)*, endemic, largest in the world length 3 mtr. or more

Kankhajura / Centipedes
(Scolopendra spp.)
17 varieties known

Bichhoo / Scorpion
(Scorpiones Spp.)
5 varieties known

Robber Crab *(Birgus latro)*, largest land crab

Butterflies & Moths

With about 290 species, the A & N Islands house some of the largest and most spectacular butterflies of the world. Ten species are endemic to these islands. Mount Harriet National Park alone is the home to six species of wild endemic silk moths. The largest of them is the Andaman Atlas moth and most spectacular is the Moon moth.

Other endemic species are Andaman Clubtail, Andaman Mormon, Andaman Swordtail, Andaman Cruiser, Andaman Crow, Dark Glassy Tiger, Leopard Lacewing, Lime Butterfly, Painted Jezebel. The Nicobar Yeoman and the Nicobar Map are endemic to the Nicobar Islands.

Andaman Clubtail
(Pachliopta rhodifer)

Andaman Mormon
(Papilio mayo)

Andaman Swordtail
(Graphium antiphates)

Andaman Cruiser
(Vindula erota)

Dark Glassy Tiger
(Parantica ageleoides)

Andaman Crow
(Euploea arolaranensis)

Leopard Lacewing
(Cethosia cyane)

Lime Butterfly
(Papilio demoleus)

Painted Jezebel
(Delias hyparete)

Source : ZSI, ICARI

Birds

More than 270 varieties of birds have been identified out of which 106 are endemic. These islands are one among the 218 declared endemic bird areas of the World.

Nicobar Megapode (Megapodius freycinet nicobariensis) They frequent the open jungle area near the shore where soil is light to build its mound easily. Each egg weighs around 150g., which is approx. 1/6th of the birds own weight. The Megapode does not take part in the incubation of its eggs. Decaying leaves of mound give enough humidity and heat for incubation.

Narconduam Hornbill *(Rhyticeros narcodami)*

This bird is restricted to Narcondum island ,it lives in small flock of three or four.

Andaman Teal *(Anasgibberifrons arbgularis)* This duck like migratory bird measuring upto 47 cm. lives in fresh water ponds, tidal creeks and swampy grasslands abundant with weeds and vegetation. Present dense population is seen in North Reef island.

Olive-Backed Sunbird
(Nectarinai jugularis)

Swiftlets *(Collocalia fuciphaga inexpectata)* There are two varieties of 'Hawabill'. This bird is capable of flying to a speed of 150 kmph., this versatile flyer is capable of spending much of its life flying. They build nest using their saliva, which is said to have medicinal and aphrodisiac value.

Wild Life

BIRDS

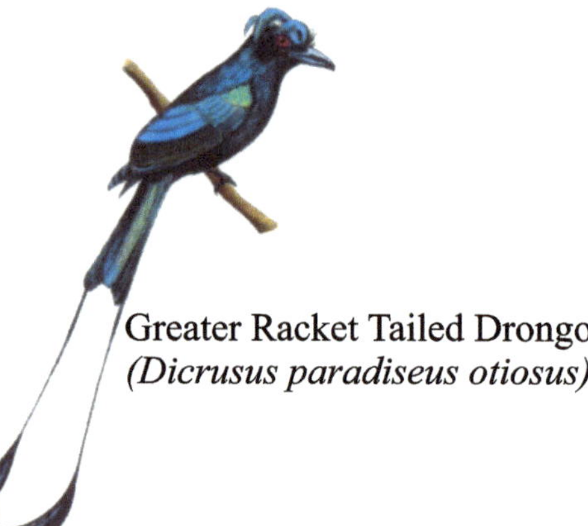

Greater Racket Tailed Drongo
(Dicrusus paradiseus otiosus)

Black Napped Tern
(Sterna sumatrana)

Andaman Drongo
(Dicrurus andamanensis)

Long Tailed Parakeet
(Psitacula longicauda)

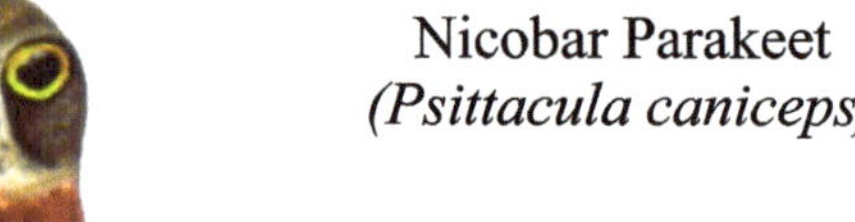

Nicobar Parakeet
(Psittacula caniceps)

Green Imperial Pigeon
(Ducula aenea)

Andaman Hawk Owl
(Ninox affinis)

Nicobar Pigeon
(Caloenas nicobarica)

Andaman Dark Serpent Eagle
(Spilornis elgini)

Andaman Black Woodpecker
(Drycopus hodgei)

Black Baza
(Aviceda leuphotes)

BIRDS

Ruddy Turnstone
(Arenaria interpres)

Brown Coucal/ Andaman Crow
(Centropus andamanensis)

Asian Fairy Bluebird
(Irena puella)

Andaman Treepie
(Dendrocitta bayleyi)

Scarlet Minivet
(Pericrocotus flammeus)

Hill Myna
(Gracula religiosa)

Collared Kingfisher
(Alcedo atthis)

Forest Wagtail
(Motacilla indica)

Whimberel
(Numenius phaeopus)

Nicobari Fowl
(Gallus domesticus)
Locally called '*Thakniet*'. Best Indigenous egg layer and hardy economical bird.

50

Sea

The Union of Inida by virtue of A&N Islands possess about six lakh (6,00,000) sq. km. of sea area as Exclusive Economic Zone which is about 30% of Indian EEZ thus all the living & non-living resources within this zone belongs to India. A part of this maritime zone on Eastern front falls upon Andaman Sea which although is comparatively less productive in terms of bio-mass production holds deposits of Maganese nodules, Hydrocarbons, Methane Clathrate/Fire ice etc. possibly due to its closed nature and intense Geo-thermal activity whereas larger part of A&N Island's EEZ is on the Western side covering Bay of Bengal which shows better and higher rate of bio-mass production. Harvestable Fish production of A&N Islands EEZ is postulated to be about one lakh forty eight thousand (1,48,000) tonnes per annum based upon catch effort method of estimation comprising of Oceanic resources like Tunas etc. 60,000 tons, shallow water surface fishes like Carax, Mackerel, Sardines etc. 56,000 tons and Bottom dwelling fishes like Perches and Groupers etc. 32,000 tons.

6.1 Fishes

Fishes the master of water world have lived in it for more than 360 million years. Today we have about 40,000 varieties of fishes known to science. They range in size from 10 mm (Philippine Gobie) to 21 m. (whale shark). Some are flattened, other inflated, many spindle shaped, a few snakelike, still others are compressed, depending on the environment in which they live or particular way of life.

From around A & N Islands more than 1200 species are recorded, out of them 240 are edible and 250 ornamental. Fishes are broadly classified as:-

1. ***Jawless fishes*** - *Agnatha* are snake like primitive fishes, which include Hagfishes and lampreys. Hagfish lives imbedded in deep waters and are excellent scavengers of the ocean floor. Lampreys stay near to land but could migrate to very deep water and feed mostly on crustaceans (prawn, lobster etc.). They enter into the body of large fishes and consume them from inside out.

2. Cartilaginous fishes - *Chondrichthyes*. They are devoid of true skeleton; this group includes Sharks, Skates and Rays. Shark, the most fearsome sea fish keeps replacing its long teeth. The female conceive through coitus. It appears that they don't perceive pain. they have a buit-in-immunity to cancer. The largest living fish belongs to this group.

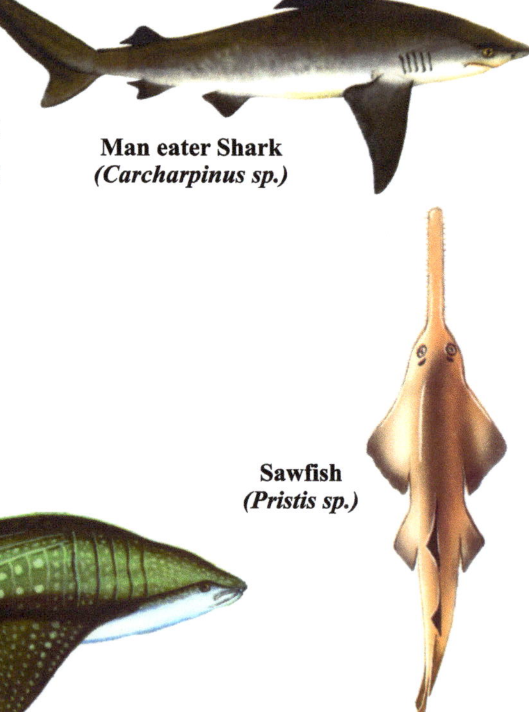

Man eater Shark
(Carcharpinus sp.)

Spoted eagle ray
(Aetobatus narinari)

Hammer head Shark
(Sphyrna sp.)

Sawfish
(Pristis sp.)

Whale Shark
(Rhincodon typus)

Largest fish of the World growing more than 20m.

3. Bony fishes *- Teleostomi / Osteichthyes.* The true bony fishes are the most modern and the largest group of fish. They are ray finned fishes, where fin rays or spines support the body of each fin. They constitute the vast majority of bony fishes. Fertilization is generally external. They posses a specialised buoyancy regulation device called Gas bladder. Many of them are capable of sex reversion as per demand. Some have bio-luminous organs to light up the deepest

Edible Fishes

Sword Fish
(*Xiphias gladius*)

Sail Fish
(*Istiophorus indicus*)

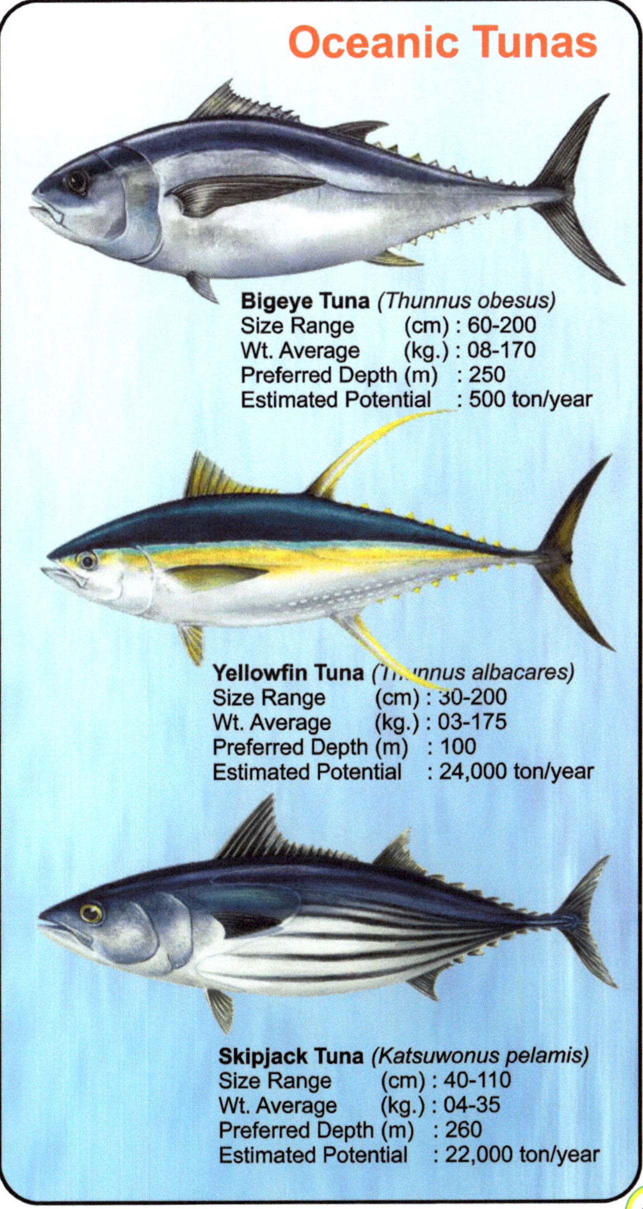

(Overall Hooking Rate around A&N Islands by Tuna Long liner is reported to be - 36.3%)

Sea Life

*Occurrenece

Lutjanus argentimaculatus
Occu.-FAO-1974
Photo FAO-1983 Plate I

Lutjanus bohar
Occu.-FAO-1974
Photo FAO-1983 Plate I

Lutjanus sebae
Occu.-FAO-1974
Photo FAO-1983 Plate I

Aphareus rutilans
Occu.-Rajan P.T.-2003
Photo FAO-1983 Plate IV

Lutjanus malabaricus
Occu.-FAO-1974
Photo FAO-1983 Plate I

Lutjanus gibbus
Occu.-FAO-1974
Photo FAO-1983 Plate I

Snappers

Sea Life

Ornamental Fishes

Moorish idol
(Zanclus sp.)

Powder blue surgeon
(Acanthurus sp.)

Bat fish
(Platax)

Maroon clown
(Premnas sp.)

Lionfish
(Pterois sp.)

Filefish
(Oxymonocanthus sp.)

Yellow tang
(Zebrasoma sp.)

Trigger
(Balistapus sp.)

Saddle back butterlfy
(Chaetodon sp.)

6.2 Shells

SHELLS

The principles of shell formation are still mysterious. They are formed by a soft bodied creature known as 'Mollusca'. A shell is loosely defined as an external skeleton made up of calcareous matter exhibiting a wide variety of shapes and colours specific to the animals that creates it through complex physiological process. This group of animals includes a seemingly infinite variety of forms in their evolution. Many authorities believe this group to be second largest in the animal kingdom. Shells are commonly divided as Univalve and Bivalve depending on the number of calcareous valves or shells they possess.

These islands are traditionally endowed with wide variety of shells specially Turbo, Trocus, Murex and Nautilus. Earliest recorded commercial exploitation began during 1929. Shells are important to us because many of them are being used as novelties supporting many cottage industries producing a wide range of decorative items and ornaments. Shells such as Giant clam, Green mussel and Oyster support edible shell fishery, a few like Scallop, Clam and Cockle are burnt in kiln to produce edible lime.

Deep Sea *Nautilus sp.*

Cassis sp.

Turbo sp.

Cassis sp.

Sea Life

Lambis sp.

Xancus sp.

Cypraea argus

Conus. aulicus

Trochus

Conus textile

Cypraea moneta

Harpa major

Architectonia sp

Coral and the Coral Reef

Survey conducted during 1986 - 87 estimated an area of 2000sq.km under coral coverage in A&N Islands which amounts to 6% of the 34965 sq.km of total continental shelf area. The human induced degradation was estimated to be about 360 sq.km. However the post Tsunami coral coverage is bound to be less due to large scale uprooting in southern islands and receding of sea level in northern islands beside wide spread burying of reef under silt. So far more than 200 varieties of corals have been documented from A&N Islands, which is the highest in this part of World. There are hundreds of fringing reefs scattered throughout the territory, and three luxuriant coral banks on Western side.

The renowned marine naturalist Cap. Jacques Cousteau and his team during 1989 explored 12 dive sites for 70 hours around Andamans and recorded "Indian waters are preserved, rich and virgin".

Under optimum environmental conditions many coral colonies flourish at a suitable site creating coral reef which under favourable ecological condition keeps growing, which in fact is an assemblage of more than 3,000 living organisms in perfect harmony; a magnificent manifestation of nature's ability to create, thread and balance various life forms in space and time. Coral reef ecosystem is the most intricate, diversified and aesthetically appealing ecosystems of this planet.

The noun coral is believed to have been derived its origin from an Arabic word '*garal*', which means small stone, or Hebrew '*goral*' which means pebble. Later, the Greeks adopted it as *'korallion'* and in Latin it appeared as *'Coralium'*. The present day English version means 'the hard stony skeleton secreted by certain marine polyps'. The animal, which secrets and builds this skeleton originated some 570 million years ago. This tiny boneless, fragile creature is genetically endowed with exceptionally high architectural skill. Polyps create multispectral & multidimensional skeleton know to us as corals. A close look at a dead bleached coral piece will reveal its porousness. Millions of pores are found in a small piece of coral. Each 'pore' was the home for a 'polyp' and the piece of this calcareous *'garal'* held for observation is the outcome of cumulative effort put in by billions of polyps for a long time.

Fringing reef

Off Wandoor

Coral Polyps

Benefits of Coral and Coral Reef

- Excellent nursery grounds for innumerable marine fishes.
- Yields many varieties of fishes, shells, lobsters, sea urchin, sea cucumber and sea grasses etc. which play a significant economic role. In A & N islands about 17% of annual fish catch is from coral reef on which thrive the export industry.
- They are the first guardians of coastline, lessening erosion.
- Coral reefs hold complex bio chemicals for the future.
- Anti-tumour, anti leukemia, anti-microbial & ultraviolet blocker chemicals of high medicinal value are being extracted from reef inhabitants.
- Coral reef provides support and sustenance to adjoining coastal terrestrial ecosystem on which people depend.
- Coral colonies are excellent record keepers of the past meteorological events.
- Coral reef glamour support tourism.

***Gorgonia* - Red & Yellow Colony**

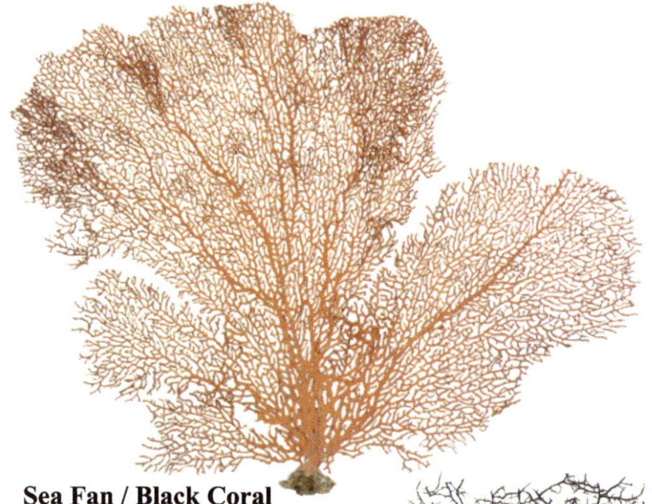

Sea Fan / Black Coral

Red Coral, *Tubipura*

Black Coral, *Antipatherian Soft Corals*

Blue Coral, *Heliopora*

Healthy Stony corals

Sea Life

Healthy Soft corals

Red Gorgonia - Melithaeid sp.

Red Sea whip - Ellisella andamanensis

Red Coral - Tubipora musica

Marble coral - Isis hippuris.

Feather worm - Sabella sp.

Sea anemone - Heteractis sp.

6.4 Coral & Coral Reef - M.G.Marine National Park, Wandoor

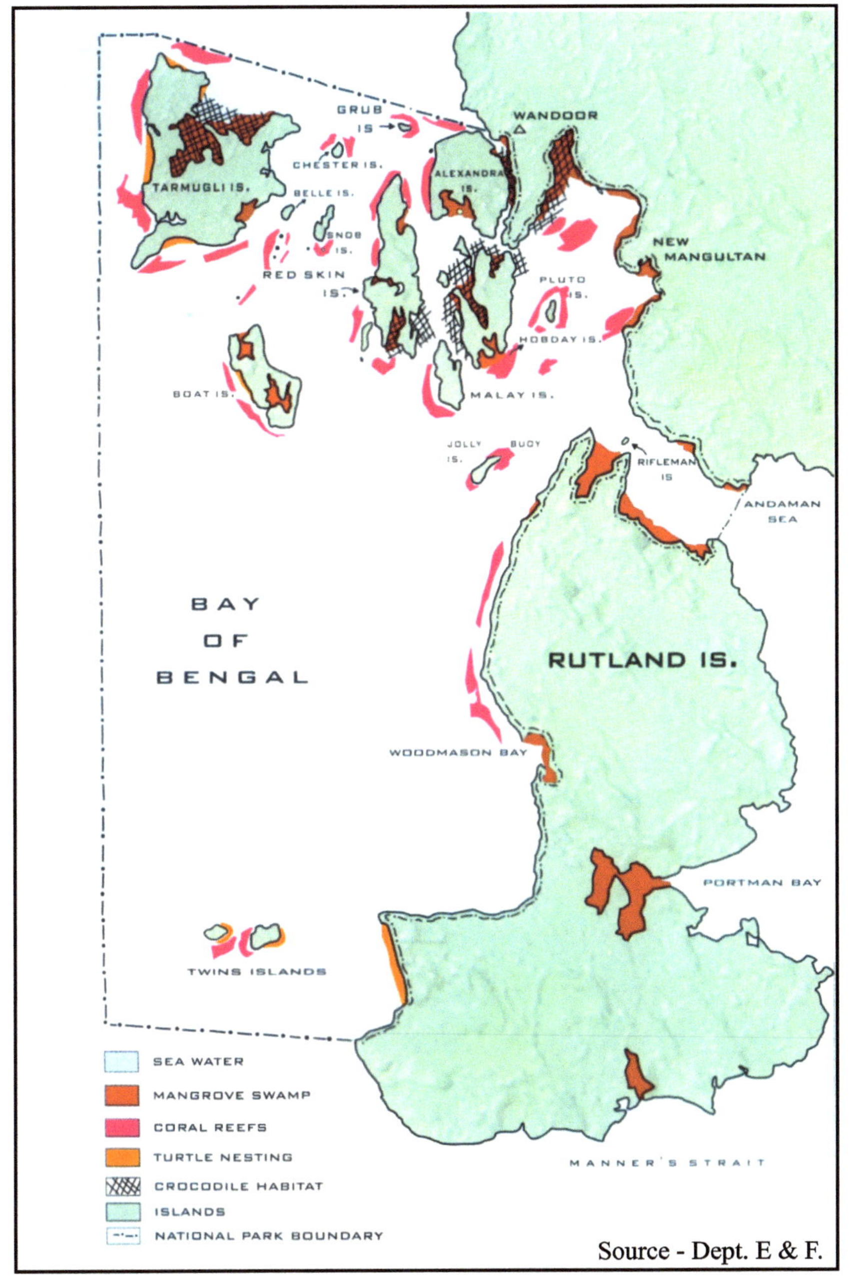

Source - Dept. E & F.

Cluster coral -*Acropora*

Brown stem coral -*Acropora*

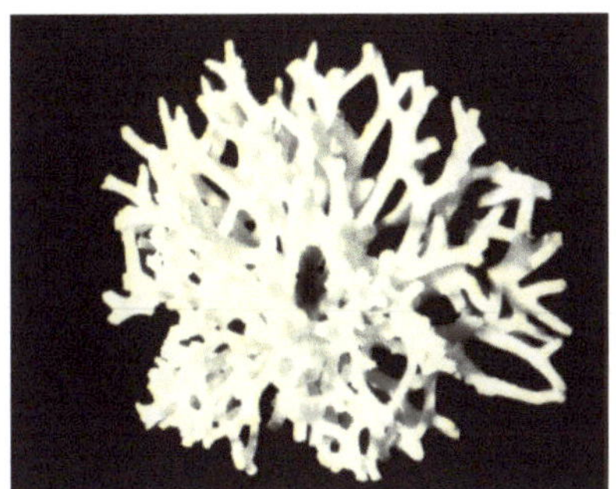
Birds nest coral - *Seriatopora*

Mushroom coral - *Fungia*

Branch coral - *Acropora*

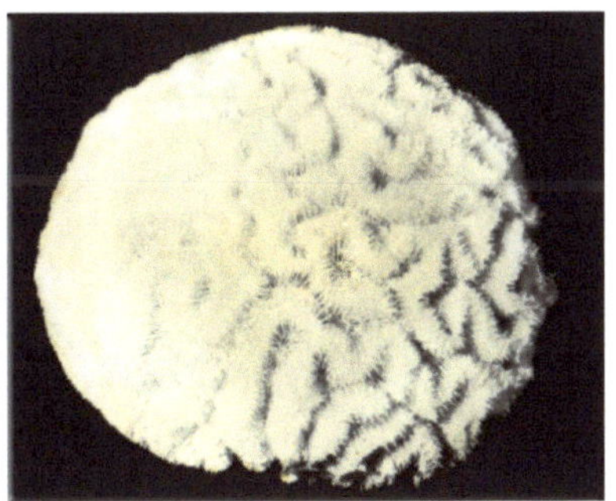
Brain coral - *Diploria*

Marine Venom

Common Shallow Water Venom Alert

	Animal	Type of Injury	Symptoms	First Aid
1	Scorpion/Lion fish	*Prick through fin spine*	*Intense pain, burning sensation, breathing difficulty, rarely, nausea & weakness*	❖ *Calm down the victim.* ❖ *Immobilise the inflicted extremity.* ❖ *Soak the injured area with tolerable hot water.* ❖ *Under medical care a local anaesthesia or Emetine HC1 may be injected at injured site.*
2	Stone fish (World's most venomous fish)	*Spine prick (Lethal)*	*Lethal myotoxin causes muscular paralysis & unconsciousness associated with breathing difficulty & intense pain*	❖ *Calm down the victim.* ❖ *Immobilise the inflicted extremity.* ❖ *Soak the injured area with tolerable hot water.* ❖ *RUSH to hospital for appropriate anti venom treatment.*
3	Coral cat fish	*Prick through fin spine*	*Burning, itching & allergic manifestation.*	❖ *Soak the injured area in hot water.*
4	Sting ray	*Spine stab*	*Feeling of prick & Lacerated injury.*	❖ *Calm down the victim.* ❖ *Stop bleeding.* ❖ *Soak in hot water.* ❖ *RUSH to hospital.*
5	Cone shell	*Sting*	*Numbness, weakness, breathing difficulty, heart failure*	❖ *Artificial respiration CPR.* ❖ *Hospitalisation.*
6	Portuguese Man-of war Box jelly fish	*Sting by almost invisible tentacles*	*Intense burning in skin with throbbing pain, area becomes red, brown or purple with blistering & swelling*	❖ *Calm down the patient.* ❖ *Douse the injured area with vinegar.* ❖ *Seek medical assistance.*

Protected Area in Andaman and Nicobar Islands

8

Sl. No.	National Pask	Area (Sq.Km.)
1.	Mahatma Gandhi Marine National Park	281.50
2.	Rani Jhansi Marine National Park	0.44
3.	Campbell Bay National Park	426.23
4.	Galathea National Park	110.00
5.	Mount Harriet	32.536
6.	North Button	0.44
7.	Middle Button	256.142
8.	South Button	46.62
9.	Saddle Peak	0.03
	Total	**1153.938**

Sl. No.	Sanctuaries	Area (Sq.Km.)	Sl. No.	Sanctuaries	Area (Sq.Km.)
1.	Arial	0.05	49.	North Brother	0.75
2.	Baltimaliv Island	2.07	50.	North Reef	3.484
3.	Bamboo	0.05	51.	Oliver	0.16
4.	Barren	8.100	52.	Orchid	0.10
5.	Belle	0.08	53.	Ox	0.13
6.	Benette	3.46	54.	Oyster	0.08
7.	Bingham	0.08	55.	Oyster	0.21
8.	Blister	0.26	56.	Paget	7.36
9.	Bluff	1.14	57.	Parkinson	0.34
10.	Bondaville	2.55	58.	Passage	0.62
11.	Brush	0.23	59.	Peacock	.62
12.	Buchanan	9.33	60.	Petric	0.13
13.	Channel	0.13	61.	Pitman	1.37
14.	Cinque	9.53	62.	Point	3.07
15.	Clyde	0.54	63.	Potanma	0.16
16.	Cone	0.65	64.	Ranger	4.26
17.	Curlew	0.03	65.	Reef	1.74
18.	Curlew (B.P.)	0.16	66.	Roper	1.46
19.	Cuthbert Bay	5.82	67.	Rose	1.01
20.	Defence	10.49	68.	Rowe	0.01
21.	Dot	0.18	69.	Sandy	1.58
22.	Dottrill	0.13	70.	Sea Serpent	0.78
23.	Duncan	0.73	71.	Shearme	7.85
24.	East	6.11	72.	Sir Hugh Rose	1.06
25.	East or Inglis	3.55	73.	Sisters	0.36
26.	Egg	0.05	74.	Snake	0.03
27.	Enterance	0.96	75.	Snake	0.73
28.	Flat	9.36	76.	Snark	0.60
29.	Galathea Bay	11.44	77.	South Brother	1.24
30.	Gander	0.05	78.	South Reef	1.17
31.	Goose	0.01	79.	South Sentinel	1.612
32.	Gurjan	0.16	80.	Spike	11.70
33.	Hump	0.47	81.	Spike	0.42
34.	Interview Island	133.00	82.	Stoet	0.44
35.	James	2.10	83.	Surat	0.31
36.	Jungle	0.52	84.	Swamp	4.09
37.	Kwangtang	0.57	85.	Table (Dalgarne)	2.29
38.	Kyd	8.00	86.	Table (Excelsior)	1.69
39.	Landfall	29.48	87.	Talabaicha	3.21
40.	Latauche	0.96	88.	Temple	1.04
41.	Lohabarrack Crocodile Sanctuary	100.00	89.	Tillongchang Island	16.83
42.	Mangrove	0.39	90.	Tree	0.03
43.	Mask	0.78	91.	Trilby	0.96
44.	Mayo	0.10	92.	Tuft	0.29
45.	Megapode Island	0.12	93.	Turtle	0.39
46.	Montogomery	0.21	94.	West	6.40
47.	Narcandum	6.812	95.	Wharf	0.11
48.	North	0.49	96.	White Cliff	0.47
				Total	**466.218**

Source - DE&F

References

1. **Majumdar. R.** - 1975 - ***Penal settlement in Andamans***. Pub., Gazetters Unit, GOI., p-21., pp. - 339.

2. **Nunn. P. D.** - 1994 - ***Oceanic Islands.,*** Blackwell publishers., Oxford. pp. 413.

3. **Stephen oppenheimer** - 2004 - ***Out of Eden*** : The peopling of the World., Pub. Constable., pp., 400.

4. **Ramachandra. R.** - 2005 - Waves from the past., ***Frontline.,*** 22 (2)., 116 - 120.

5. **Anon** - 2005 - ***Wizard.,*** XIV (3)., 22

6. **Anon** - 2005 - ***The Hindu.***, Chennai., 11thFeb 2005.

7. **Chakraborty. S.** - 2005 - The Day of Judgement.,***CNN***., New Delhi., 3 (2)., p.34 quoting USGS.

8. **Ramachandran. R.** - 2005 - Of Plates Tectonics., Quakes & Tsunamis., ***Frontline.,*** 22 (2)., 118.

9. **Revkin. C. Andrew** - 2005 - Missed signals.,***Frontline.***, 22(2)., 115

10. **Anon** - 2005 - ***The Daily Telegrams.***, 11th May 2005.

11. **Anon.** - 2005 - ***The Daily Telegrams.,*** 25th April 2005

12.**Mudur. G. S.** - 2005 - First man's Children on Andamans., ***The Telegraph***., Calcutta., 13th May 2005.

13. **Mukherjee. M.** - 1999 - Out of Africa into Asia., ***Scientific American***. 20(1)., p.14.

14. **Shekhar Singh *et.al.*** - 1991 - ***Directory of National Parks and Sanctuaries in A & N Islands.,*** IIPA, New Delhi., pp. 171.

15. **Kloss, C. Boden** - 1902 - ***Andamans & Niobars,*** Vivek, Publishing House, New Delhi., Reprint 1971., p . 178., pp. 373.

16. **Dhingra Kiran** - 2005 - ***The A & N Islands in 20th century.***, A gazetteer., Oxford Press., p. 29., pp. 398.

17. **Anon.** - 2001 - 2021 - ***A & N Islands. Forest and Environment.***,Dept. of Environment and Forest.

18. **Anon.** - 2003 - ANET., ***Status paper.***

19. **Sekhsaria. Pankaj** - 2003 - ***Troubled Islands.,*** Mudra printers., Pune., p. 3., pp. 89.

20. **Oberai, C.P.** - 2000 - ***Eco Tourism Paradise Andaman & Nicobar Islands.*** B.R. Publishing Corporation, Delhi., p. 177., pp 218.

21. **Anon.** - 2005 - ***The Daily Telegrams*** 31st August 2005.

22. **FAO** - 1974 - 1983 - ***Identification Sheet, Area 51, 57 & 71***

❇ Andaman Wood Pigeon - State Bird

Andaman Wood Pigeon is an endemic bird, which is found only in Andaman and Nicobar group of Islands. This bird is of the size of a domestic pigeon with longer tail. This bird has whitish head with checkerboard pattern on neck. The upper parts are dark slate grey in colour and underparts are pale blue grey Mettalic green sheen on upper side and reddish bill with yellowish tip and purplish red orbital skin are identification characters. The bird lives in dense broad leaved evergreen forest.

❇ Andaman Padauk - State Tree

Andaman Padauk is a tall deciduous tree found only in Andaman. it grows upto height of 120 feet/36.6 mtr. The timber is highly prized for making furniture. Burr and Buttress formation add charm to the tree and used in making unique furniture.

❇ Dugong - State Animal

Dugong, an endangered marine mammal, also knows as Sea Cow, is only strictly marine mammal, which is herbivorous. It mainly feeds on sea-grass and other aquatic vegetation. Dugong is distributed in shallow tropical waters in Indo-Pacific Region. The animal is about three-metre length and weighs about 400 kg. In India Dugong is reported from Gulf of Kutch, Gulf of Mannar, Palk Bay and Andaman and Nicobar Islands. Within A&N Islands Dugong has been reported from Ritchie's Archipelago, North Reef, Little Andaman and parts of Nicobars.

चंद ऐसे भी कंवल होते हैं
खिलते नहीं और वक्फ़े अजल होते हैं
ये बात जुदा है के तामीर ना हो पाए
हर ज़हन में चंद ताजमहल होते हैं ।

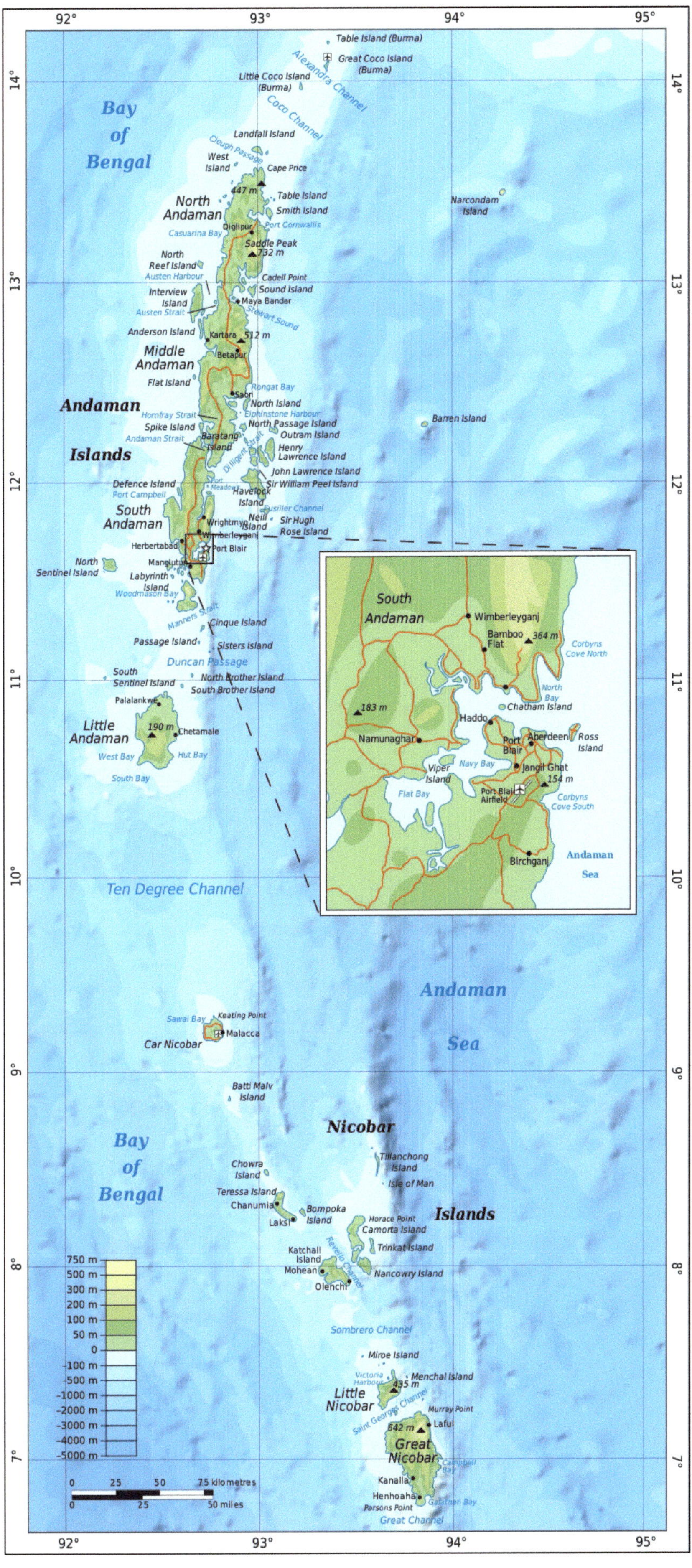
Bay of Bengal
Andaman Islands
North Andaman
Middle Andaman
South Andaman
Little Andaman
Ten Degree Channel
Andaman Sea
Nicobar Islands
Car Nicobar
Little Nicobar
Great Nicobar
Table Island (Burma)
Great Coco Island (Burma)
Little Coco Island (Burma)
Coco Channel
Alexandra Channel
Landfall Island
West Island
Cape Price
Table Island
Smith Island
Port Cornwallis
Narcondam Island
Diglipur
Saddle Peak 732 m
Cadell Point
Sound Island
Maya Bandar
Interview Island
Austen Strait
Anderson Island
Kartara
512 m
Betapur
Flat Island
Rongat Bay
North Island
North Passage Island
Outram Island
Henry Lawrence Island
John Lawrence Island
Sir William Peel Island
Barren Island
Spike Island
Baratang Island
Defence Island
Havelock Island
Neill Island
Sir Hugh Rose Island
Herbertabad
Port Blair
North Sentinel Island
Labyrinth Island
Cinque Island
Passage Island
Sisters Island
Duncan Passage
South Sentinel Island
North Brother Island
South Brother Island
Chetamale
190 m
West Bay
Hut Bay
South Bay
Wimberleyganj
Bamboo Flat
364 m
Corbyns Cove North
North Bay
Chatham Island
Haddo
183 m
Namunaghar
Aberdeen
Ross Island
Viper Island
Navy Bay
Jangli Ghat
154 m
Flat Bay
Port Blair Airfield
Corbyns Cove South
Birchganj
Andaman Sea
Keating Point
Sawai Bay
Malacca
Batti Malv Island
Tillanchong Island
Isle of Man
Chowra Island
Teressa Island
Chanumla
Bompoka Island
Laksi
Camorta Island
Trinkat Island
Katchall Island
Mohean
Olenchi
Nancowry Island
Sombrero Channel
Menchal Island
Murray Point
Lafui
642 m
Kanalla
Henhoaha
Great Channel
750 m
500 m
300 m
200 m
100 m
50 m
0
-100 m
-500 m
-1000 m
-2000 m
-3000 m
-4000 m
-5000 m
0 25 50 75 kilometres
0 25 50 miles
92°
93°
94°
95°
14°
13°
12°
11°
10°
9°
8°
7°

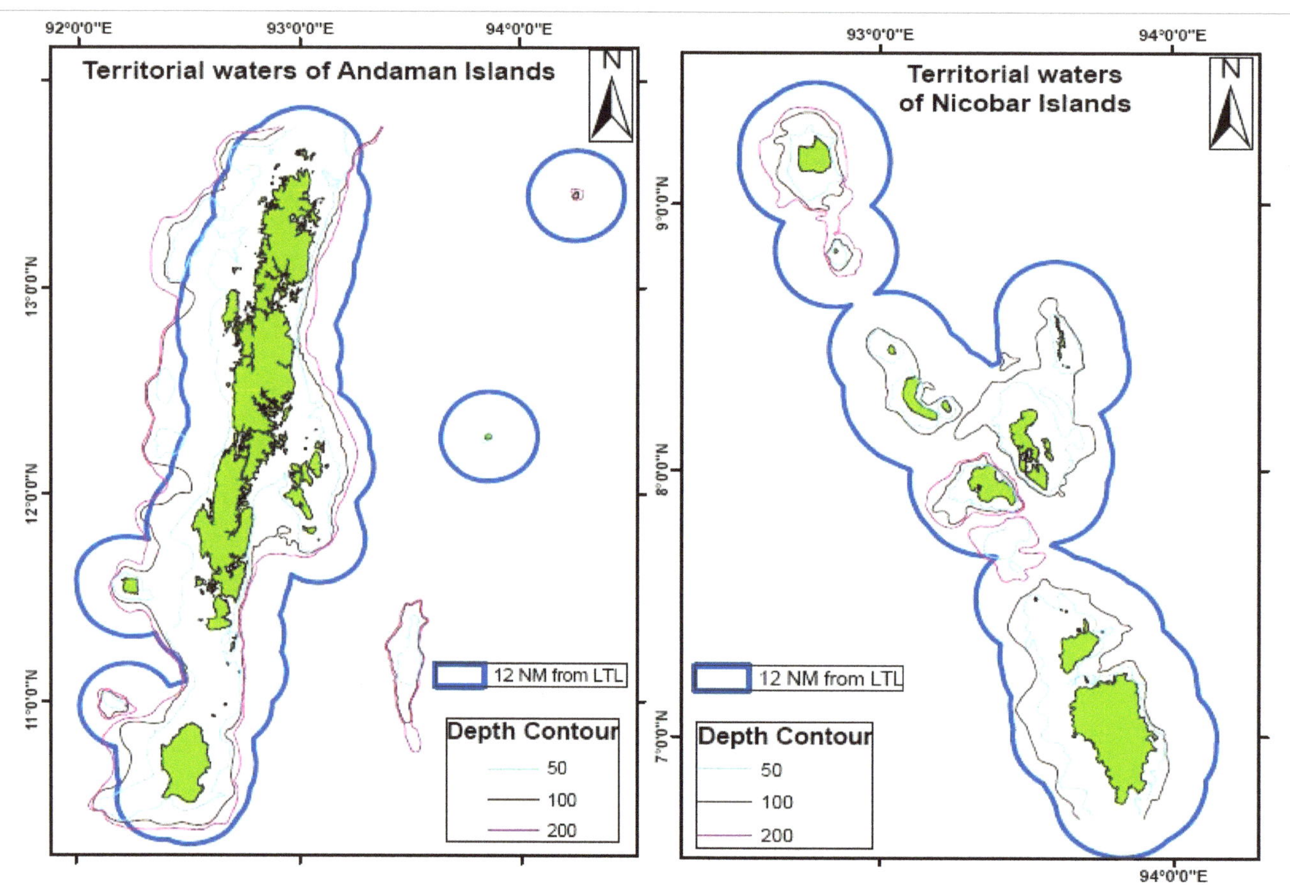

Once unauthorizedly created wrong map

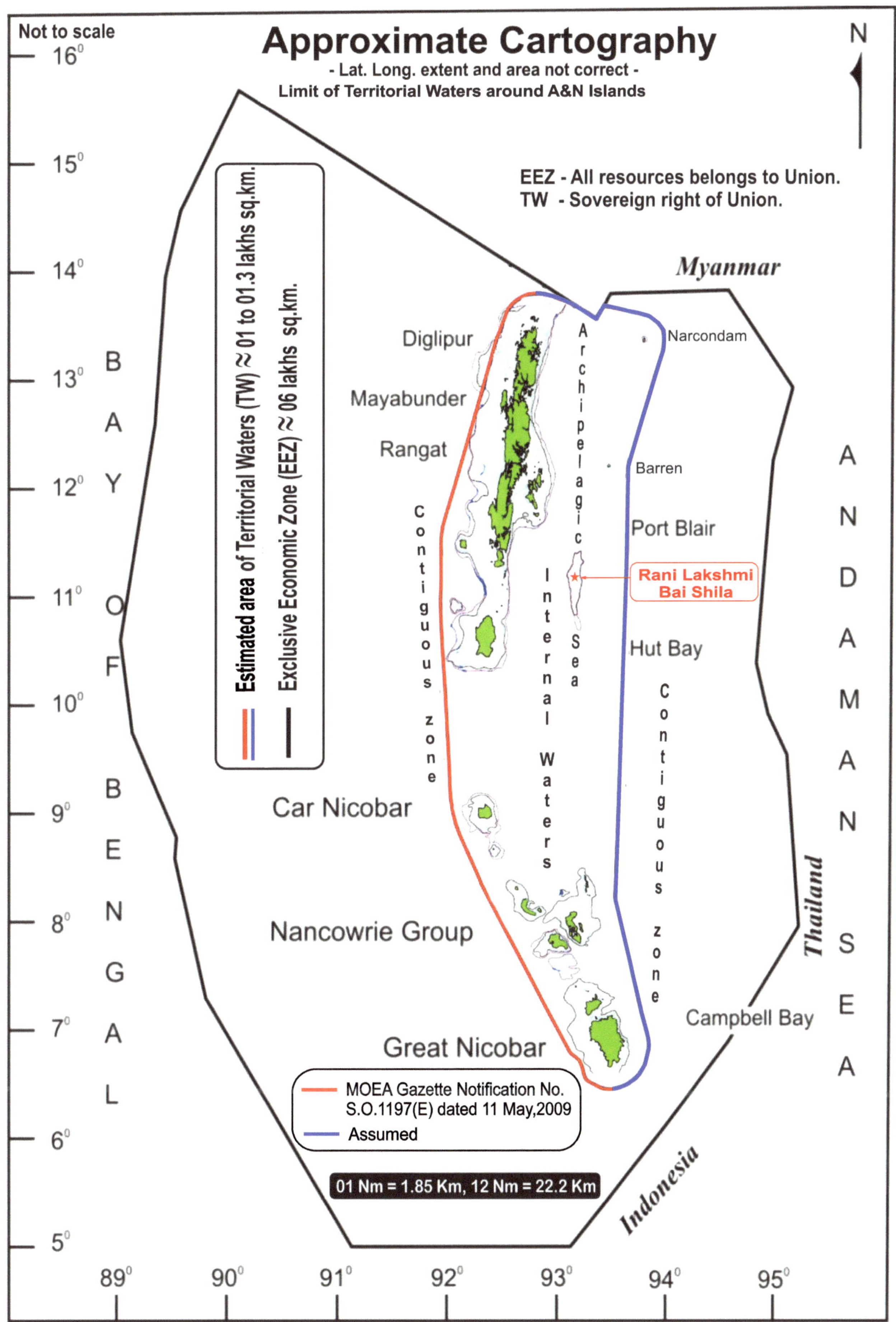

Not to scale
Approximate Cartography
- Lat. Long. extent and area not correct -
Limit of Territorial Waters around A&N Islands
N
EEZ - All resources belongs to Union.
TW - Sovereign right of Union.
Estimated area of Territorial Waters (TW) ≈ 01 to 01.3 lakhs sq.km.
Exclusive Economic Zone (EEZ) ≈ 06 lakhs sq.km.
Myanmar
Narcondam
Diglipur
Mayabunder
Rangat
Barren
Port Blair
Rani Lakshmi Bai Shila
Hut Bay
Archipelagic Sea
Internal Waters
Contiguous zone
Contiguous zone
Car Nicobar
Nancowrie Group
Great Nicobar
Campbell Bay
Thailand
Indonesia
BAY OF BENGAL
ANDAMAN SEA
MOEA Gazette Notification No. S.O.1197(E) dated 11 May,2009
Assumed
01 Nm = 1.85 Km, 12 Nm = 22.2 Km
16°
15°
14°
13°
12°
11°
10°
9°
8°
7°
6°
5°
89°
90°
91°
92°
93°
94°
95°

Dr. Pankaj Sharma
Joint Secretary (D&ISA)

विदेश मंत्रालय, नई दिल्ली
MINISTRY OF EXTERNAL AFFAIRS
NEW DELHI

office 1948
26/07/17

REGD BY SDS

AE-1/1061/4/2017 26 July 2017

Subject: REVIEW OF SEA AREAS AROUND ANDAMAN & NICOBAR ISLANDS

Dear Sir,

Please refer to your letter No PMO ID No. PMOPG/D/2017/0032667 dated 20 January 2017.

2. Your concern for the benefit of the country based on your rich and professional experience in the maritime domain is highly appreciated.

3. This is to kindly inform you that the issue has been examined in details by means of inter-ministerial mechanism and the provisions of UNCLOS 1982 have been pursued whilst promulgating the India's Baseline System under Indian Maritime Act vide Ministry of External Affairs Gazette Notification (MEA Notification No.S.O.1197 (E), dated 11 May 2009). Also, the National Hydrographic Office has not published any navigational chart depicting the extent of the territorial waters.

4. We trust this clarifies the situation.

Kind regards

Yours sincerely,

(Pankaj Sharma)

Shri Arif M Mustafa
Gafoor Manzil, Gafoor Lane
24/2- Shastri Road Aberdeen bazaar
Port Blair
Andaman and Nicobar Islands – 744101.

ArSquare Development Pvt. Ltd.

GoExplore Andaman

November, 2021 ₹ 100

Landings at Indian Subcontinent's Lone

Active volcano at Barren Island

Tale of Two Cellular Jails:

Resemblance is Uncanny

When the Jarawas came

across the so-called civilized people

When The Hostile Sentinelese Tribe

Showed Friendly Gesture

Travel

Magazine

Landings at Indian Subcontinent's Lone *Active* Volcano *at* Barren Island

By Arif M. Mustafa

"Amidst this catastrophic scenario of erupting volcano at Barren Island, a weak and trembling cry of a baby goat attracted the attention of the team. The starving goat was rescued and sent on board the vessel."

A team of ten explorers on the forenoon of 15th May 1991 landed at Barren Island and went up the hill adjacent to the volcano. The team comprised of two Geologists, three marine scientists, and others, along with the then Lt. Governor, Lt. Gen. Ranjit Singh Dayal. The team perched upon the peak of the adjacent 250 m. high hill. They observed from the seaward sidewall of the blown away large cauldron located just one and a half km away from the Indian subcontinent's lone active volcano.

Feeble fuming was seen in the crater, whereas the lateral vent was continuously ejective basaltic lava that flowed down the dome. The volcano was active after a lull of about a hundred years.

Awakening was noticed during March 1991 by the Indian Coast Guard. The National Remote Sensing Agency recorded about a km long vertical column of smoke in the sky above Barren Island.

A river of lava, about 200 m wide, was formed by the time we landed. The island was trembling with the intermittent rumbling boom. The volcano was spitting fire, ash, and gas like an angry mythological dragon. The sky was overcast, and it suddenly started raining, but to our astonishment, we were drenched in black rain.

Meanwhile, gusts of monsoon-like winds increased in intensity, strong lightning paired with screaming thunder. Down the hill, the lava front was a mix of glowing cinders and molten rock that was engulfing vegetation. The awful powers of nature were unleashed, and the environment I found myself in was one of the most adventurous moments of my life.

Another unforgettable opportunity on the same amphitheater came 12 years later, on 16th March 2003 when the volcano relaxed a little.
This provided me the opportunity to walk over the rim of the widened crater with a team of foreign journalists, including National Geographic. Among the 18 landings I have made at Barren Island, the memory of 15th May 1991 and 6th March 2003 encounters with the mystic mood of nature will ever remain afresh.

India has two volcanic islands in the Bay of Bengal namely Barren Island Volcano and Narcondum Island Volcano. The former awoke during 1991, while the latter remains extinct. Barren Island is located about 135 km northeast of Port Blair. The uninhabited island is almost circular in shape, about 3 km in diameter, 10 sq km in area. It is quite steep with a sloping surface rising from a depth of 2.2 km in the sea to a peak at about 250 meters above mean sea level.

There are a couple of theories about the evolution of this island. The most widely accepted one is that initially, an eruption took place in the sea about 1 to 2 million years back during the late to post Pleistocene period which resulted in the emergence of this island.

After hearing the news of the resurgence of a volcano, an unsuccessful attempt to land at Barren Island was made on 11th May 1991. A long streak of volcanic smoke became visible a few minutes after crossing Inglish Island, the smoke was drifting towards the southeast. The team anchored in the vicinity of Barre Island by 5 p.m. and moved landward in a small boat. About 100 m. away from the shore intense showers of volcanic glass/dust/ash particles started, which scared some of the team members, and the party returned without landing. The dormant volcano was actively spitting lava from an auxiliary vent located on the northern side about 50 m. below the main crater. The discharge was pulsating with a rumbling noise and ejected boulders were rolling down the slope.

The thick and dark smoke column was about 900 m. high from the top of the volcano. The surrounding forest was burning and there was a charred smell accompanied by dancing sparks throughout.

A second attempt by the 10-member team was made on 15th May 1991. The team left Port Blair by M.V. Tarmugli at 6 p.m. The vessel anchored the next day at around 6 a.m, about 500 m. away. Suddenly a Sperm whale appeared in the water and greeted the party with a fascinating jet spray and disappeared.

The volcano that the team ultimately landed on was dangerously violent and continuously throwing out glowing boulders and dense fumes with intermittent rumbling. The surrounding vegetation was partially charred and the entire area was buried under thick volcanic ash. The edge of the advancing super-heated lava river was about 500 m. away from the seashore. The lava front was amazing, slowly advancing, producing cracking noise, fuming, releasing sparklets expanding latterly.

Amidst this catastrophic scenario, a weak and trembling cry of a baby goat attracted the attention of the team. The starving goat was rescued and sent on board the vessel. To get a better view, we went up the charred hill on the left side and climbed to a height of 250 m.

Contrary to the name, the island is fairly rich in flora and fauna. On 15th May 1991, we found a few live animals such as feral goat, Rock skipper fish, rock mussel, etc. Live crabs were seen vacating the western shore, moving up the ash-covered mountain range, and escaping towards the southeastern side.

I made many more attempts to land at the volcano. On 26th June 1991 landed with the support of the Coast Guard. This time the volcano was dangerously eruptive. The 12 m. tall light beacon was completely engulfed by the advancing lava river, while the sea-lava interface was producing enormous amounts of steam. The main crater collapsed to the level of the lateral vent, considerably enlarging its circumference.

On 10th July 1991, volcanism appeared to be at its climax. On 1st Nov 1991, I landed with the support of A&N Police along with well-known German volcanologists Dr. Ing Peter Halbach & Dr. Margret Halbach. The volcano was silent, coastal topography totally changed. The next landing was on 4th January 1992 when we found the volcano silent with the exception of intermittent gas emission.

Volcano
Puluga laka bang
at
Barren Island

Volcano at Barren Island

Dedicated with love to 'Ishu'/Ishrat for her deep compassion, patience, understanding and sacrifices during my innumerable adventures around the island. The strength in me, from within, came through her everlasting affectionate support.

Arif M.Mustafa

The Sponsor

WE OPERATE LIVEABOARD CRUISE GAME FISHING TRIPS TO FAR OFF ISLANDS LIKE BARREN ISLAND ACTIVE VOLCANO, NARCONDAM ISLAND EXTINCT VOLCANO AND MANY MORE ISLANDS IN ANDAMANS WHERE DAILY BOATS CANT REACH. LIVEABOARD IS THE BEST WAY TO FISH AND EXPLORE THE ISLANDS OF ANDAMANS AS THERE IS NO NEED TO COME BACK TO THE SHORE FOR 2-3 DAYS. OUR LUXURY LIVEABOARD HAS TWO SUITE CABINS FOR COMFORTABLE SLEEP FOR 4 PERSONS, FULL FLEDGE GALLEY, BATHROOM, LIVING AREA, FULL AIR CONDITIONED. ANDAMAN ISLANDS ARE OFTEN SAID TO BE THE FIRST GAME FISHING DESTINATION NOT ONLY IN SOUTHEAST ASIA BUT ALSO IN THE WORLD. INDEED YOU CAN CATCH BLACK & BLUE MARLIN, BROADBILL SWORDFISH, SAILFISH, TUNA MONSTER, WAHOO, DOGTOOTH TUNA, BARRACUDA, DOLPHIN FISH, RED SNAPPER, GIANT TREVALLY, GROUPER'S AND SHARKS TO NAME JUST A FEW THAT TOO ALL XXXL SIZE.

Puluga laka bang

at

Barren Island

First Edition - 2014

Authors

Arif M. Mustafa
Asst Director of Fisheries
Department of Fisheries,
A&N Administration,
Mobile - 09434261302

Altamash Mustafa
Vocational Instructor DC
ITI Dollygunj
A&N Administration
Mobile - 09476089909

Acknowledged

Type set
Ms. **Mohini Kumari**
South Point, Port Blair.

Layout & Graphics
Mr. **CHS Srinivas Rao**
RGT Road, Port Blair.

Photographs
(1) IG,VSR Murthy PTM, TM, Commander CGR A&N
(2) Shri Akshay Malavi
(3) Google Images

Special thanks & gratitude to Indian Coast Guard, A&N Islands

Sl.No.		Topic	Page
1	-	Index	2
2	-	Message	
2.1	-	Hon'ble Lt. Governor, A&N Islands	3
2.2	-	Chief Secretary, A&N Administration	4
2.3	-	Sponsor	5
3	-	Intro Remarks	6
4	-	Encounter	7
5	-	Background	10
6	-	Location	11
7	-	Physiography	12
8	-	Genesis	13
9	-	Geology	14
10	-	Volcanic History	15
11	-	About Valcanoes	23
12	-	Threat	24
13	-	Glossary of Geological terms	25

For bibliographic purpose this document may please be quoted as follows :-
***Mustafa A.M; Mustafa A** - 2014, Volcano at Barren Island.*

Volcano *Puluga laka bang* at **Barren Island**

Lt. Gen (Retd) A.K.Singh
PVSM, AVSM, SM, VSM
Lieutenant Governor
Andaman and Nicobar Islands

Raj Niwas,
Port Blair - 744101
Tel : (O) 03192-233333
(R) 03192-233300
Fax : 03192-230372

Message

I am happy to see that a booklet on the Volcano at Barren Island is being published for the first time. This booklet reflects the recent resurgence of research interest with reference to this geological phenomenon. The authors have sought to bring together amazing geological facts and beautiful pictures of the volcano. I am confident that this publication will not only serve in educating us but will help generate a lot of interest about the volcano, both for tourists as well as scientists engaged in serious research.

I congratulate Shri Arif M. Mustafa and his son, Shri Altamash Mustafa for this commendable effort. I also congratulate the sponsor, Shri Zeeshan Ghani of Andaman Sea Game Fishing.com for his valuable support for this effort.

My best wishes.

Lt. Gen (Retd) A K Singh
Lieutenant Governor
Andaman & Nicobar Islands

आनन्द प्रकाश
Anand Prakash
मुख्य सचिव
Chief Secretary

फोन / Ph. No. 03192-233110/234087
फैक्स सं. / Fax No. 03192-232656
ई–मेइल / E-mail : cs-andaman@nic.in

अण्डमान तथा निकोबार प्रशासन
ANDAMAN AND NICOBAR ADMINISTRATION
सचिवालय / SECRETARIAT

Port Blair, dated the 06th December 2013.

Message

I am deeply impressed to see that a Fisheries Department Officer of the Andaman Government and his son so brilliantly attempted to unravel the geological marvels & mysteries of our lone active volcano at Barren Island. The chronological documentation in lucid language is factual and brings out all the salient features.

This prime publication will definitely go a long way in educating the people around. This publication in the days ahead may become a milestone for adventure tourism in Andaman & Nicobar Islands. The chapter on 'Threat' based upon geological facts deserve due attention of our planners and administrators especially those dealing with Disaster Management.

I congratulate both the authors and also the sponsor for this valuable contribution at a very timely juncture.

I wish them all success in all such fruitful endeavours ahead.

(Anand Prakash)

ZEESHAN GHANI
Director
Andaman Sea Game Fishing.com
Foreshore Road, Haddo
Port Blair, A&N Islands - 744102.

I feel great to be a part of this publication as a sponsor. The author Mr. Arif a renowned fisheries expert who has authored number of publications should be highly appreciated for his work on barren island. He is the only person in India who has done research to the hilt on barren island with maximum number of landings on it.

India is fortunate to have everything from Deserts to Snowfall, Mountains, Sea, Islands and the most rare among all is a live Volcano. We should be on the Top 10 country as a Travel Destination.

I am happy to say that my company has come out with an economically priced luxury cruise fishing tour to Barren Island where one gets the opportunity to see the active volcano.

I wish Mr. Arif to keep the good work going.

The human race has named everything from a sub atomic particle to distant galaxies, but somehow the civilized world forgot to name the volcano at Barren Island and has always associated it with the host island, and its therefore commonly known as the "Barren Island Volcano".

Let us rechristen it as *Puluga laka bang (PL)* meaning *'Creator his mouth'* as it was called by the original inhabitants of these islands, the Great Andamanese[1].

Volcano / Puluga laka bang at Barren island - 1960

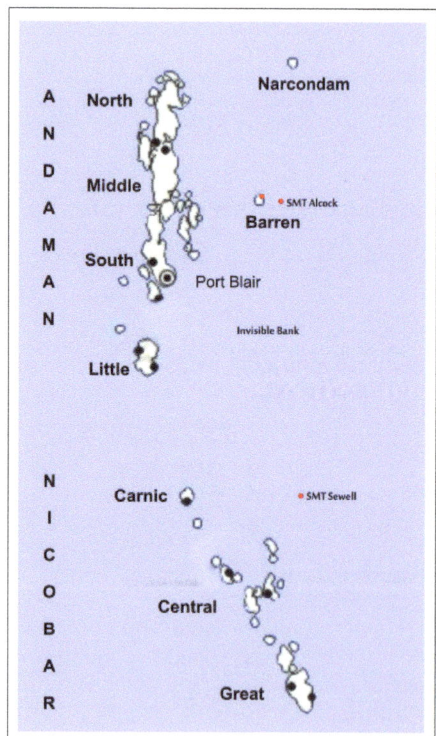

Whereas the name of extinct volcano Narcondum/*'Narkandam'* appears to be a corrupt version of Sanskrit word *'Narak'* (Hell) because "*the Coromandal Brahmins used to say that the Rakshasas (Demons) had their abode on the island of Andaman lying on the route from Pulicut to Pegu*".... "*can it be that in old times, but still contemporary with Hindu navigation, this Volcano was active*"[1].

● *Reported sites of active volcanism in Andaman and Nicobar Islands*

1. *Man. E.H. 1883 - 'The Aboriginal Inhabitants of Andaman Islands'.*

Looking at the inferno in the crater of Barren Island, a strato /composite type volcano, *Puluga laka bang* reminds us of the geological hypothesis of volcanism for the past 4.6 billion years ago. Its remnants still rage beneath us.

Awakening during March 1991

A team of ten explorers on the forenoon of 15th May 1991 landed at Barren Island and went up the hill adjacent to volcano. The team comprised of two Geologists, three marine science scientists and others, alongwith Lt. Gen. (Rtd.) Ranjit Singh Dayal, PVSM. ADVC. Hon'ble Lt. Governor of A&N Islands. The team perched upon the peak of the adjacent 250 m. high hill. They observed from the seaward side wall of the blown away large cauldron located just one and a half kilometre away from the active volcano. The most noticeable feature was the monstrous lateral vent of the least studied, lone, active volcano of India.

Fuming & advancing lava river, 12m Signal tower also seen

Feeble fuming was seen in the crater, whereas the lateral vent was continuously ejecting basaltic lava that flowed down the dome and mixed with pumice to form a frothy lava blown aloft alongwith a swirling cloud of ash mixed with noxious gases. The volcano was active after a lull of about a hundred years.

Team members drenched in black rain

Fomer Hon'ble Lt. Governor Lt. Gen. (Retd.) Ranjit Singh Dayal, PVSM, ADVC, A&N Islands, Secretary to LG Shri Manoj Parida, IAS, Shri Vidya Sagar Pokriyal, ADC & Others

Advancing lava river on 15th May,1991.

Awakening was noticed during March1991 by the Indian Coast Guard. Oceanographic Research vessels of the Department of Ocean Development, reported an abrupt jump in Magnetometer reading and the National Remote Sensing Agency recorded about a kilometre long vertical column of smoke in the sky above Barren Island. A smoke cloud drifting south easterly was also noticed and an abnormal increase in temperature radiating from the heaps of solidified ejected material tephra was also reported.

A river of lava, about 200 m. wide, was formed by the time we landed. The island was trembling with intermittent rumbling boom. The volcano was spitting fire, ash and gas like an angry mythological dragon. The explosive volcano emitted sticky, thick magma and gases under pressure, that exploded when released. The sky was overcast and it suddenly started raining, but to our astonishment we were drenched in black rain. Minutes later we were losing hold on the ground - the rain had triggered the formation of a cement like slurry, by mixing with the ash blanket that was spread all over. This slurry formed all around began to descend in wave from all the heigh rises around us.

Meanwhile, gusts of monsoon-like winds increased in intensity, strong lightening paired with screaming thunder, and the smoke column was momentarily impregnated with hues of red, pink, indigo, blue and bright sparkles. Down the hill, the lava front was a mix of glowing cinders and molten rock that was engulfing vegetation. The awesome powers of nature were unleashed, and the environment I found myself in was definitely one of the most adventurous moment of my life.

Another unforgettable opportunity on the same amphitheatre came 12 years later, on 26th March 2003 when the volcano relaxed a little. This provided me the opportunity to walk over the rim of the widened crater with a team of foreign journalist including National Geographic. Among the 18 landings I have made at Barren Island, the memory of 15th May 1991 and 6th March 2003 encounter with mystic mood of nature will ever remain afresh.

Lava river expanding from sides

Ash deposit over southern vegetation about 20m in height

Exposed tips of the buried vegetation

Background

India has two volcanic Islands in the Bay of Bengal namely Barren Island volcano (*PL)*and Narcondum Island Volcano. The former awoke during 1991, while the latter continues to remain extinct. Geologically these two volcanoes belong to the general Sunda group lying on the Neogene Inner Volcanic area, and evolved as a result of Eastward subduction of the Indian Ocean lithosphere below the South East Asian plate.

The *Puluga laka bang* is known for resurgent eruptive volcanism and the recent activity of strombolian nature started during March 1991 at a time when Mt. Pinatubo of Phillipine and Mt. Unzen of Japan were also active. This provided an opportunity for the scientific community of India to study an active volcano of their own.

In addition to the two volcanoes mentioned above we also have two prominent submerged volcanic sea mounts sitting on the floor of Andaman sea basin, the Alcock SMT is located just 60 km. East of Barren Island and the Sewell SMT 200 km. East of Car Nicobar.

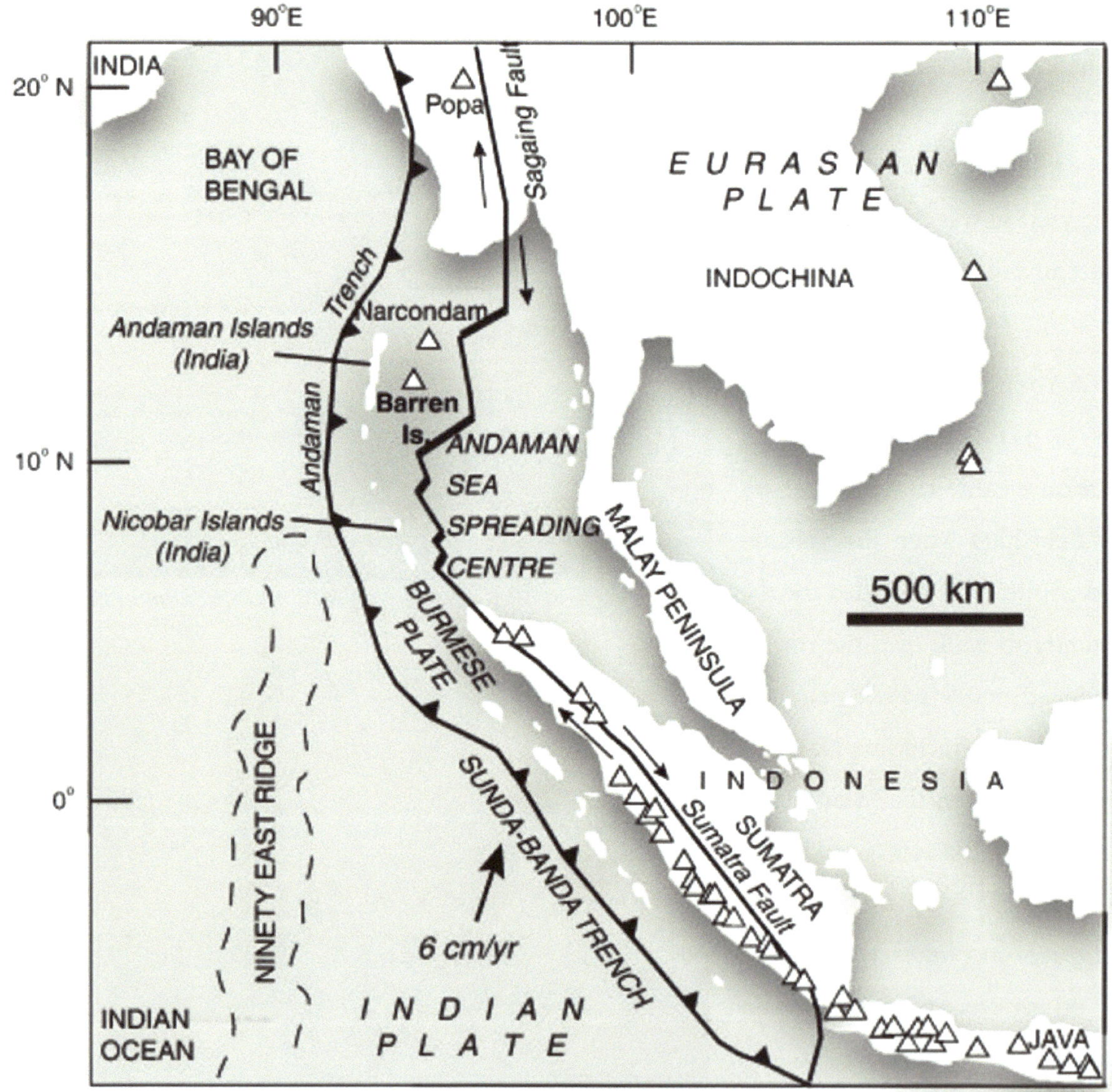

Resultant Geological shift

6 Location

Barren Island is located at 12° 17' 30" N latitude and 93° 50' 30" E longitude in the northern part of Andaman Sea in the Bay of Bengal. It is about 135 km North east of Port Blair.

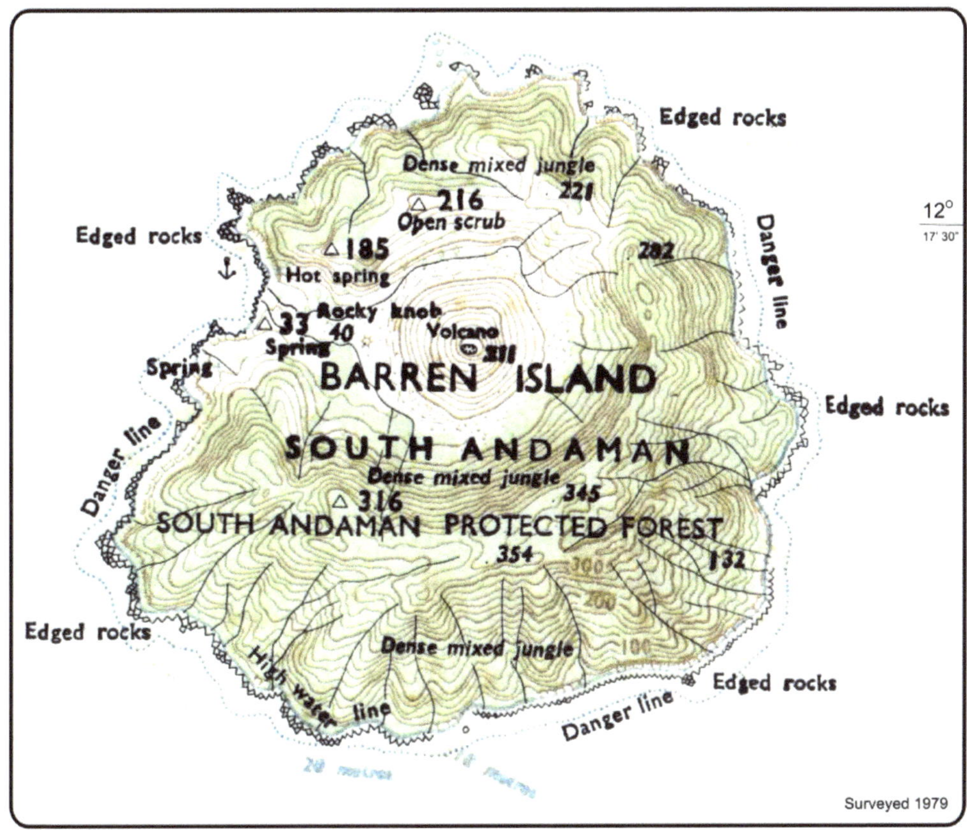

Altered topography 2013

The uninhabited Barren Island is almost circular in shape, about 3 km. in diameter, 10 sq km. in area, and slightly tapering towards West. It is quite steep with sloping surface rising from a depth of 2.2 km in sea to a peak at about 250 m. above mean sea level. It looks like a truncated cone from a distance. Due to its steep topography, any conspicuous rocky, sandy or muddy beach is absent. The volcano, *Puluga laka bang*, is located towards Northern tip of Barren Island.

Changing Topography

There are couples of theories about the evolution of this island. The most widely accepted one is that, initially an eruption took place in the sea about 1 to 2 million years back during late to post Pleistocene period which resulted in the emergence of this island. This phenomenon is considered to have occurred in the late Tertiary period and a giant volcanic cone developed that encompassed the whole island. Subsequently, this giant volcanic cone was blown out and the ejected debris and ash material was deposited on the relict cauldron as pyroclastic.

Thus *Puluga laka bang* is the resultant of Burma Micro plate subduction beneath Sunda plate. Argon dating confirms 42,000 years old ash deposit.

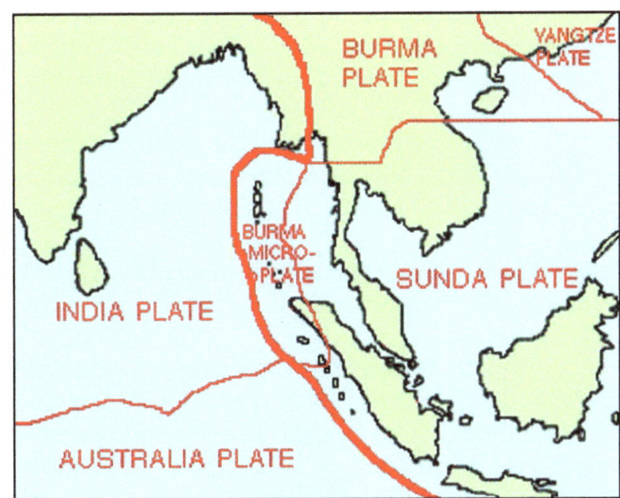

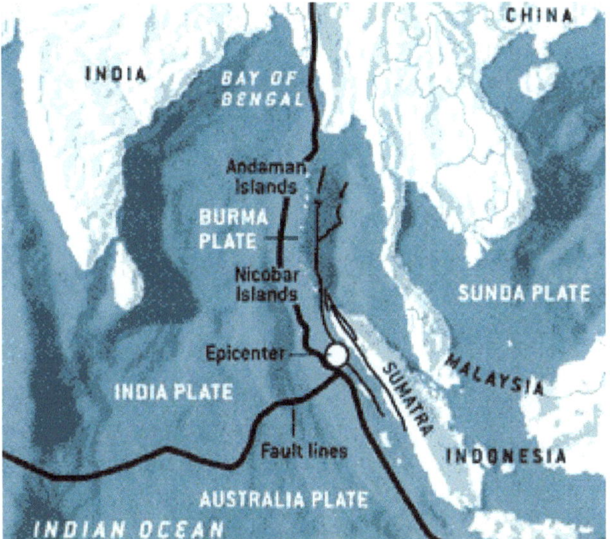

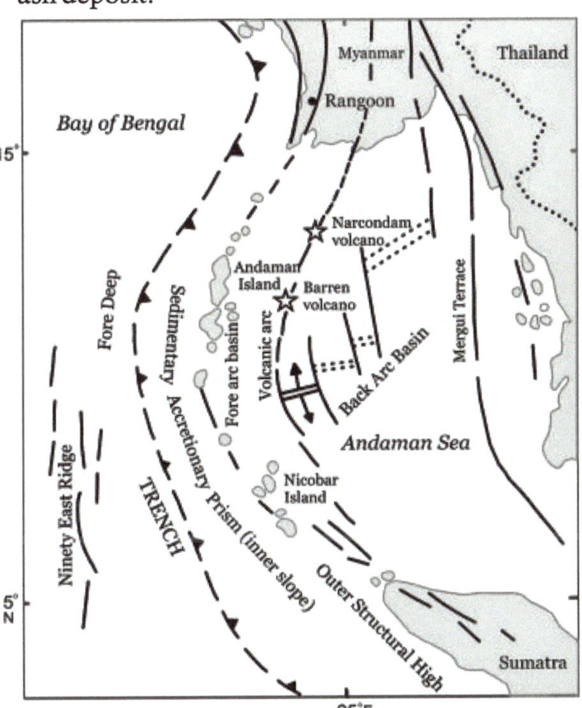

Maps showing continuing Geological turbulence in the area

Geo tectonically both the islands of Barren and Narcondumlie on the Miocene to recent volcanic arc, which proceeds from Mount Popa and Wuntho of Burma in the North and swerving South easterly into the active volcanoes of Indonesia and Sunda region.

It is believed that, during the geological time the existing volcanic cone was developed near the central part of the blown-off cauldron and lava pile of high-alumina olivine basalt erupted out from the main crater as well as from the three subsidiary vents that developed in the North eastern, North western and Southern faces. The lava coming from these vents and the then existing volcanic cone about 50m. down the main crater filled up the valley of the amphitheatre. The lava pile finally flowed down into the sea towards the Western side of the island.

The eruption of 1803 was on the relict part of the older volcano and deposited pyroclastic matter. The SiO_2content was 45 % to 53.5 % (basalt). The older deposited lava sediment is olivine basalt followed by high alumina olivine basalt whereas the present lava is olivine bearing basaltic andesite. Prior to 1991 activity, the circumference of the crater was about 60 m. at top and 25 m. at base, the sealed cup depression was 15 m. in depth. The surface temperature of the vent prior to 1991 eruption was approximately 100 °C.*(Dr. D. Haldar, Scientist, GSI, Personnel Communication)*.

The resurgent activity lasted for about nine months from March 1991 to November 1991.It considerably altered the topography. Reduction in the height of volcanic cone from 305 m. to 250 m. above mean sea level accompanied by an enlargement of the crater was most prominent. The earlier landing site is totally engulfed by advancing lava and at that point a 30m. high lava bed formed in the sea, subsequently a new crescent cove about 60 m. long emerged some 200 m. north of previous landing site during 2005 which was ultimately occupied by lava river during 2010..

'aa' type of lava

Recorded eruptive activity of *Puluga laka bang*	
....	1787
....	1789
....	1795
....	1803
....	1857
....	1890
March	1991
December	1994
January	1995
May	2005
April	2006
January	2009
September	2010
February	2013
October	2013

Landing with the help of Coast Guard

According to the records available, during 1795 Capt. Blair observed enormous volumes of smoke and frequent showers of red-hot stones. "*Somewhere of a size to weigh three or four tons, and had been thrown some hundred yards past the foot of the cone. There were two or three eruptions while we were close to it; several of the red-hot stones rolled down the sides of the cone and bounded a considerably way beyond us .... those parts of the island that are distant from the volcano are thinly covered with withered shrubs and blasted trees*".

Collapse of main crater during June 1991

Ash wrapped coastal terrain

Buried vegetation

Un-oxidized tephra

A few years later Horsburgh records an explosion every ten minutes and a fire of considerable extent burning on the Eastern side of the crater. In the next thirty years subterranean forces had considerably diminished in activity and at the end of that period only volumes of white smokes with no flames were to be seen. Drs. Mouat and Liebig, who visited the island within few months of each other in 1857 wrote respectively of volumes of dark smoke and clouds of hot watery vapour. In 1866, a whitish vapour was emitted from several deep fissures and during 1890 steam was seen to be gushing from a sulphur-bed which was liquid and pasty and a new jet was coming from a lump on the sloping side of the cone; while the sole evidence of activity to be observed was the deposition of Sulphur and escape of steam that often condensed on the surface rocks. There is no authentic record available thereafter, till 1991.

After hearing the news of resurgence of *Puluga laka bang* an unsuccessful attempt to land at Barren Island was made on 11th May 1991, a long streak of volcanic smoke became visible few minutes after crossing Inglish Island, the smoke was drifting towards the south east. The team anchored in the vicinity of Barren Island by 5:00 p.m., opposite the 12 m. high signal tower and moved landward in a small boat. About 100 m. away from the shore intense showers of volcanic glass / dust / ash particles started, which scared some of the team members, and the party returned without landing. The dormant volcano was actively spitting lava from an auxiliary vent located in the Northern side about 50 m. below the main crater. The discharge was pulsating with a rumbling noise and ejected boulders were rolling down the slope.

The smoke was thick and dark, with white patches at regular interval. The smoke column was about 900 m. high from the top of the volcano. The surrounding forest was burning, and there was a charred smell accompanied by dancing sparks throughout.

A second attempt by the team, comprising of 10 persons, left Port Blair by M.V. Tarmugli on 15th May 1991 at 6:00 p.m. The sea was calm with a little swell throughout. Visibility was poor at dusk. The vessel anchored at 6.10 am. about 500 m. away from the lone light beacon on the North Western coast at a depth of 800 m. On the left, the island terrain was almost vertical, rising straight up from the sea. The landing point on the right side, next to the 12 m. high navigational light beacon, was a small sandy patch. On the other side the sloping terrain was highly uneven rising upto 250 m. Suddenly a Sperm whale (*Physeter catodon?*) appeared in the black water and greeted the party with a fascinating jet spray and disappeared.

Participants of 15th May 1991 expedition		
1	Hon'ble Lt. Governor Andaman & Nicobar Island.	Lt. Genl. Ranjit Singh Dayal, PVSM. ADVC (Retd.)
2	Secretary to Hon'ble Lt. Governor	Shri Manoj Parida, IAS
3	Director of Shipping Services.	Cdr. T.R. Dewan, IN
4	Incharge, Andaman Division of Geological Survey of India.	Dr. D.R. Haldar.
5	Incharge Andaman & Nicobar Center for Ocean Development	Shri Arif Mohd. Mustafa.
6	ADC to Hon'ble Lt. Governor.	Shri Vidya Sagar Pokriyal
7	Asst. Geologist from GSI.	Shri J.K. Biswas.
8	Zoological Asst. from Zoological Survey of India.	Shri P.T. Rajan.
9	Research Scholar from Oxford University.	Ms.Zohra Fatima.
10	Diving Assistant from ANCOD.	Shri Raju

Fissure exhuming noxious gases

Driblet cone formation

Weathering of un-oxidized ejected rock

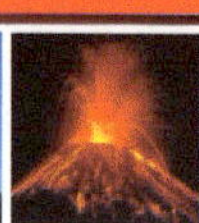

Marginal green colour is nickel and red is iron

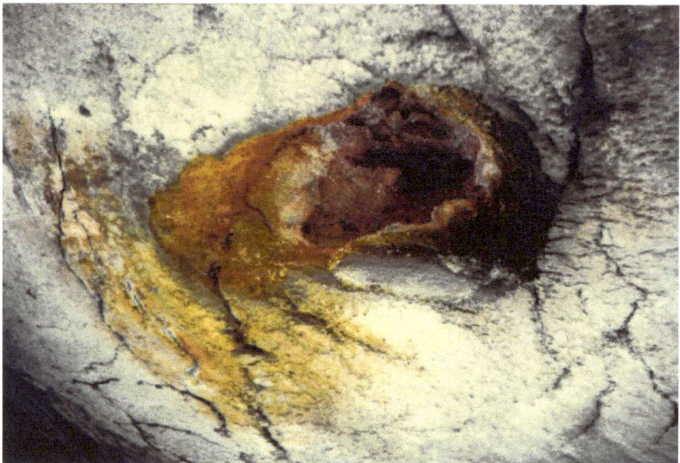
Heated sulphur emitting vent

Fumaroles

The volcano that the team ultimately landed on was dangerously violent and continuously throwing out glowing boulders and dense fumes with intermittent rumbling. The surrounding vegetation was partially charred. The entire area was buried under thick volcanic ash. The edge of the advancing super-heated lava river was about 500 m. away from the sea shore. The lava was of 'aa', 'ropy' and 'scoriaceous' type. The lava front was amazing, slowly advancing, producing cracking noise, fragmenting solidified lava pieces, fuming, releasing sparklets expanding laterally, radiating heat, at times unbearable, and had a red hot glowing interior. Amidst this catastrophic scenario, a weak and trembling cry of a baby goat *(Capra hircus)* attracted the attention of the team. The starving goat was rescued and sent onboard the vessel. To get a better view, we went up the charred hill on the left side and climbed to a height of 250 m.

Sea surface slick marks indicating the sub-surface drift were fast changing position, appearing and disappearing in quick succession. Strong coastal drift was felt moving from sandy patch and beyond towards the vertical land mass further North. A whirlpool fed by coastal conventional current was also seen churning near the landing beach.

On the sandy patch, the fine sand deposit was more than 60 cm. The top layer was thoroughly mixed and stirred with the black dust cinder giving it an unusual blackish appearance. The submerged area was totally covered upto about 18 cm. with loose lapilli

/dust cinder / ash particles frequently raining therein. Average dust cinder / volcanic glass particle size was 3 mm. The particles were found to be sensitive to magnetism.

The coastal salinity ranged from 34 to 36 ppm. whereas 32 ppm. is the average for Andaman Sea. The specific gravity was in the range of 1.022 to 1.027 and the pH value was 6 to 6.2. The pH of regular sea water is relatively constant around 8.0, due to the buffering action of the carbonic acid system in the photic zone. The drop in pH value is attributed to the high sulphur emission from volcano which disturbed the coastal ionic balance.

Contrary to name the island is fairly rich in flora and fauna. On 15th May 1991 we found a few live animals such as feral goat, the Rock skipper fish *Andamia sp.* rock mussel *Crassostrea sp.* and *Saccostrea sp.* the shell *Cellana sp.* and *Nerita sp.* Alongwith *Chiton sp.* and *Patella sp.*Large number of dead shore crab *Pelocarcinus humei* of family *Geocarcinidae* were found all along the coast. Live crabs were seen vacating the western shore, moving up the ash covered mountain range and escaping towards south eastern side. In order to observe this unusual phenomenon of landward migration, we followed the migrating crabs uphill, and found that they had reached upto 225 m. than descending eastward through forest cover and entering the sea. The coral community was restricted to a small area comprising of 9 species belonging to 5 genera namely *Acroporidae, Pocilloporidae, Favites, Favia* and *Porites*.

Fumaroles

Author and the German volcanologist Prof. Dr. Ing. Peter Halbach

Condensed lava river

Driblet cone over condensed lava river

On 26th June, 1991 landing with the support of the Coast Guard, the volcano was dangerously eruptive. The 12 m. tall light beacon was completely engulfed by the advancing lava river, while the sea-lava interface was producing enormous amounts of steam. The synergistic impact of steam from the sea and fiery fountains from the crater were horrifying. Main crater collapsed to the level of lateral vent, considerably enlarging it's circumference.

Sulphur matrix

On 10.7.1991 landed with support from Coast Guard; volcanism appeared to be at its climax.

On 1.11.1991 landed with support from A&N Police along with a well known German volcanologist Prof. Dr. Ing. Peter Halbach & Dr. Margret Halbach. The volcano was silent, coastal topography totally changed, 20 m. heigh lava front was found much ahead of the earlier landing point, and at the sea-lava interface, the water temperature was 72 °C & pH 6. Weathering of the condensed magma was in progress. A new crescent shape landing site about 60 m. long developed some 200 m. ahead of previous landing point. Air temperature was 38 °C. Due to the intense heat and constant steaming and emission of sulphur dominated gases, we were not able to stay longer in the absence of breathing mask.

Author at the rim of crater

Changing mood of volcano

Subsidiary vent on the rim of crater

Inside the silent crater

Resurgence

Rumbling boom

Paired vent

Precipated sulphur

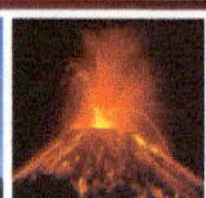

On 4th January, 1992 we landed with support from Coast Guard. The volcano was silent with the exception of intermittent gas emission, two 'driblet' lava cones were found developed on the solidified lava river, sulfataric activity continuing at places. Sea surface water temperature at interface was 29 °C & pH 7.6. Three Fumaroles were found developed on ash covered slope near the southern foot of *Puluga laka bang*. Fumaroles are small holes formed in the vicinity of an active volcano through which hot air is seldom expelled. A few, call it pulse of volcano because the temperature gradient and gas composition in fumaroles are the indicator of the status inside. Marine algae was found developing over certain rocks, no other marine or terrestrial life form sighted with the exception of the crab *Pelocacinus humei*. These crabs were in large numbers from the edge up to the foot of Volcano.

% Composition – 1994 Magma		
Si O_2	52.78	52.36
Al_2 O_2	20.69	21.42
Fe_2 O3	7.29	8.78
P_2 O_5	0.17	0.37
Ca O	8.12	8.45
MgO	3.02	
K_2 O	0.51	0.81
Na_2 O	5.15	5.27
LOl	1.52	1.87
%o Important Elements		
Gold	Au.	< 0.02
Palladium	Pd.	0.51
Platinum	Pt.	< 0.02
Rhodium	Rh.	< 0.02
Source :Gimpex Ltd., Chennai		

Expanding island

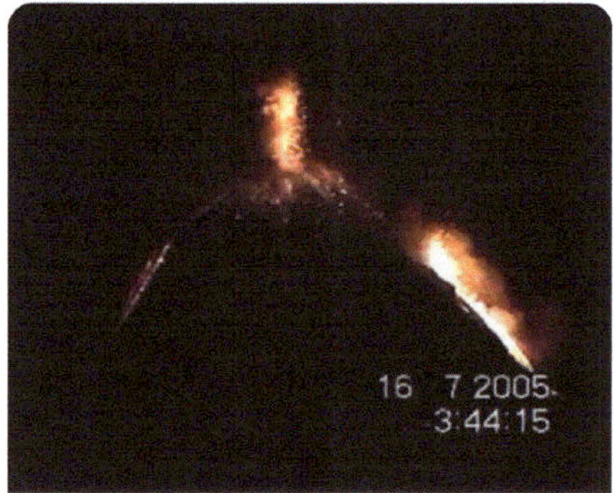

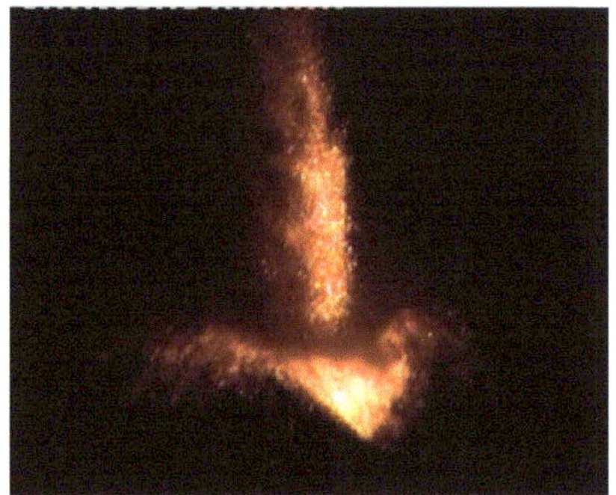
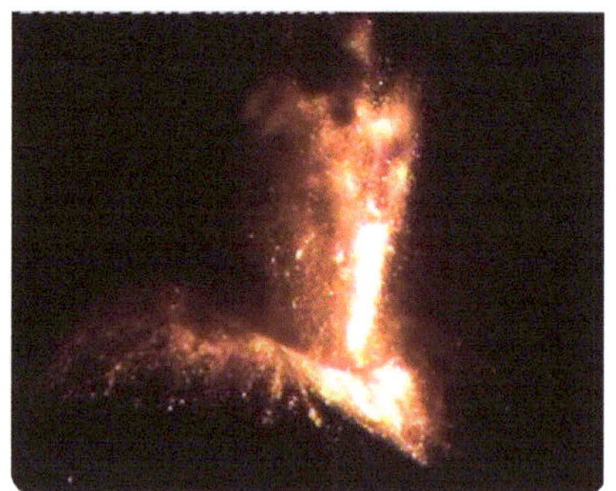
Night vision

There are about 550 volcanoes in the World today, they are generally found where tectonic plates are diverging or converging; rarely due to stretching or thinning of Earth's Crust or rises from near the core mantle boundary 3500 km. deep. Radioactive decay of elements keep the inner core of our planet heated up to 5,000°C, and this temperature gradually drops to 4,000°C at outer core, and then 3,000°C & 1,500°C in inner mantle and upper mantle, respectively. The mantle is finally wrapped in 100 to 170 km. thick crust; the upper layer is 8 km deep and the lower 96 to 161 km. This crust is divided into sections called Tectonic plates.

The inner heat moves the upper tectonic plates slowly but continuously on which rest the continents. At places ascending heat currents force gas-charged molten fluid through the crust as lava.

Volcanoes rejuvenate the soil through replenishment of nutrients directly or indirectly as circulating blood does for us.

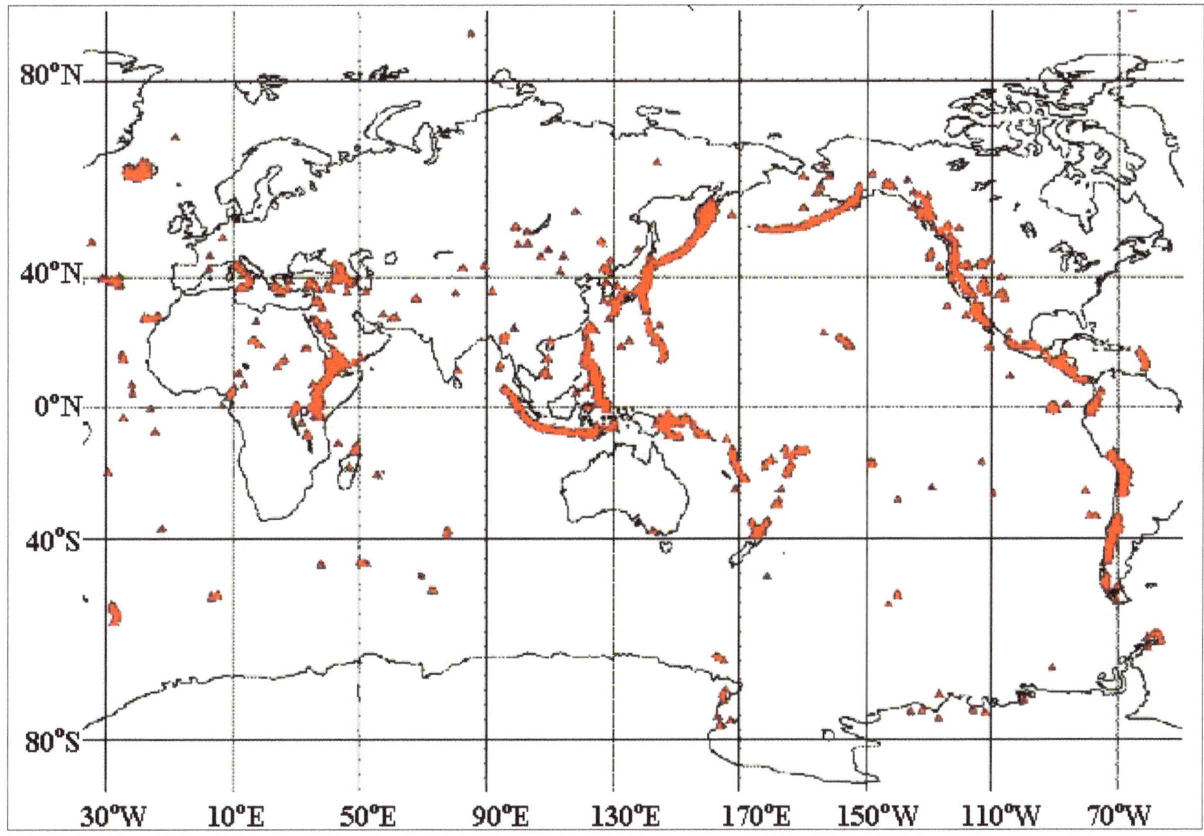

Map showing distribution of valcanoes

Fortunately *Puluga laka bang* is isolated, and thus there is no immediate threat to the civilization around. However, one has to be extremely cautious while landing at Barren Island. On 1st November, 1991 Inspector Pabla of A&N Police stepped down on the sea floor from small motor boat despite warnings from German volcanologists. The water was around 1.20 m deep. The Inspector instantly cried, jumped and hung on to the boat dangerously tilting it. The floor temperature upto 2 m. depth was recorded to be 80°C while the volcano was apparently in silence. On 4th January 1992 Shri Raju the skin diver of ANCOD fainted in the vicinity of a lava river and was instantly moved away. He revived within 3-4 minutes but developed acute cough associated with burning sensation in trachea and lung due to inhalation of noxious gasses. On 26th March 2003 shoe sole of Shri Rustum of IP&T Department melted while climbing up the volcano; he for a moment inadvertently had stepped on a sulphur matrix.

Super heated cloud of ash and noxious gasses such as Sulphur dioxide, Hydrogen chloride and Carbon dioxide could kill all living creatures on land and in the sky, on mixing with moisture, they form an aerosol cloud containing Sulphuric acid. This aerosol cloud deflects sunlight and causes acid rain. However, dry and thin Sulphur dioxide layer may stay in stratosphere for five years. Volcanic clouds also impair jet engines, which is an aviation hazard.

What else? The worst thing that could be imagined would be the collapse of aged *Puluga laka bang* in line with what happened in the small island of Krakatau in Sunda Strait between Java and Sumatra. The volcano walls collapsed allowing sea to stream in; creating a water, steam and super heated magma mix. This pressure bomb exploded at 10:02 AM on 27th August 1883 producing the loudest explosion ever heard in modern time. The sound of the explosion was heard about 4,800 km away, a 37 m. high Tsunami formed, wiping 300 towns and killing 36,000 people.

Smoke / ash darkened the sky for 450 km. In the United States, the fire brigade was called to extinguish a fire approaching from the eastern horizon which was glowing as if on fire.

Similarly volcanic cone of Mt. St. Helens in Washington State collapsed during 1980, the most destructive behaviour of a volcano.

Let us pray our *Puluga laka bang* may remain intact as it is.

Aa Lava with blocky structure *(Hawaiian)*.

Andesite The term used for a fine-grained intermediate volcanic igneous rock which gets characterized by the presence of oligoclase of andesine. Porphyritic varieties are almost common, both ferromagnesian minerals and feldspars occur as phenocrysts the latter commonly showing zoning. Hypersthene and enstatite have been found to be more common in andesites than in diorites. Pure glassy andesites are rare, but glass is of wide occurrence in the groundmass of andesites, commonly in a devitrified state.

Basalt The term used for a crystalline, sometimes glassy, basic igneous rock of volcanic origin. The essential minerals include a calcic plagioclase and pyroxene (usually augite), with or without olivine. Magnetite forms an important accessory, where as quartz, hornblende, and hypersthene sometimes occur in significant amounts. Glassy basalt is known as tachylite; basalt glass having olivine and augite phenocrysts is known as limburgite, but analysis reveals that limburgites and commonly undersaturated. Basalts are having low SiO_2 content (45 50%), tholeities having the higher silica content. FeO, MgO, and CaO are generally high, whereas Na_2O and K_2O are low, especially in the olivine basalts.

Driblet Cones Small cones which are formed by lavas which are very viscous and usually acidic.

Fumarole A hole which is usually found in volcanic areas, from which volatile vapours or gases escape out.

Lapilli Fragments of pyroclastic rock which measure between 4 and 32 mm in diameter.

Lava The material which is extruded by a volcano. It consists of molten or partially molten silicate material and reaches the earth's surface through volcanic vents and fissures. Their composition ranges from acidic to ultrabasic, although basic lavas account for more than 90% of the whole. Acid lavas have been highly viscous and rarely cover very large areas, while basic lavas have a much lower viscosity and flow readily.

Lithosphere Refers to the outer part of the earth's crust which has a high strength as compared with the asthenosphere. It consists of the surface recks, the sial and the upper part of the sima.

Magma A naturally occurring molten rock material which is formed within the crust or upper mantle of the earth and may consolidate to form an igneous rock. Magna is having a complex system of molten silicates with water and other gaseous material in solution.

Olivine $(Mg,Fe_2)SiO_4$. A neositicate group of olive-green or brown mineral. It has the isomorphous solid-sotution series forsterite-fayatite. Olivine is a common rock forming mineral of basic, ultrabasic, and low-silica, igneous rocks (gabbro, basalt, periodotite, dunite). It gets crystallized early from a magma, weathers readily at earth's surface, and gets metamorphosed to serpentine. It is having vitreous luster and its hardness on Mohs scale has been 3.27-3.37. The essential structure involves a series of isolated SiO_4 tetrahedra that are linked by means of metal cations. The olivines are orthorhombic, show no cleavage and display conchoidal fracture. Many names are assigned to the intermediate compounds between these pure end members:-

Forsterite	0 10% Fa
Chrysolite	10 30% Fa
Hyalosiderite	30 50% Fa
Hortonolite	50 70% Fa
Ferrohortonolite	70 90% Fa
Fayalite	90 100% Fa

Pumice A light coloured cellular glassy rock which is commonly having the composition of rhyolite. It is usually sufficiently buoyant to float on water. It is economically useful as light-weight aggregate and as an abrasive.

Pyroclastic Rocks This type of rocks contain fragmental volcanic material which is blown into the atmosphere because of explosive activity. They are generally formed from volcanoes whose lava is of a more viscous type. The coarsest types of pyroclastic material, the agglomerates or volcanic breccias, generally occur in very close proximity tow or actually within, the volcanic tent.

Ropey lava It is a synonym of pahoehoe.

Scoriaceous A term used for describing a lava or pyroclastic rock having empty cavities, or amygdales.

Strombolian Eruption A type of eruption which has been characterized by jetting of clots or fountains of fluid basaltic lava from a central crater.

Sulphur A non-metallic native element, which occurs in areas of recent volcanic activity and around hot springs and in sedimentary rocks having gypsum and limestone. It is found as orthorhombic crystals.

Tectonic Refers to an adjective which is used to relate a particular phenomenon to a structural or orogenic concept, e.g., 'tectonic control of sedimentation' means that the process of sedimentation has been controlled by orogenic activity; a ' tectonic map' implies a map which is designed to demonstrate structural features rather than stratigraphical or lithological ones.

Tectonics Refers to the study of the major structural features of the earth's crust of the broad structure of a region.

Tephra A term used for all fragmental volcanic products which get ejected through the vent, e.g., ash, cinders, lapilli, scoriae, pumice bombs etc.

Vent Refers to the opening through which a volcano ejects igneous material.

Volcanic Refers to one of the three groups into which rocks have been divided. The volcanic assemblage are including all extrusive ricks and the associated intrusive ones. The group has been dominantly basic strictly magmatic, and usually but necessarily associated with orogeny.

Volcanism The term used for the movement of magma and its associated gases from the interior into the crust and to the surface of the earth. It is also known as volcanicity.

Volcano Refers to a vent or fissure in the earth's crust through which molten magma, hot gases, and other fluids tend to escape to the surface of the land or, in certain cases, the bottom of the sea. Volcanoes have been classified broadly into central types where the products are escaping via a single pipe (vent) and fissure types, where the products are escaping from a linear vent or rock.

Many volcanoes get developed parasitic or daughter cones on the flanks of the main mountain, and these may become so extensive that the main crater has become extinct; this has actually happened with Etna, which is probably the wreck of a shield or Hawaiian-type volcano.

Know the Authors

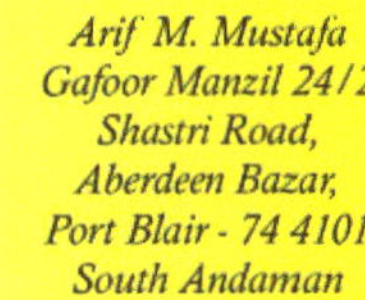

Arif M. Mustafa
Gafoor Manzil 24/2
Shastri Road,
Aberdeen Bazar,
Port Blair - 74 4101
South Andaman

An environmental friendly Fisheries professional, author of three popular books and many scientific publications, representing second progeny of those Indians who were exiled to Andamans from mainland India during British Raj.

Endowed with the instinct to scientifically explore the natural heritage around A&N Islands.

Sponsor

Altamash Mustafa,
B.Tech. Civil; M.Tech. Env. Civil; MBA
Gafoor Manzil 24/2
Shastri Road,
Aberdeen Bazar,
Port Blair - 74 4101

Young energetic, environment conscious, enthusiastic and adventurous Engineer. Substantially contributed in post Tsunami power restoration and disaster management. An orator good enough, to ignite the care for environmental protection among his students.

Sponsor

Sport Fishing in Andaman Islands

This publication is dedicated to **Er. Yameen Md. Murtaza,** *B.E.E. Hons.,* Engineer Emeritus formerly Superintending Engineer, Electricity Department, A&N Islands to acknowledge his endeavoring logistic support towards publications made by first author since 1980.

Sport Fishing

A boon for Tourism
Attractive to Economy
Adventurer for Fisheries
Asset for Environment
Additive to Coastal Security

Mike

Mike's Fishing Adventures

is based at Havelock Island in the Andaman Islands, where the pristine waters surrounding us hold record breaking fish. Our itineraries range from half day trips, day trips, multiple day packages and bespoke exploration excursions.

With so many near virgin reef systems you are likely to hook in to some real monsters. Be prepared for a fishing adventure that will certainly put you and your gear to the ultimate test.

Mike has over 17 years of experience fishing tropical waters and has developed a 6th sense in fish location and will put you in the best possible areas. He came to the Andamans for the first time in 2008, fell in love with the pure beauty of the islands and the amazing fishing.

David began fishing canals and ponds during his formative years and soon graduated to spinning for golden masher. While posted at Chennai he became an avid surfcasting angler but soon began venturing offshore. A 2009 visit to the Andamans ended up in him relocating.

David

Beach No.5, Havelock Island, Andaman and Nicobar Islands - 744211
Email : drive@barefootindia.com or visit : www.mikefishing.com
Phone : +91-9474263120
Join us on www.facebook.com/MikesFishingAdventures

Fishing Rates & Packages

1/2 day's fishing	: Rs. 15,000 per trip
1-2 day's fishing	: Rs. 26.000 per day
3-4 day's fishing	: Rs. 24.750 per day
5-6 day's fishing	: Rs. 24.000 per day
7+ day's fishing	: Rs. 23.500 per day

Equipment Hire Rates

Trolling gear is free of charge	
Jigging gear	: Rs. 2,000 per day
Popping gear	: Rs. 3,000 per day

SURMAI
fishing club
Le Rover 230, un stickbait qui a intéressé tous les poissons, dont des carangues à points bleus.
SURMAI FISHING PACKAGES
PACKAGES ONE: 2 or 3 anglers
With SURAMI I – open 28 feet powered with twin Yamaha 4 stroke 100 hp.
7 days of fishing
1150 € per angler if 3 anglers share the boat.
1650 € per angler if 2 anglers share the boat.
Sponsor
PACKAGES TWO: 4 or 5 anglers
With SURAMI II – open catamaran 32 feet powered with twin Yamaha 4 stroke 200 hp.
7 days of fishing
1200 € per angler if 5 anglers shares the catamaran
1450 € per angler if 4 anglers share the catamaran
Package Include:
• 7 Days of fishing (8hours a day)
• Transfers to guest house or hotel to jetty (up and down)
• Assistance of one skipper plus one sail fisherman helper on board
• Midday pack lunch with 2 bottles mineral water / per angler
Package does not Include:
• Air fair to Andaman Islands.
• Hotel, Personal and Food expenses (except the midday meal on board).
FOR BOOKING
Contact : 91-9434269839, 91-9734411668
Email : surmaifishing@hotmail.com
contact@andamanadventurefishing.com
Surmai Fishing Club Pvt. Ltd
www.andamanadventurefishing.com

WWW.ISLANDGAMEFISHING.COM
ISLAND FISHING
Sponsor
Beach #1, Havelock Island, Andaman and Nicobar Islands, India
Phone: +91 947426381 | Email Id: biswajit@islandgamefishing.com
Website: www.islandgamefishing.com
Fishing Packages & Rates
1/2 Day's Fishing Rs. 13,000/- Per Trip
1-2 Day's Fishing Rs. 25,000/- Per Day
3-4 Day's Fishing Rs. 23,500/- Per Day
5-6 Day's Fishing Rs. 22,000/- Per Day
We offer charter services like Sport Fishing, Boat Charter, Snorkeling and Lodging in the Ritchie's Archipelago based from beautiful Havelock Island. Our boat is equipped with all equipments like Fish-finder, GPS, and VHF along with lifejackets, also equipments for trolling, popping and jigging. All our staffs are from Andaman Islands who have very good knowledge of the islands and the ways of the sea.

Sport Fishing in Andaman Islands

First Edition - 2013

Authors

Arif M. Mustafa
Asst Director of Fisheries
Department of Fisheries,
A&N Administration,
Mobile - 09434261302

Akshay Malavi
Sport Fishing Expert
Director
Sea Fishing India Pvt. Ltd.
Mobile - 09933235622

Acknowledged

Type set
Ms. **Mohini Kumari**
South Point, Port Blair.

Layout & Graphics
Mr. **CHS Srinivas Rao**
RGT Road, Port Blair.

Photographs
(1) M/s Sea Fishing India Pvt. Ltd.
(2) Google Images

Index

Sl.No		Topic		Page
1	—	**Index**	—	02
2	—	**Message**		
2.1	—	Chief Secretary, A&N Administration	—	03
2.2	—	Secretary (Fisheries), A&N Administration	—	04
2.3	—	Head, A&N Environmental Team	—	05
3	—	**Introduction**	—	06
4	—	**History**	—	09
4.1	—	Land Based Recreational Fishing	—	10
4.2	—	Salt Water Big Game Fishing	—	10
5	—	**Important Sport Fishes of Andamans**	—	11
5.1	—	Grouper	—	11
5.2	—	Snapper	—	11
5.3	—	Trevally, Jack, etc.	—	14
5.4	—	Sail & Sword Fish, etc.	—	17
6	—	**Technique**	—	21
6.1	—	Trolling	—	21
6.2	—	Popping	—	24
6.3	—	Jigging	—	25
6.4	—	Fly Fishing	—	26
6.5	—	Bait Fishing	—	27
7	—	**Equipment**	—	28
7.1	—	Rod	—	28
7.2	—	Reel	—	29
7.3	—	Fishing Line	—	29
7.4	—	Hook	—	30
7.5	—	Gaff	—	32
7.6	—	Artificial Bait & Natural Bait	—	32
8	—	**Types of Sport Fishing Vessels**	—	34
8.1	—	Center Console FPR Boat. 5.5 – 7.3 m.	—	34
8.2	—	Center Console FPR Boat. 8.5 – 12.1 m.	—	34
8.3	—	Metal Live-aboard Boat. 12.1 – 18.3 m.	—	35
9	—	**Ethics of Sport Fishing**	—	36
9.1	—	Deep Sea Fish Handling and Release	—	37
10	—	**Recommendations**	—	39

2.1 Message

आनन्द प्रकाश
Anand Prakash
मुख्य सचिव
Chief Secretary

फोन / Ph. No. 03192-233110/234087
फैक्स सं. / Fax No. 03192-232656
ई–मेइल / E-mail : cs-andaman@nic.in

अण्डमान तथा निकोबार प्रशासन
ANDAMAN AND NICOBAR ADMINISTRATION
सचिवालय / SECRETARIAT

Port Blair, dated the 11th December 2012.

I am delighted to pen the message for a book which perhaps is the prime publication on the subject in the country.

Fisheries and Tourism are two equally important core sectors identified for development in A&N Islands, Sport fishing offer a new horizon combining and complementing each other, sustainably.

Sport fishing keeps the phobia of depletion away and liberates the state from the worry of resources replenishment, which is a major Global concern.

This emerging amalgamated aspect of Fisheries deserves logical consideration by all those involved in the development of this territory.

I deeply congratulate both the authors and all the sponsors for this valuable initiative, effort and contribution in introducing us all to a new chapter.

(Anand Prakash)

2.2 Message

एस.एन.मिश्रा
S.N.MISRA
सचिव (मत्स्य)
Secretary (Fisheries)

फोन / Ph. No. 03192-233205 (O)
फैक्स सं. / Fax No. 03192-232479 (F)
ई–मेइल / E-mail : devcom@and.nic.in

अण्डमान तथा निकोबार प्रशासन
ANDAMAN AND NICOBAR ADMINISTRATION
सचिवालय / SECRETARIAT

Port Blair, dated the 20th December 2012

I am impressed to see such a publication on **Sport Fishing in Andaman Islands**, highlighting various aspects of this sport.

It is a delight to go through the presentations in the book, and I am sure that this will go a long way in popularizing Sport Fishing in the Andaman Islands, thereby contributing greatly to tourism also.

I congratulate the authors and the sponsors as well on their work and extend my best wishes to them for their future endeavors.

(S.N.Misra)

For bibliographic purpose this document may please be quoted as follows :-
***Mustafa A.M; Malavi Akshay** - 2013 Sport Fishing in Andaman Islands*

Game Fishing
at
Havelock Island
SPORTS FISHING
HAVELOCK * ANDAMAN INDIA
Captain hook's
Pop Jigg Troll
Contact Qutub : +91 9434280543/ 9933210052
www.AndamanSportsFishing.com

Sponsor
GAMEFISHINGINDIA.COM
PREDATOR FISHING
Sea Fishing India Pvt. Ltd.
2nd Floor, 03 Foreshore Road,
Haddo P.O.
Port Blair
Contact: +91 94342 80117
+91 99332 35622
+91 3192 241610
gamefishingindia@gmail.com
www.gamefishingindia.com

Sponsor

Tasneem Khan
Assistant Director
Andaman and Nicobar Islands Environmental Team
Centre for Island Ecology
North Wandoor
(A division of the Madras Crocodile Bank Trust)

There is much to be learned with regard to the fish and fisheries in the waters of the Andaman Islands. Global fish stock are considered to be declining rapidly and it is indeed a refreshing approach that the fisheries sector is now looking beyond, solely the exploitation of marine resources. Sport fishing is essentially a recreational activity that involves, catch and safe release of fish. Sport fishing is gaining increasing popularity in the waters of the Andaman Islands, an island group where boat fishing and tourism are critical aspects of the economy. Sport fishing has the scope of function in favour of both revenue generation as well as to promote conservation. There is also tremendous scope for this activity, which generally caters to a niche market, to involve local communities as well.

Well regulated catch and release operations can result in a better understanding of seasonal patterns, keeping track of fish stock and an active involvements of tourists, local communities and researchers, in conservation. However, it is essential to promote large-scale sport fishing in India with prudent guidelines and monitoring mechanisms.

I am pleased to know that this publication is a collaborative initiative between a government professional and a private entrepreneur. Joint associations such as this can often have far-reaching and positive impacts and will hopefully set a good example for similar efforts within the island group.

3 Introduction

Art of fishing is as old as the human civilization itself this is an evolved ancient skill of primary production basically to harvest natural food - the Fish, one of the staple food, one of the richest of all sources of animal protein, however, with the advent of modernization this art of reaping transformed itself in an act of glamour, thrill and pleasure during the leisure without damaging the ecosystem and is termed 'Sport Fishing', less known in this part of the World.

'Sport' as defined in Webster's dictionary is "an activity or experience that gives enjoyment or recreation; part time; diversion".

Ethical protocol of Sport Fishing demands safe release of prey. In sport Fishing the Fish caught is released back to sea in live condition, it is only the thrill and pleasure drive out of it, a classical eco-friendly aspect of Fishing.

In India Sport Fishing has just begun in Andamans. Today in South Andaman we have 28 Registered Sport Fishing boats owned by 13 entrepreneurs out of which 18 are based in Ritchie archipelago (Havelock & Neil Island) and remaining 10 in Port Blair. 99% of Sport Fishing clients are foreign tourist. The pristine tropical sea around Andaman & Nicobar Islands fortunately still remains untouched by the ghost of depletion whereas in very many proximal and distal Ocean and Sea World over the shadow of depletion has crept in, however, our coastal as well as oceanic regime still abound with quality sport fish varieties sought by tourist from far and near.

These Islands are annually visited by about hundred or hundred and fifty thousand tourists a year out of which ten to twelve thousand are foreign nationals, rationally 5% of them indulge in Sport Fishing trips varying from a day to a week.

M/s Surmai Fishing Club, Dignabad, Port Blair is the pioneer of Sport Fishing in Andamans, the Company brought in the first specialized Sport Fishing vessel during 2001 for exploratory fishing and started a full-fledged commercial operation in 2006. Since then this aspect of fishing is developing at a rapid pace, conservative estimate indicate seasonal turn over of about ₹ 2.5 to 3 crore through five to six hundred customers. Andamans in the days ahead could be the star tourist spot of continental India for Sport Fishing. Unlike mainland India where there are millions of square kilometers that are suitable grounds for various fishing activities the Andaman and Nicobar chain offer limited areas suitable for viable exploitation. Diversifications of fishing methods and gear have evolved to be able to exploit the potential of these tropical Island ecosystems. Hand lines and vertical lines have been an effective method of harvesting fish. This peculiar natural underwater topography also makes these grounds a perfect setting for Sport Fishing. Most common and much sought after sport fishing species are demersal and semi pelagic. Also the virtually non existent continental shelf and relative close distance to the drop offs make these Islands an ideal location for targeting pelagic species of fish as well. Port Blair located on the Eastern coast of the South Andaman Island is the Capital of these beautiful tropical Islands and is the center of the local fishing industry. As the territory is experiencing rapid enhancement in growth, the Fisheries sector plays a pivotal role in exploitation of fisheries in the Indian EEZ. This Island territory is emerging as a major fish landing, processing and seafood exporting center, but now a new form of fisheries is being born. Sport Fishing still in its infancy if encouraged and guided has the capability of emerging as the fourth front of Fishing and Port Blair could well be the Sport Fishing Capital of the nation.

This Island territory stretches for about a thousand kilometers from North to south and has a vast oceanic regime of 6 lakh sq. km. From its highest peak to its deepest depth this territory exhibits unique Geological, Topographical, Meteorological and Oceanographic features. Wild and Intense volcanic activity is believed to have led to the creation of this archipelago, which is endowed with an extremely shrunken pseudo continental shelf of about 35,000 sq. km. which restricts the scope of benthic production. There are un-chartered depths around the islands some of which reach a depth of 4187m to the east of Car Nicobar Island. The territory experiences two monsoons namely the South-west and the North-east each year, with an average down pour of 3000 mm and experiences innumerable moderate to intense gales and cyclones which frequently wreck havoc in the upper Bay of Bengal and Arabian Sea. An above normal deep sea temperature due to submarine tectonic activity and innumerable thermal vents makes the Andaman Sea that lies to the east of this island territory Oligotrophic i.e. less productive in nature due to the natural limitations imposed upon it. However the Western sector is comparatively better due to upwelling. The insufficient food stock in the form of phyto and zoo plankton is to a large extent supplemented by detritus load brought in through associated ecosystems.

SPORT FISHING FLEET OF ANDAMANS *(January 2013)*		
	Port Blair	
Sl. No	***Name & address of Owner / Company***	***No. of Boats***
1	M/s Surmai Fishing Club, Mob - 09434269839	04
2	Shri Darren Davis, M/s Sea Fishing India Pvt. Ltd., Mob - 09434280117	01
	Shri Akshay Malavi, M/s Sea Fishing India Pvt. Ltd., Mob - 09933235622	01
3	Shri Anshul Jaiswal / Shri Shakir Ahmed M/s Andaman Marine Club Pvt. Ltd., Mob - 09933203707	01
4	Shri Biswajit Mondal, M/s Island Fishing, Mob - 09474286381	01
5	Shri Vincent C. Aluvilla, M/s New Rich Marine Pvt. Ltd., Mob - 09476046244	01
6	Shri Sumer Verma, M/s Nature Hotels & Resorts Pvt. Ltd., Mob - 09820419813	01
	Total	10
	Neil Island	
7	Shri Robin Sill, Ram Nagar, Mob - 09476013348	01
8	Shri Gurudas Das, Mob - 09434270454	01
	Total	02
	Havelock Island	
9	Shri Qutubuddin Taher, Captain Hook's Sport Fishing, Mob - 09434280543	04
10	Shri Deepak Govind, M/s Barefoot Resort, Mob - 09434280687	09
11	Shri Sushil Dixit, V3 Restaurant, Mob - 09434260666	01
12	M/s Andaman Bubbles Pvt. Ltd., Mob - 09476076387	01
13	Shri Bikash Samadder, Mob – 09531606021	01
	Total	16
	Grand Total	**28**

The richest of the auxiliary ecosystem in terms of productivity are the mangroves and coral reefs eco systems. Since time immemorial the shallow coastal marine areas have been exclusively exploited by the prehistoric Negrito aboriginal population of the Andaman and the Mongloid race of the Nicobar chain. Many of the shallow coastal areas in the Nicobar chain are still pristine and in the words of renowned marine naturalist Captain Jacques Cousteau after his visit in 1989 *"Indian waters are preserved, rich and virgin"*. As recent as 2002 the UNDP under the GEF programme had inferred that *"The A&N Archipelago hosts probably the healthiest and least impacted expanse of coral reefs within the Indian Ocean"*.

Fishing domain of Andaman and Nicobar Island

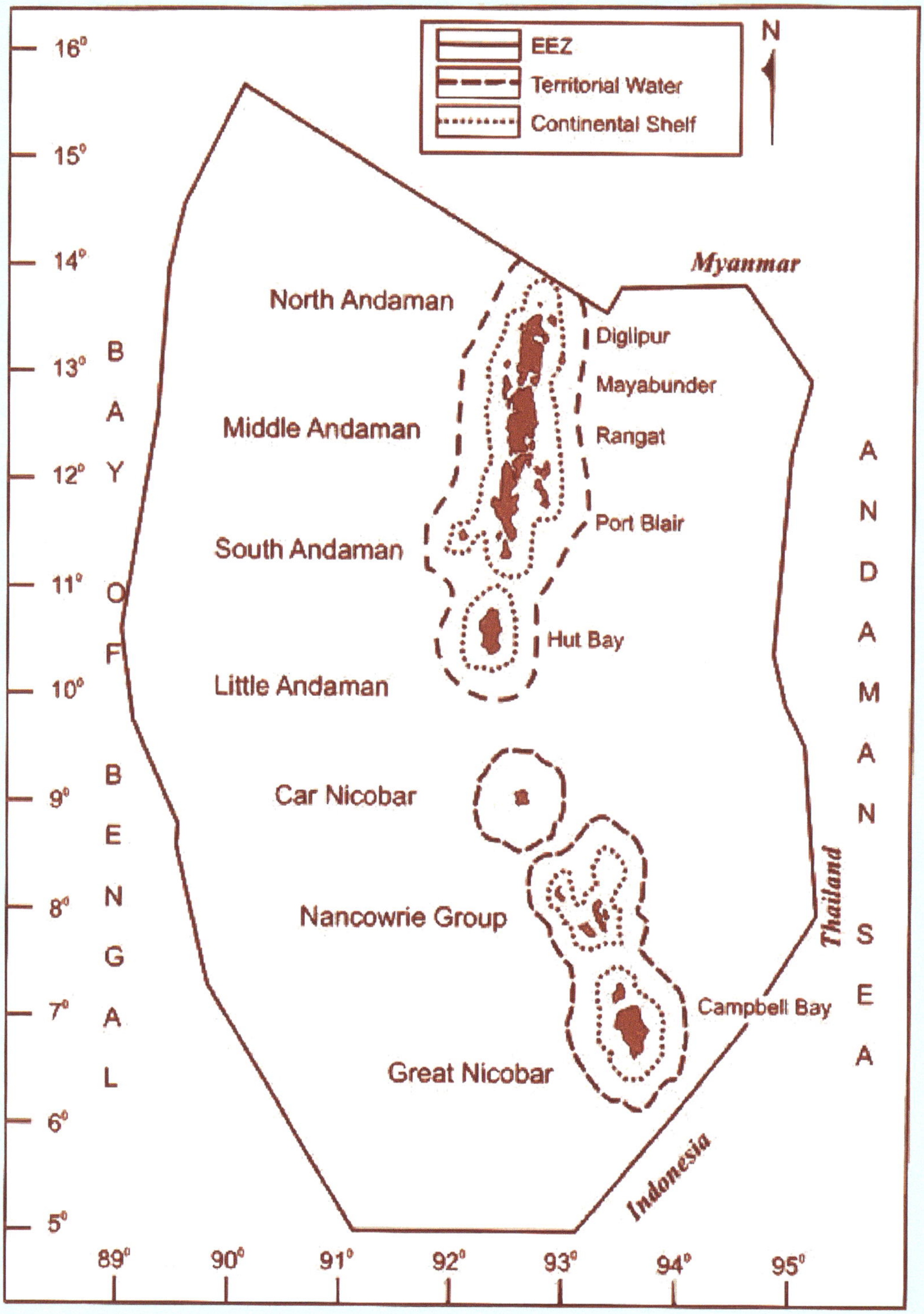

EEZ
Territorial Water
Continental Shelf
N
Myanmar
North Andaman
Diglipur
Mayabunder
Middle Andaman
Rangat
South Andaman
Port Blair
Hut Bay
Little Andaman
Car Nicobar
Nancowrie Group
Campbell Bay
Great Nicobar
Thailand
Indonesia
BAY OF BENGAL
ANDAMAN SEA
16° 15° 14° 13° 12° 11° 10° 9° 8° 7° 6° 5°
89° 90° 91° 92° 93° 94° 95°

History

Fishing is an ancient practice that dates back at least to 40,000 years ago during the Upper Paleolithic Period Skeletal remains of Taiyuan man, a 40,000 year old modern human from Eastern Asia, indicates that he regularly consumed freshwater fish. Spear fishing with barbed poles (Harpoons) was widespread in Paleolithic times.

In Neolithic times, culture and technology spread worldwide between 4,000 and 8,000 years ago. With the new technologies of farming came basic forms of fishing methods that are still used today.

From 7500 to 3000 years ago, Native Americans of the Western coast were known to engage in fishing with gorge hook and line.

Copper harpoons were known to the seafaring Harappans. Early hunters in India include the Mincopie people, aboriginal inhabitants of Andaman and Nicobar Islands, who have used harpoons with long cords for fishing since early times.

Fishing scenes are rarely represented in ancient Greek culture, but a cup, dating from c. 500 BC, that shows a boy crouched with a fishing-rod in his right hand and a basket in his left hand.

By the 12th dynasty, metal hooks with barbs were being used. Nile perch, catfish and eels were among the most important fish. Some representations indicate fishing was being pursued as a pastime.

Oppian of Corycus, a Greek author wrote a major treatise on sea fishing, the 'Halieulica' composed between 177 and 180. This is the earliest such work to have survived intact to the modern day. Oppian describes various means of fishing including the use of nets cast from boats, scoop nets held open by a hoop, spears and tridents, and various traps " which work while their masters sleep".

The Greek historian Polybius (ca 203 BC 12 BC), in his Histories, describes hunting for Swordfish by using a harpoon with a barbed and detachable head.

Pictorial evidence of Roman fishing comes from mosaics which show fishing from boats with rod and line as well as nets. The Greco-Roman sea god Neptune is depicted as wielding a fishing trident.

In India, the Pandyas, a classical Dravidian Tamil kingdom, were known for the Pearl fishery as early as the 1st century BC. Their seaport Tuticorin was known for deep sea Pearl fishing. The Paravas, a Tamil caste centered in Tuticorin, developed into a rich community because of their pearl trade, navigation knowledge and fisheries from ancient representations and literature it is clear that fishing boats were typically small and used for fishing not far from land.

Drift gillnets and setnets have been widely adapted in cultures around the world. The antiquity of gillnet technology is documented by a number of sources from many countries and cultures. Japanese records trace fisheries exploitation, including gillnetting, for over 3,000 years.

During World War II, navigation and communication devices, as well as many other forms of maritime equipments (ex. Depth-sounding and radar) were improved and made more compact. These devices became much more accessible to the average fisherman, thus making their range and mobility increasingly larger.

The introduction of fine synthetic fibers such as nylon in the construction of fishing gear during the 1960s marked an expansion in the commercial use of gillnets. The new materials were cheaper, lasted longer and required less maintenance than natural fibers.

4.1 Land Based Recreational Fishing

The earliest English literature on recreational fishing, which was well ahead of its time, was published in 1496, by Dame Juliana Berners. The essay was titled ' Treatyse of Fysshynge with an Angle'. Recreational fishing or angling went on to become one of the major interests of the nobility. During the 16th century the essay became very popular and was reprinted many times. ' Treatyse of Fysshynge wyth an Angle' included detailed information on fishing waters, the construction of rods and lines, and the use of natural and artificial baits. Considering the age when the essay was published it is also astonishing to find it also includes modern concept of concerns about conservation.

With the publishing of Izaak Walton's ' The Compleat Angler, or Contemplative Man's Recreation' in1653 the concept of recreational angling gained mass appeal. This book is one of the finest examples of the position of the angler who loves ' fishing for the sake of fishing'. More than 300 editions of The Compleat Angler have been published, which makes it one of the most frequently reprinted books in English literature.

4.2 Salt Water Big Game Fishing

The history of sport fishing took another leap forward with the invention of the motor boat. This allowed the development of big game fishing around the start of the twentieth century. Now, the sport is well developed and increasingly popular, although it does face new challenges. The dwindling stocks of fish in many waterways have led to new practices being developed where fish are no longer killed as part of the catch, but are measured and released back into the water. In some cases, electronic tags are used to identify catches, and even to send data back to government agencies.

Sport fishing will doubtless face many challenges in the years ahead, with the ever growing awareness of the need to protect the environment and the species which inhabit it. The sport has made an excellent start in dealing with any criticisms which may have been leveled against it, by adopting the methods of releasing fish back into the water once caught, and sport fishing is now actively contributing to conservation by reporting data back to government agencies. The next few decades are going to be a crucially important time in the history of sport fishing.

Important sport fishes of Andamans

5.1 Grouper : Members of Serranidae family are referred to as Coral Groupers. The word "Grouper" comes from the word for the fish, most widely believed to be from the Portuguese name, Garoupa. The Grouper is a highly prized sport fishing species and can be a difficult fish to catch based on the fact it dwells close to structure. Common methods utilized are use of natural and artificial baits, use of surface lures and also the use of lead jigs in deeper locations. Care has to be taken while reeling in this species as they are susceptible to barotrauma thus making the task of effectively releasing them a highly specialized one that comes with experience. Grouper are commonly found around the Andaman Islands. They are commonly known as Gobra in Andamans.

Mr. Goh Yu Shen Singaporian; Off Passage Island with a Grouper.

5.2 Snapper : Fishes of family, Lutjanidae, mainly marine, some inhabiting estuaries. Some are important food fish. One of the best known is the Red snapper. Snappers are a widely popular fish to catch based on the ease to locate this species. They are also an ideal fish to catch for novices as they are hardy and a easy fish to release. They can be caught in a number of ways the most popular being use of natural baits using hook.

5.2.1 Two-spot Red Snapper (Lutjanus bohar) is a species of snapper belonging to the genus Lutjanus. It is also known as Twinspot or Bohar snapper. It reaches a length of up to 80 cm. The deep

Mr. Stewart Newnham of U.K; off Neil Island with a Two-spot Red Snapper

red orange colour of this fish also makes it another popular fish in the Lutjanid family. As previously mentioned these are an easy species to catch especially on natural baits. This fish is common in the waters of the Andaman Islands and is referred to as the Kutta Bhetki.

5.2.2 Green Jobfish *(Aprion Virescens)*, is a fish of the Lutjanidae family, and is the only member of the genus Aprion. This fish is considered one of the 'brutes' of the reef often seen snatching baits from other slower fish. They can be effectively caught using surface lures and artificial flies. These fish are commonly known as *Rui Macchi*.

Singaporean Angler with a Green Jobfish ; off Rutland

5.2.3 Rusty Jobfish *(Aphareus rutilans)* (Cuvier, 1830) - The Rusty Jobfish is an elongate species with a large lunate tail. It has a large mouth with a protruding lower jaw and minute teeth. The long-based dorsal fin is not deeply incised at the junction of the spinous and soft-rayed portions. The last ray of both the dorsal and anal fins is elongated. Body colour varies from blue-grey to mauve or reddish. These are specialized fish to catch and are out of the realm of the novice angler. Often found in depths ranging from 60 to 120 m they can only be caught on natural baits and lead jigs. This species is referred to as the *Mrigal Macchi* by local fishermen.

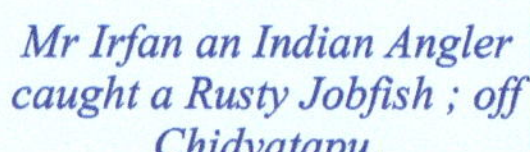

Mr Irfan an Indian Angler caught a Rusty Jobfish ; off Chidyatapu.

5.2.4 Rosy Snapper Jobfish *(Pristipomoides filamentosus).* - Often mistaken for the Rusty Jobfish the Rosy Snapper again is found in the same locations. Rosy Snapper are not a popular Sport fishing species as they are seldom caught. They can be targeted on lead jigs. They are locally known as *Mrigal Macchi.*

A Malaysian Angler with a Rosy Snapper ; off Passage Island

5.2.5 Mangrove Red Snapper *(Lutjanus argentimaculatus)* - This is an explosive and powerful sportfish renowned for its superb eating and fighting qualities. Coloration of the mangrove red snapper ranges from burnt orange, to copper, to bronze and dark reddish-brown, depending on its age and environment.

Younger fish caught in estuarine areas are often darker than older fish taken from offshore reef areas and exhibit lighter vertical bands down their flanks. The mangrove red snapper is a highly regarded table fish with firm, sweet-tasting, white flesh. It is a popular fish locally known as the *Khadi Bhekti*.

Mr. Kanta Rao an Indian with a Mangrove Red Snapper; off Baratang

5.3 The Carangidae are a family of fish which includes the Jacks, Pompanos, Jack mackerels, and Scads. They are common in tropical to subtropical waters of the Indo-Pacific, ranging from South Africa in the West to Japan in the East, typically inhabiting

inshore reefs and bays. They are a prolific species in the Andaman Islands and are commonly known as Kokari.

5.3.1 Giant trevally, (Caranx ignobilis) also known as the Giant kingfish, Lowly Trevally, Barrier Trevally, or GT, The giant trevally is distributed throughout from South Africa in the West to Hawaii in the East.

Mr Casal Roman Jose Manuel of Spain with a Giant Trevally caught off North Brother

This is a very common fish of the Andaman Island. It is distinguished by its steep head profile, strong tail scutes, and a variety of other more detailed anatomical features. It is the largest fish in the genus Caranx, growing to a maximum known size of 170 cm and a weight of 80 kg. The giant trevally is a powerful apex predator in most of its habitats, and is known to hunt individually and in schools. The fish grows relatively fast, reaching sexual maturity at a length of around 60 cm at three years of age.

This species of fish has a dedicated following of anglers who travel the world in search of pristine areas where these fish can be caught. Current hotspots in the world for targeting this species are Oman, Djibuti, Socotra Island in Yemen, Maldives, Andaman Islands, Northern Indonesia, Japan, Australia and accessible Polynesian Islands. The preferred method of catching this species is by the use of surface lures, but they can also be caught using a wide array of methods.

5.3.2 Yellow Spotted Trevally, *(Carangoides fulvoguttatus)* also known as the Goldspotted trevally and Tarrum is a widespread species of large inshore marine fish. The Yellow Spotted Trevally inhabits the tropical and subtropical waters of the Indo-west Pacific region, commonly found in large shoals off deep reefs in the Andaman Islands. They are predatory, taking fish, cephalopods and crustaceans. It is considered an excellent sport fish by anglers and is a good table fish.

Mr.Dilip Ekka with a Yellow Spotted Tevally ; off Neil Island

5.3.3 Bluefin Trevally, *(Caranx melampygus)* also known as the Bluefin Jack, Bluefin Kingfish, Bluefinned Crevalle, Blue ulua, Omilu and Spotted Trevally, is a species of large, widely distributed from Eastern Africa in the West to Central America in the East, including Japan in the North and Australia in the South. Bluefin trevally is easily recognised by their electric blue fins, tapered snout and numerous blue and black spots on their sides. Bluefin trevally is a popular target for both commercial and recreational fishermen. This species is targeted with the use of artificial surface baits and natural baits. They are commonly referred to as *Neela Kokari.*

French Angler with a Blue fin Trevally; off Passage Island

5.3.4 Almaco Jack *(Seriola Rivoliana)* is a game fish of the family Carangidae. They feed, both day and night, on other smaller fish such as baitfish and small squid. This species is more widely caught than its larger counterpart the Greater Amberjack. They can be effectively caught on deep sea mounts with the use of jigs. Also known by local fishermen as the *Pathar Kokari.*

Mr Eric Belmondo of France with a Almaco Jack ; off Ross Island

5.4 Scombridae is the family of fast swimming oceanic fishes comprising of sail, Sword, Seer & Tunas etc.

5.4.1 Indo Pacific Spanish Mackerel *(Scomberomorus commerson)* is a mackerel of the Scombridae family. It is found in a wide ranging area centering in South-east Asia but as far west as the east coast of Africa and from the Middle East and along the northern coastal areas of the Indian Ocean. They are easily caught by trolling, which by far is the most effective way to catch this species. They are much sought after fish in the Tourist trade as a food fish and known as *Surmai.*

Indian Anglers with a Spanish Mackeral ; off Baratang.

5.4.2 Sailfish : There are two species of fish in the genus Istiophorus, living in warmer sections of all the oceans of the world. They are predominately blue to gray in color and have a characteristic erectile dorsal fin known as a sail, which often stretches the entire length of the back. Another notable characteristic is the elongated bill, resembling that of the swordfish and other marlins. They are therefore described as billfish in sport fishing circles. Individuals have been clocked at speeds of up to 110 Km. per hour (68 mph), which is the highest speed reliably reported in any water creature. Generally, sailfish do not grow to more than 3 meters (9.8 ft) in length and rarely weigh over 90 kilograms (200 lb). Sailfish are highly prized game fish and are known for their incredible jumps and great speed. They can appear in a startling array of colors, from subdued browns and grays to vibrant purples and even silver.

A group from England with a Sail fish; off Sister Island.

5.4.3 Swordfish (Xiphias gladius) from Greek ξίφος: sword, and Latin gladius: sword), also known as Broadbill in some countries, are large, highly migratory, predatory fish characterized by a long, flat bill. They are a popular sport fish of the billfish category, though elusive. Swordfish are elongated, round-bodied, and lose all teeth and scales by adulthood. These fish are found widely in tropical and temperate parts of the Atlantic, Pacific and Indian oceans, and can typically be found from near the surface to a depth of 550 m (1,800 ft). They are a prized gamefish that have been caught by commercial fishing vessels in the Andaman Sea. To date there is no record of these fish having been caught on rod and line. In a world where there are very few achievements that can be accomplished for the first time ever. Catching a Swordfish on rod and line in the Andaman Islands would be one of those achievements.

5.4.4 Indo-Pacific Blue Marlin *(Makaira mazara)* as a separate species is under debate. Genetic data suggests that, although the two groups are isolated from each other, they are both the same species. These fish are migratory and not much is known about their movements in the Andaman waters. It is however noticed that they are caught in larger numbers just before the onset of the S.W. Monsoon. They are also commonly referred to as the *Hawabill.*

Mr Wilson Grant, Mr Bryce James, Mr Bowman Damian from Australia and Mr Lim Heng Fong from Malaysia caught a Blue Marlin; off Neil Island.

5.4.5 Black Marlin *(Makaira indica)* is a species of marlin found in tropical and subtropical Indo-Pacific and East Pacific oceans from near the surface to depths of 915 m (3,002 ft).It is a large commercial game fish with a maximum published length of 4.65 m (15.3 ft) and weight of 750 kg (1,700 lb). It is one of the largest marlins and bony fish. This fish is highly prized if caught. They are also known as the *Hawabill* in Andaman Islands.

A Spanish couple with a Black Marlin ; off Burmanala, South Andaman.

5.4.6 Dogtooth Tuna *(Gymnosarda unicolor)* This is a large fast-swimming fish of the family Scombridae. It is placed in its own genus, Gymnosarda. It has the large teeth and straight edged first dorsal fin characteristic of the smaller bonitos (Genus Sarda) and although it far exceeds them in size, reaching weights of over 150 Kg. is considered more closely related to them than to the true Tunas. South Korea is where the long-standing world record of 130.6 Kg. / 288 lbs was caught.

Dogtooth tuna frequent reef environments, with smaller fish being more commonly found near shallow reef areas and larger ones haunting deep reef drop off areas, seamounts and steep underwater walls. The Dogtooth Tuna is one of the apex non-pelagic predators in its environment, sharing that position with Giant Trevally, Napoleon Wrasse, and large Groupers, as well as Reef, Bull and Tiger sharks. The Dogtooth Tuna is appreciated in most of its range as a fine food fish and also as a game fish sought by both rod and reel anglers and Dogtooth Tuna used to be mostly taken as an incidental catch by anglers trolling for other game fish - with natural baits for black marlin, for instance, or with lures for wahoo and Spanish (narrowbarred) mackerel.

In the last 10 to 15 years there has been more dedicated effort directed at this species because of its rarity and sporting qualities. Dogtooth Tuna are now a highly coveted prize by many European and Asian sports anglers and are one of the hardest gamefish to catch. High speed jigging with a variety of metal lures has increased tremendously in popularity in the last several years as advancements in tackle technology have resulted in lightweight rods and reels that are capable of handling heavy spectra-type braided lines. Some of the more popular destinations for anglers seeking this species include Okinawa and other islands of southern Japan, Rodrigues and other Indian Ocean islands such as the Maldives, Andaman & Nicobar Islands, Bali and the Great Barrier Reef. Though seen rarely they are known as the *Surmai Bangdi*.

Mr Michael Bailey from Canada with a Dogtooth Tuna; off Outram Island.

5.4.7 Great Barracuda (*Sphyraena barracuda*) also known as the giant barracuda is a species of barracuda. Great barracudas often grow over (1.8 m. / 6ft) long and is a type of ray-finned fish. Great barracudas are large fish. Mature specimens are usually around 60100 cm record-sized specimen caught on rod-and-reel weighed 46.72 kg (103.0 lb) and measured 1.7 m (5.6 ft), Barracudas possess strong, fang-like teeth that are unequal in size and set in sockets in the jaws and on the roof of the mouth. The head is quite large and is pointed. These fish are common in the Andaman waters and are referred to as the *Dandoos Macchi.*

Javed Salim Siddique an Indian Angler with a Great Barracuda ; off Inglis Island.

5.5 Yellow Fin Tuna (*Thunnus albacares*) is a species of tuna found in pelagic waters of tropical and subtropical oceans worldwide. The yellow fin tuna is among the larger tuna species, reaching weights of over 181 Kg / 400 pounds. The second dorsal fin and the anal fin, as well as the finlets between those fins and the tail, are bright yellow, giving this fish its common name. The main body is very dark metallic blue, changing to silver on the belly, which has about 20 vertical lines. These fish are pelagic hence are not residents in the Andaman Sea. We see the first of the large aggregations passing through our waters in early March. They can be targeted off shore as they pass. Yellow Fin Tuna can be seen in large aggregations accompanied by pods of Dolphin, Whales and_Oceanic birds in search of smaller baitfish and shrimp. It locally is a popular fish with most of the catch being exported. They are known by fishermen as *Tuna.*

Javed Salim Siddique an Indian Angler with A Yellow fin Tuna; off Port Blair.

6.1. Trolling

Trolling is one of the most basic forms of fishing from a boat and is the most popular way of catching fish throughout the world. Trolling is a method of fishing where one or more fishing lines, baited with lures or rigged bait-fish, are drawn through the water at varying depths and distances behind a boat. Trolling is used to catch pelagic fish such as mackerel and kingfish.

Trolling is commonly confused with trawling, a different method of fishing where a net (trawl) is drawn through the water instead of lines. Trolling is used primarily for recreational whereas trawling is used mainly for commercial fishing.

Trolling from a moving boat involves moving at various speeds through the water depending on species being targeted. Trolling speeds can vary from 1-2 knots to 15-16 knots. Multiple lines are often used, and outriggers can be used to spread the lines more widely and reduce their chances of tangling. Downriggers can also be used to keep the lures or baits trailing at a desired depth.

Outriggers are poles which allow a boat to troll multiple lines in the water without tangling by increasing the 'spread' or distance between lures. Effective use of outriggers can simulate a school of bait fish. Downriggers are devices used while trolling to keep bait or lure at predetermined and desired depth. Fish swim at different depths according to factors such as temperature and the speed and direction of water currents.

A downrigger consists of a one or two meter horizontal pole which supports a weight, typically about three kg of lead, on a steel cable. A line with a 'release clip' attaches the fishing line to the weight, and the bait or lure is attached to the release clip. Upon a 'strike' the main line releases from the release clip and the angler is left with a fish on his line caught at a depth that would otherwise have been impossible for him to have reached.

6.1.1 Big Game Trolling for Marlin

Marlin fishing is considered by some game fishermen to be the pinnacle of offshore game fishing, due to the size and power of marlin and the relative rareness of this species.

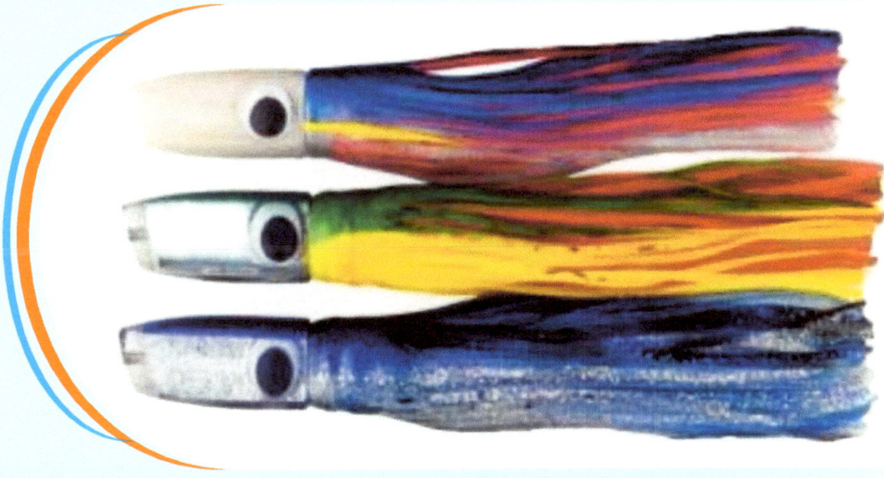

Marlin Fishing Techniques

Fishing styles and gear used in the pursuit of blue marlin vary, depending on the size of blue marlin common to the area, the size of fish being targeted, local sea conditions, and often local tradition. The main methods use artificial lures, rigged natural baits, or live bait.

A typical Marlin trolling setup in the Andaman Sea, off Neil Island.

Artificial Lure Fishing

Blue marlin are aggressive fish that respond well to the splash, bubble trail and action of a well presented artificial lure probably the most popular technique used by Blue marlin Hunters worldwide. Artificial lure fishing has spread from its Hawaiian origins. Today, marlin lures are produced in a huge variety of shapes, sizes and colours, mass-produced by large manufacturers and individually crafted by small-scale custom makers.

Artificial Marlin lures

Typical marlin lures come in various sizes, small (7-8 inch), medium (10-12 inch) to large (14 inches or more) with a shaped plastic or metal head to which a plastic skirt is attached. The design of the lure head, particularly its face, gives the lure its individual action when trolled through the water. Lure actions range from an active side-to-side swimming pattern to pushing water aggressively on the surface to, most commonly, tracking along in a straight line with a regular surface pop and bubble trail. Experienced anglers can fine tune their lures to get the action they want.

Lures are normally fished at speeds of between 7.5 to 9 knots; faster speeds in the 10 to 15 knot range. The speeds allow substantial areas to be effectively worked in a day's fishing. A pattern of four or more lures is trolled at varying distances behind the boat. Lures may be fished either straight from the rod tip "flat lines" or from outriggers.

Natural Bait Fishing

Rigged natural baits have been used by sport fishermen seeking Blue Marlin. Rigged Spanish mackerel and Horse Ballyhoo are widely used for Atlantic blue marlin. Rigged natural baits are sometimes combined with an artificial lure or skirt to make 'skirted baits' or 'bait/lure combinations'.

Rigged Natural Bait

6.2 Popping

Popping is one of the newest disciplines of sport fishing pioneered by the Japanese. Over the last decade there have been vast changes and developments in the sport fishing industry especially with fishing equipment. Popping is primarily the use of surface lures or 'poppers'. These lures are 'cup faced' and float on the surface and when pulled or dragged through the water they make a 'popping' sound/action. They resemble either a bird diving down to feed on the surface or the splashing of an injured fish and are attacked by larger fish.

Poppers are available in various sizes and often range from 50grms to 200grms in weight. There are various shapes of poppers as well that have various actions to resemble various bait fishes in the area.

Also common is the use of 'stick baits'. Stick baits are primarily made out of wood or resin and resemble wounded or injured fish. Most commonly stick baits are used just under the surface of the water.

6.3 Jigging

Jigging is the practice of fishing with a jig, a type of fishing lure. A jig consists of a lead sinker with a hook molded into it and usually covered by a soft body to attract fish. Jigs are intended to create a jerky, vertical motion.

The jig is very versatile and can be used at varying depths and many species are attracted to the lure which has made it popular amongst anglers for years. For successful jigging, the ' jigger ' needs to use a rod which is good for feeling a strike, and needs to stay in contact with the lure and get it to where the fish are. Most fish caught by jigs are on or near the bottom.

The more modern version of jigging is 'speed jigging' where jigs are reeled up at a much faster speed and flutter as they swim. Jigs are used with a single hook attached by a separate cord often made of Dyneema or Kevlar so it is not cut by fish with sharp teeth.

The separate single hook is called an 'assist hook' attached to the jig by an 'assist cord'.

6.4 Fly Fishing

Saltwater fly fishing is typically done with heavier tackle than that which is used for freshwater trout fishing, both to handle the larger, more powerful fish, and to accommodate the casting of larger and heavier flies. Salt water fly fishing typically employs the use of wet flies resembling baitfish, crabs, shrimp and other forage. Many saltwater species, particularly large, fast and powerful fish, are not easily slowed down by "palming" the hand on the reel. Instead, a purpose-made saltwater reel for these species must have a powerful drag system. Furthermore, saltwater reels purpose-made for larger fish must be larger, heavier, and corrosion-resistant. Corrosion-resistant equipment is key to durability in all types of saltwater fishing, regardless of the size and power of the target species.

Saltwater fly fishing is most often done from a boat, for pursuing Sailfish, Tuna, Dorado, Marlin and other pelagic. Ocean fish are usually harder to catch. Hooks for saltwater flies must also be extremely durable and corrosion resistant.

Wet Flies

6.5 Bait Fishing

The use of natural baits like small sardines and mackerel either whole or cut into chunks is the most basic form of sport fishing requiring nothing more than a rod, reel line and hook. Typical 'rigs' have a lead weight right at the bottom to get them down to the sea bed. Hooks are rigged on short lines above the lead weight and baited with cut fish. Upon feeling a 'bite' a fish is then reeled up to the surface. Most anglers start fishing this way as it is the cheapest form of fishing and also the most effective.

Various methods of rigging bait fish.

Fishing Tackle is the equipment used by fishermen when fishing. Almost any equipment or gear used for fishing can be called fishing tackle. Some examples are Hooks, Lines, Sinkers, Floats, Rods, Reels, Baits, Lures, Nets, Gaffs and Tackle boxes.

Gear that is attached to the end of a fishing line is called Terminal tackle. This includes hooks, leaders, swivels, sinkers, floats, split rings and wire, snaps, beads, spoons, blades, spinners and clevises to attach spinner blades to fishing lures.

Fishing tackle can be contrasted with fishing techniques. Fishing tackle refers to the physical equipment that is used when fishing, whereas fishing techniques refers to the manner in which the tackle is used when fishing. However the term fishing gear is more usually used in the context of commercial fishing, whereas fishing tackle is more often used in the context of recreational fishing. Below are listed some common tools that make up fishing tackle.

7.1 Rod

A fishing rod in its most basic sense is an additional tool used with a line and a hook at one end that is baited. Early fishing rods are depicted on inscriptions in ancient Egypt, China and Rome. In Medieval England they were called Angles (hence the term angling). As they evolved they were made from materials such as split bamboo, Calcutta reed, or ash wood, which were light, tough, and pliable. Handles and grips were made of cork, wood, or wrapped cane. Guides were simple wire loops.

Modern rods are sophisticated casting tools fitted with line guides and a reel for line stowage. They are most commonly made of solid fiberglass. With the advent of technology the use of carbon fiber became common. Fishing rods vary in action as well as length, and can be found in sizes between 24 inches and 20 feet. The longer the rod, the greater the mechanical advantage in casting. There are many different types of rods, such as Fly rods, Spin and Bait casting rods, Spinning rods, Surf rods and Trolling rods.

7.2 Reel

A fishing reel is a device used for the deployment and retrieval of a fishing line using a spool mounted on an axle. Fishing reels are traditionally used in angling. They are most often used in conjunction with a fishing rod. Some of the various kinds of reels commonly used are Spinning reels, Multiplier reels, Trolling reels and Fly reels. Reels are built to hold a long length of line wrapped around the 'spool' and also have a 'drag' that allows line to come off the spool at a predetermined pressure. The use of a reel with drag is what 'tires' a fish before it can be reeled in. Reels have complex gearing and gear ratios are based on the application the reel is built for.

7.3 Fishing line

A fishing line is a cord used or made for fishing. The earliest fishing lines were made from leaves or plant stalk. Later lines were constructed from horse hair or silk thread, with catgut leaders. From the 1850s, modern industrial machinery was employed to fashion fishing lines in quantity. Modern lines are made from artificial substances, including Nylon, Polyethylene, Dacron and Dyneema. The most common type is monofilament made of a single strand. Fishermen often use monofilament because of its buoyant characteristics and its ability to stretch under load. Recently, other alternatives to standard nylon monofilament lines have been introduced made of copolymers or fluorocarbon, or a combination of the two materials. There are also braided fishing lines, cofilament and thermally fused lines, also known as 'superlines' for their small diameter, lack of stretch, and great strength relative to standard nylon monofilament lines.

Multifilament line, also referred to as The Super Lines, is a type of fishing line. It is a braided line which is made up of a type of polyethylene, an extremely thin line for its strength. By weight, polyethylene strands are five to ten times sturdier than steel. Multifilament line is almost similar to braided dacron in terms of sensitivity but a diameter about one-third that of monofilament. Multifilament works best on conventional and bait casting reels. On spinning and spin casting reels, the line's limpness can make sure for awkward manipulation, as it doesn't "spring" off the reel like monofilament. Consequently, knot-tying is more difficult with multifilaments. Certain knots work better with super line, like the palomar knot. Applying a type of super glue will help to prevent other types of knots from slipping. This type of fishing line is expensive sometimes four times the cost of equivalent monofilament. This can become a considerable expense, especially considering that the line is so thin that you need more of it to fill a reel spool. Sometimes, a backing of monofilament or other line is used under the braided line on the spool.

7.4 Hook

The use of the hook in angling is descended, historically, from what would today be called a "gorge". The word "gorge", in this context, comes from an archaic word meaning "throat". Gorges were used by ancient peoples to capture fish. A gorge was a long, thin piece of bone or stone attached by its midpoint to a thin line. The gorge would be fixed with bait so that it would rest parallel to the lay of the line. When a fish swallowed the bait, a tug on the line caused the gorge to orient itself at right angles to the line, thereby sticking in the fish's gullet.

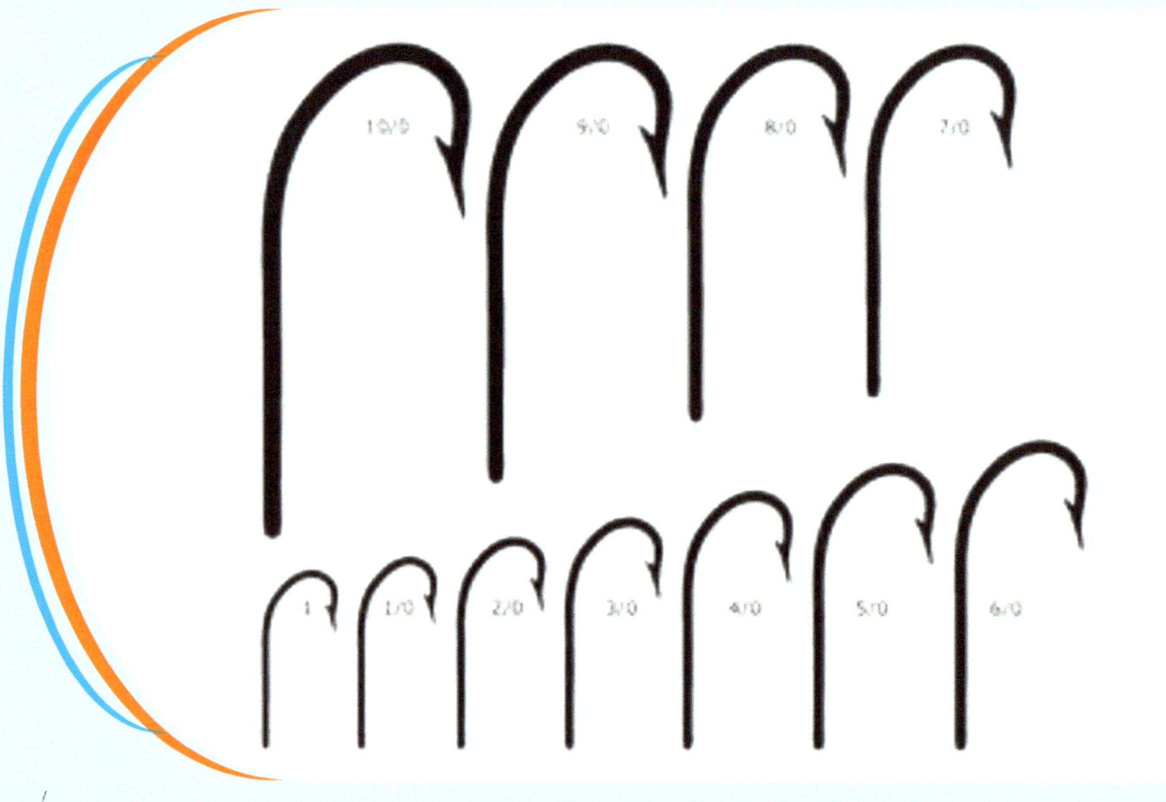

A fish hook is a device for catching fish either by impaling them in the mouth or, more rarely, by snagging the body of the fish. Fish hooks have been employed for millennia by fishermen to catch fresh and saltwater fish. Early hooks were made from the upper bills of eagles and from bones, shells, horns and thorns of plants. In 2005, the fish hook was chosen by Forbes as one of the top twenty tools in the history of man. Fish hooks are normally attached to some form of line or lure device which connects the caught fish to the fisherman. There is an enormous variety of fish hooks. Sizes, designs, shapes, and materials are all variable depending on the intended purpose of the hook. They are manufactured for a range of purposes from general fishing to extremely limited and specialized applications. Fish hooks are designed to hold various types of artificial, processed, dead or live baits (bait fishing); to act as the foundation for artificial representations of fish prey (fly fishing); or to be attached to or integrated into other devices that represent fish prey (lure fishing).

Circle Hook

A circle hook is a type of fish hook which is sharply curved back in a circular shape. It has become widely used among anglers in recent years because the hook generally catches more fish and is rarely swallowed. Since the circle hook catches the fish on the lips at the corner of its mouth, it usually decreases the mortality rates of released fish as compared to J-hook (like O'Shaughnessy or Octopus hooks) which are often swallowed by the fish, causing damage to the gills or vital organs. The circle hook's unique shape allows it to only hook onto an exposed surface, which in the case of a fish means the corner of its mouth. The fish takes the baited hook and swallows it, and as the hook is reeled in, it is safely pulled out of the fish until it reaches the mouth. At this point it will catch the corner of the mouth of the fish, resulting in fewer gut-hooked fish. In terms of technique, it is important to not strike (or set the hook) when the fish bites, but rather just reel in. The act of striking while using a circle hook often results in the hook being pulled out of the fish altogether.

7.5 Gaff

In fishing, a gaff is a pole with a sharp hook on the end that is used to stab a large fish and then lift the fish into the boat or onto shore. Ideally, the hook is placed under the backbone. Gaffs are used when the weight of the fish exceeds the breaking point of the fishing line or the fishing pole. A gaff cannot be used if it is intended to release the fish unharmed after capture, unless the fish is skillfully gaffed in the lip, jaw, or lower gill using a thin gaff hook.

"Flying gaff" is a specialized type of gaff used for securing and controlling very large fish. The hook part of the gaff (the head) detaches when sufficient force is used, somewhat like a harpoon's dart. The head is secured to the boat with a length of heavy rope or cable.

7.6 Artificial Bait and Natural Bait

7.6.1 Artificial Lures or Plugs

Many people prefer to fish solely with lures, which are artificial baits designed to entice fish to strike. There is a wide array of sizes of these lures and they can be used from catching the smallest of popular game fishing species to some of the largest.

7.6.2 Spoons

Casting spoons are made of metal and are usually have a high sheen and are bright silver/gold. They are cast out and retrieved at high speed especially at a school of fish. They look identical to small fleeing fish and often entice larger predatory fish into striking.

7.6.3 Natural Bait

Natural baits are often the first choice of a novice angler and are widely available at any seaside location. They are often bought before boarding a sport fishing vessel from a market close by or also can be caught while en route to the fishing grounds. Natural bait can be used whole or can be cut and used in bite sized chunks based on hook size and target species. There are numerous rigs and methods that are used. Most common being the paternoster rig where the weight/sinker is right at the bottom and the hook or hooks are located on branches off the mainline. A minimum of 2 hooks and a maximum of 4 hooks are used.

8 Types of Sport Fishing Vessels

Sport Fishing is best carried out from boats as it gives anglers a novel experience, its easier to find fish as enthusiasts are not restricted to land.

Center console is a type of single-decked open hull boat where the console of the boat is in the center of the boat. There is a cabin on some models, within which there might be a toilet facility. The boat deck surrounds the console so that a person can walk all around the boat from stern to bow with ease.

The console of a boat is where all the controls are located, including steering, ignition, trim control, radio and other electronic devices, switches etc. It may have a small storage space and/or ahead. In general there is no weather protection or berths, making the design ill-suited to cruising. The console may have a T-Top cover to provide limited relief from the sun and rain, but this is not universal because it interferes with fishermen's casting. Most center consoles are powered by outboard motors.

8.1 Center Console FRP Boats 5.5 m. / 18 ft to 7.3 m. / 24 ft.

18 to 24ft craft are ideal for tourists that are first time fishers or would like to experience angling for the first time. Typically suited for a group of 2 to 3 anglers and meant for 3 to 4 hours of fishing from close by locations. These boats will give newcomers a glimpse of an exciting sport.

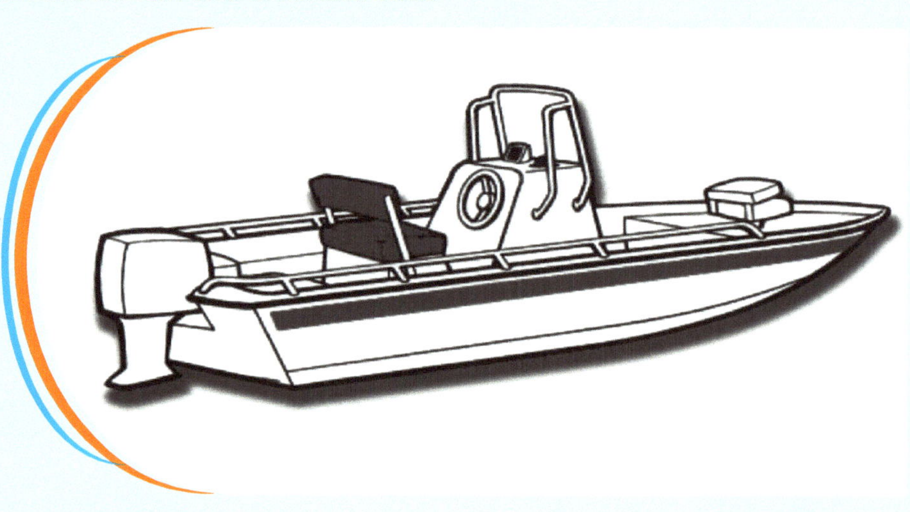

8.2 Center Console FRP Boats 8.5 m. / 28 ft to 12.1 m. /40 ft

Sport fishing craft of these size are meant for the most serious angler and for long range fishing trips craft should be fitted with modern electronics like GPS, Fish finder, Digital Maps, VHF, EPIRB, Fire fighting and life saving appliances. Typically these boats will sail in day light hours with 2 to 6 anglers on board apart from crew.

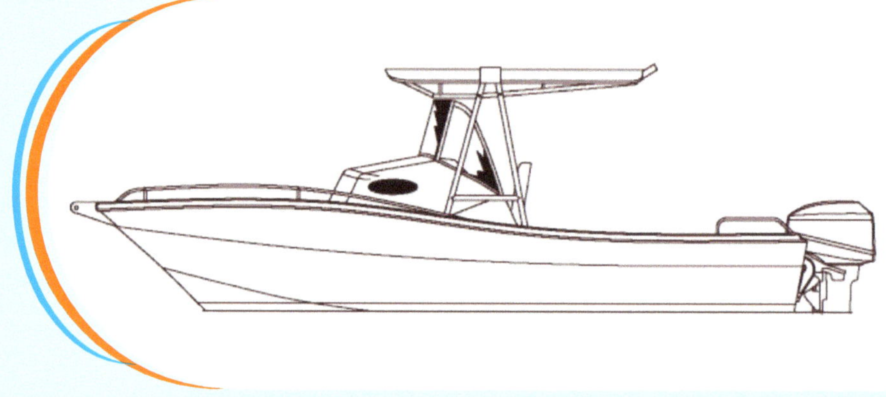

8.3 Metal Live-aboard Boats 12.1 m. / 40ft to 18.3 m. /60ft FRP

These boats are meant for Marlin fishing which involves fishing off shore. Tropically a group of four anglers would go out to see fishing from 3 to 7 days on one of these boats without the necessity of touching land. This configuration of fishing is not cheap and can be afforded by a small and delegated group of anglers who travel world wide.

These are expensive purpose-built offshore vessels with powerfully driven deep sea hulls. They are often built to luxury standards and equipped with many technologies to ease the life of the deep sea recreational fisherman, including outriggers, flying bridges and fighting chairs, and state of the art electronic aids, such as fish-finders and navigation electronics Fish finders, also known as bottom machines or echo sounders, are now commonplace. Other electronics used to narrow down the search for fish may include radar, forward or side-scanning sonar, water temperature sensors and sea surface temperature imagery obtained from satellites. The cost of a suitable boat, electronics, tackle and the operating costs (fuels and other consumables, insurance, mooring fees and maintenance) can be very substantial. Consequently, many big-game anglers prefer to use charter services where they hire the use of a boat and equipment, and the fish-finding expertise of a captain, in preference to maintaining their own. Either way, big-game fishing can be an expensive pursuit.

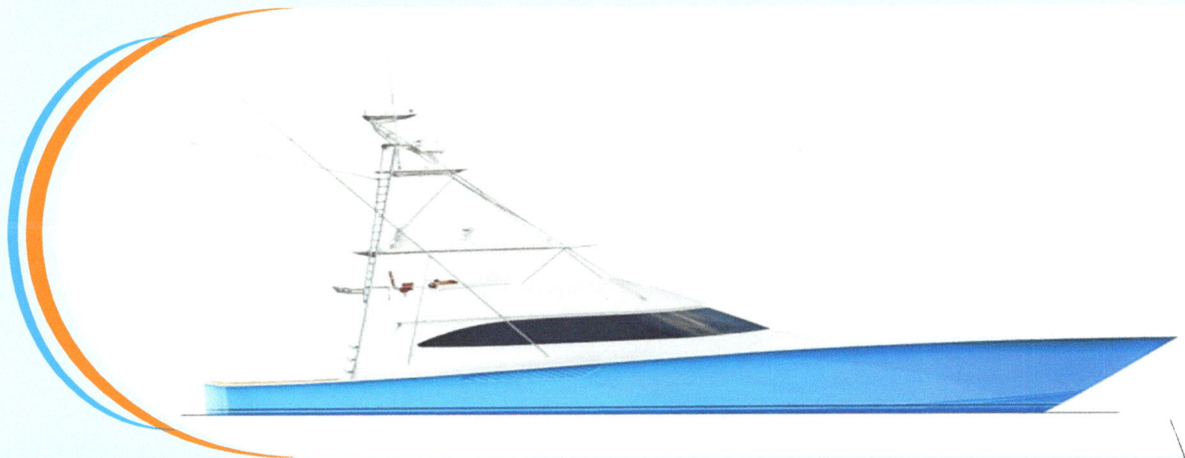

Catch and Release fundamentally is a practice within recreational fishing intended as a technique of conservation. After capture, fish are unhooked and returned to the water before experiencing serious exhaustion or injury. Using barbless hooks, it is often possible to release a fish without removing it from the water (a slack line is frequently sufficient).

Catch and Release has been performed for more than a century by fishermen in order to prevent target species from disappearing in heavily fished waters.

Anglers fishing for fun rather than for food accepted the idea of releasing the fish while fishing in so-called "no-kill" zones. Conservationists have advocated catch and release as a way to ensure sustainability and to avoid over fishing of fish stocks.

Live and Safe release

Catch and release is now widely used to conserve and indeed is critical in conserving vulnerable fish species.

Effective catch and release fishing techniques avoid excessive fish fighting and handling times, avoid damage to fish skin, scale and slime layers by nets, dry hands and dry surfaces and avoid damage to gills by poor handling techniques.

The use of barbless hooks is an important aspect of catch and release; barbless hooks reduce injury and handling time, increasing survival. Frequently, fish caught on barbless hooks can be released without being removed from the water. To make a hook barbless, the barb is simply crushed flat with a pair of needle-nosed pliers, a trivial task. Medium grit sandpaper can be further used to ensure complete removal of the barb, but this is not necessary and is rarely done.

The effects of catch and release vary from species to species. Studies of fish caught in shallow water on the Great Barrier Reef showed high survival rates (97 %+) for released fish if handled correctly and particularly if caught on artificial baits such as lures. Fish caught on lures are usually hooked cleanly in the mouth, minimizing injury and aiding release.

9.1 Deep Sea Fish Handling and Release

Some deep sea fish species suffer from the sudden pressure change when wound to the surface from great depths; these species cannot adjust their body's physiology quickly enough to follow the pressure change.

Recovery Check before release

The result is called "barotrauma". Fish with barotrauma will have their enormously swollen swim-bladder protruding from their mouth, bulging eyeballs, and often sustain other, more subtle but still very serious injuries. Upon release, fish with barotrauma will be unable to swim or dive due to the swollen swim-bladder.

The common practice has been to deflate the swim bladder by pricking it with a thin sharp object before attempting to release the fish.

Emerging research indicates both barotrauma and the practice of deflating the swim bladder are both highly damaging to fish, and that survival rate of caught-and-released deep-sea fish is extremely low. However, barotrauma requires that fish be caught at least 30 50 feet below the surface. Many surface caught fish, such as billfish, and all fish caught from shore, do not meet this criterion and thus do not suffer barotrauma.

Reviving Giant Trevally before release

10 Recommendations

Globally we are struggling to acquire that mystique mechanic of trade which brings in economical prosperity without damaging the eco-niche around us, Ah! a task so complicated! two sides of the same coin one underpinning the other.

How fortunate this Union Territory is? We possess the fish wealth and the recently inducted trade of Sport Fishing, one complementing the other, provided the resources are otherwise not subjected to overexploitation.

We the Authors through mutual perception would advocate to evolve an atmosphere conducive to hobbyist and remunerative to Union which in the days ahead could be prudent enough for the World around to come and enjoy. We personally feel that this emerging IVth dimension of Fisheries will definitely witness an explosive expansion if following aspects are taken care of :

1. BERTHING / JETTY FACILITY exclusively for berthing Sport Fishing Vessels and angling equipped with all modern facilities.
2. SPORT FISHING FESTIVAL may be organized creating awareness among locals and tourists. Holding competitions, inviting international crowd of hobbyists.
3. SPORT FISHING ZONE in-par with the standard management practices adopted elsewhere in the World, A &N Islands shall have 'NO TAKE ZONE' wherein fishing is allowed with the mandatory condition to release the catch back live and safe.
4. INCENTIVE aspiring local fishers be assisted financially through schemes for converting fishing boats to Angling Crafts and certain subsidy benefit in-line with contemporary fishery activity.
5. SPORT FISHING LICENSE to tourist on payment of fee and grant of short duration ID to each client.
6. MONITORING capabilities of the concerned surveillance agencies shall adequately be enhanced to oversee the Sport Fishing actively in tune with Coastal Security SOP.
7. RESEARCH through Fisheries /Ocenographical Universities, Institutions and Scholars shall be encouraged to utilize Sport Fishing platform.
8. SUSTAINED RATIONAL HARVEST exploitation of Territorial and Oceanic resources by Territorial and LOP fishing fleet based upon the principles of population dynamics.

Recommendations are the personal views of the Authors and NOT that of Government

Know the Authors

Arif M. Mustafa
Gafoor Manzil 24/2
Shastri Road,
Aberdeen Bazar,
Port Blair - 74 4101
South Andaman

Akshay Malavi
Sea Fishing India Pvt. Ltd.,
2nd Floor, 3 Foreshore Road,
Behind Police Station,
Chatham, Haddo (Post)
Port Blair - 744102.

A dedicated Fisheries professional possessing more than 30 years of experience in exploring, developing, expending and managing the tropical marine fisheries of A&N Islands. Island pioneer of Induced Fish breeding, Marine aquarium, Artificial reef, Coral reef survey, Cage culture, Perch fishery reporting etc. with about 30 publications to his credit in India and abroad.

Indian counterpart during Cousteau Foundation France, Russian Academy of Sciences Moscow and National Geography's Andaman expeditions.

Reported two high value sea food items a Deep sea dog fish shark and a Spear lobster from Andaman Sea.

A local inhabitant representing second generation of a 'British India' rebel *Pandit* Ayodiya Rai Sharma a *Prohit* from village Kothar, Shajahanpur exiled to Andamans during 1871, embressed Islam while in Andamans and adopted the name Nazir Mohammed.

Akshay Malavi

Sport Fishing appears to be a trait inherited, a passion so deep in his personality that makes him an expert of this art. He nurtured this desire while serving stained glass industry and travel agency after completing college education from Bangalore. He got inclined towards the art of fishing since school days while outing at lakes and reservoirs.

His visit to Andamans during 2005 took him right into the arena and he decided to convert his passion into his carrier. During 2006 in association with his close friend Mr. Darran Davis he established Sea Fishing India Pvt. Ltd. which is recognized by two apex level regulatory and conservatory authorities for Sport Fishing Worldwide namely the International Game Fish Association (IGFA) & the Bill Fish Foundation (BFF).

Presently this Company caters to dedicated anglers from around the World in its state of the art crafts. Plans are on the expansion of the Company overseas and in mainland India, using a format of sustainably managing fisheries resources and generating income by novel and new methods.

Arif M. Mustafa

Sponsor

Upcoming

Anna
Sport Fishing

Port Blair

Sporting Families

EMERGING SPORT FISHING OF ANDAMANS

Sports Fishing Boats in Operation

Fishes Caught and Released during Sport Fishing in Andamans

Stages of Sport Fishing
(CASTING, POPPING, LANDING, RELEASING)

Courtesy
SEA FISHING (P) LTD.
Port Blair

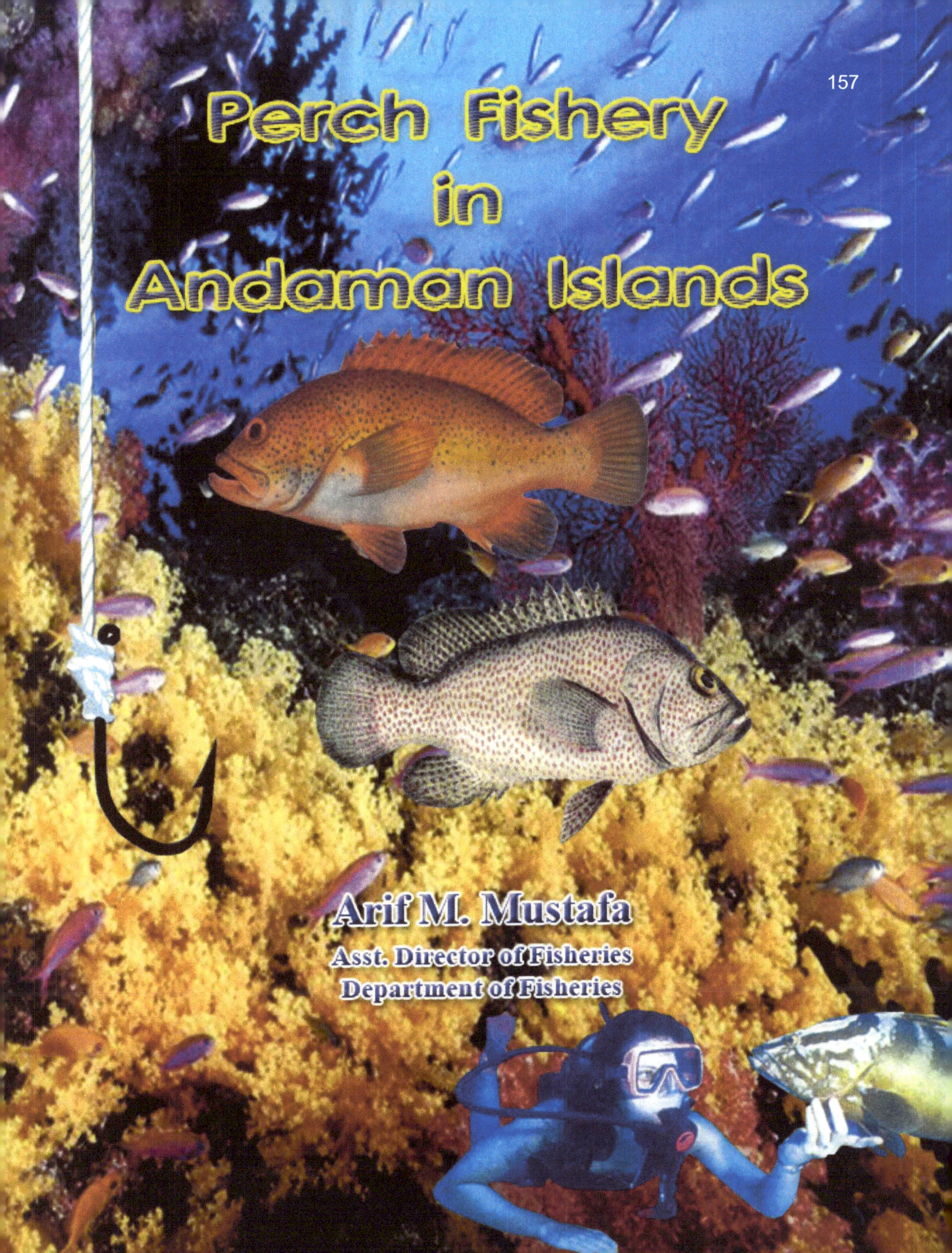
Perch Fishery
in
Andaman Islands
Arif M. Mustafa
Asst. Director of Fisheries
Department of Fisheries

Exported Snappers of Andaman Islands.

Front Cover Photo Courtesy
Jack, Migdalski, E.C., Boggs, C.,

Courtesy - FAO

PERCH FISHERY IN ANDAMAN ISLANDS

Arif M. Mustafa
Asst. Director of Fisheries
Department of Fisheries
Andaman & Nicobar Administration

2011

In the interest of promoting and developing the Marine Food Industry in sustainable and eco-friendly manner, this booklet has been published

By

The Andaman Chamber of Commerce & Industry
Port Blair

First Edition - 2011, 200 copies

Arif M. Mustafa
Asst. Director of Fisheries
Department of Fisheries
Mob. : 09434261302

For bibliographic purpose this document may please be quoted as follows:
Mustafa, A. M. 2011, Perch Fishery in Andaman Islands.

MESSAGE

शक्ति सिन्हा
Shakti Sinha
मुख्य सचिव
Chief Secretary

फोन/Ph. No. 03192-233110/234087
फैक्स सं./Fax No. 03192-232656
ई-मेइल/E-mail: cs-andaman@nic.in

अण्डमान तथा निकोबार प्रशासन
ANDAMAN AND NICOBAR ADMINISTRATION
सचिवालय / SECRETARIAT

Port Blair, dated the 03rd January, 2011

Message

It is at a very appropriate time that such a useful book has been published when the Marine Food Industry in the Andaman & Nicobar Islands is poised to grow by leaps and bounds.

The hidden Bio-Fishery facts related to Perches has been presented lucidly. This amazing group of fish deserve to be harvested sustainably.

It is my wish that our Planners and Professionals will seriously consider the recommendations made here to avoid any stock depletion or commercial extinction of Perches in the days ahead.

I congratulate both the Author and the Andaman Chamber of Commerce & Industry for this valuable initiative.

3/1/11
(Shakti Sinha)

Index

Sl. no		Topic		Page
1.	--	Message	--	03
2.	--	Introduction	--	05
3.	--	Demersal fish potential	--	06
4.	--	Perch fisheries	--	09
5.	--	Perch, the fish	--	11
6.	--	Market	--	17
7.	--	Inference	--	21
8.	--	Recommendation	--	22
9.	--	Refrences	--	23
10.	--	Perch Exporters	--	26

1. INTRODUCTION

Port Blair the insular capital of the largest Indian archipelago, the Andaman & Nicobar Islands is experiencing a rapid quantitative enhancement in the landing and export of bottom dwelling fishes collectively called Perches, comprising of three different groups of fishes namely Groupers, Snappers and Emperors.

This territory is destined to emerge as a major seafood export center during the current millennium, because it has the largest and the least exploited EEZ among all the maritime states and union territories of the country. The vast oceanic regime of about 6 lakh sq. km and the linear stretch of about a thousand km. amid Bay of Bengal in the northern Indian Ocean apparently offers tremendous scope for the development of industrial fisheries in a big way, paving way to enhance the socio-economic quality of human race settled in these islands. However, the conception of Blue revolution, still twinkle, faintly in a horizon far away.

This island territory from its highest peak to the deepest depth possess a unique and fragile eco-set-up. It exhibits a different geological, topographical, meteorological and oceanographical features. This archipelago is believed to have born of intense volcanic activity, endowed with an extremely shrunken pseudo continental shelf of about 35,000 sq.km. which drastically restrict the scope of benthic production. The depth beyond continental edge is steep and still uncharted at many places, deepest depth known is 4,198m. east of Car Nicobar island. The territory experiences two monsoons namely South-west and North-east each year, with an average downpour of 3,000mm., further, about 39,400 cubic m./second of fresh water and more than 250 million tones of sediment per year (Bhattathiri, P.M.A. & Devassy, V. P. - in anon 1980) is continuously being drained through adjoining rivers & Canals into the sea around. The vast adjoining sea- air stretch generates innumerable moderate to intense gale and cyclone, which, many a times devastates the coastal areas in upper Bay of Bengal and Arabian sea. The deep sea temperature over eastern sector is above normal due to submarine tectonic activity, presence of volcanic chain and thermal vents. Free nutrients like Phosphate and Nitrate are negligible in coastal water, deep water mass is depleted of oxygen, thus limiting productivity, however the western sector is comparatively better with reported upwelling. The whole of EEZ as such is Oligotrophic i.e. less productive in nature due to natural limitation imposed upon it. The insufficient food stock in the form of phyto and zoo plankton is to a great extent supplemented by detritus load brought in through associated ecosystems.

The richest among the auxiliary ecosystems in terms of productivity, are the 'Mangroves' and the 'Coral reef' eco-system. Since time immemorial the shallow coastal marine areas are mainly being exploited by the prehistoric Negrito aborigin population of Andaman and the Mangoloid race of Nicobar. Those areas are still pristine. The renowned marine naturalist Cap. Jacques Cousteau after exploring this area during 1989 has put on record that 'Indian waters are preserved, rich and virgin', recently during 2002 the UNDP under GEF programme after extensive expert observation has inferred that 'The A & N archipelago hosts probably the healthiest and least impacted expanse of coral reefs within the Indian Ocean'. Island's coral domine of about 2,000 sq. km. is producing about 24,600 tones. of biota per annum (Mustafa, et. al 1987) 10% of which are fishes (Rao, G.C. 1999) i.e. 2,460 tones. fish per annum. Similarly, the mangrove system is marvelous and extend substantial bio-support to coral reef system in addition to its own production.

2. DEMERSAL FISH POTENTIAL :

Various analytical and empirical postulations are afloat, significant, analytic estimate in respect of Andaman sea has been made by the National Institute of Oceanography, Goa which recorded that 4,70,000 tones fish per annum could be harvested from Andaman Sea (Qasim, S. Z. in annon 1980) a substantial part of which is under island's EEZ. Empirically the Fishery Survey of India, Mumbai, earlier reported that 2,43,500 tones fish per annum (Sudersan, D. 1989 & 1990) (Anon - 1991) comprising of 1,39,000 tones Pelagic; 22,200 tones Demersal and 82,000 tones Oceanic could be harvested. This estimate has been recently revised to 1,48,000 tones fish per annum (John, M.E, et.al. 2005) comprising of 56,000 tones Pelagic; 32,000 tones Demersal and 60,000 tones Oceanic as harvestable potential yield level for the EEZ of A & N Islands. Synthesis of literature reveals that the benthic or demersal fish stock is well under 10% of the total projected stock. The prime demersal stock assessment beyond 30m. depth was earlier made through bottom trawl swept area method using CPUE obtained in trawl survey. Based on that the potential annual yield could be :-

ESTIMATED POTENTIAL DEMERSAL YIELD PER ANNUM		
Sl. No.	Species / group	Potential yield (tonnes)
1	Elasmobranchs	4200
2	Silver bellies (Leiognathids)	5000
3	**PERCHES**	**8000**
4	Pomfrets	1900
5	Catfish	1000
6	Threadfins (Polynemids)	400
7	Croakers (Sciaenids)	1200
8	Gerrids	1400
9	Goat fishes *(Upenoids)*	900
10	Silver grunt *(Pomadasyds)*	100
11	Drift fish *(Ariomma indica)*	300
12	Threadfin breams *(Nemipterids)*	500
13	Lizard fish	150
14	Flat fish	50
15	Bulls eye (Priacanthids)	100
16	Cephalopods	100
17	Penaeid shrimps	800
18	Crabs	1000
19	Deepsea lobster	120
20	Deepsea shrimps	110
21	Deepsea fishes	1970
22	Others	2700
	TOTAL	32 000

Source - FSI

However, unlike mainland India where there are millions of square km. of suitable ground available for bottom trawling these islands offer extremely limited trawlable area due to topographical limitations mentioned above; hence it may in theory be wise to have such assumption, but to our knowledge, no tropical nation, where there are corals has assessed the demersal stock potential based on bottom trawl data. Bottom trawling is in fact lethal to any tropical island ecosystem and has been forbidden Globally in coral infested areas; hence, gear diversification was later adopted for survey introducing bottom set long line and vertical lines beside hand line. The family and genus wise composition thus observed is as follows:

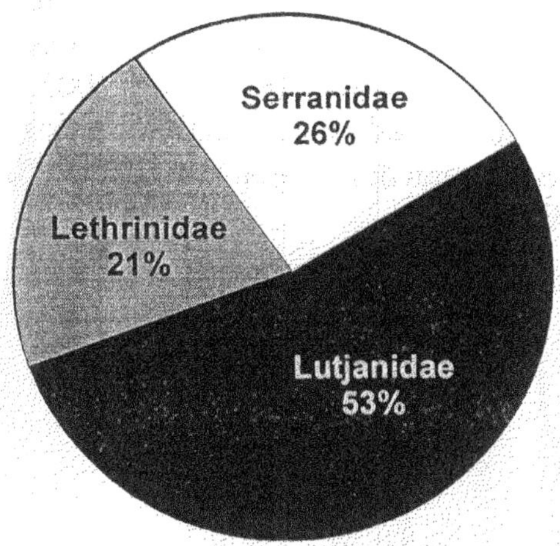

Family-wise percentage composition of perches recorded in bottom set longline survey during 2000-06

(Source - Somvanshi V.S.)

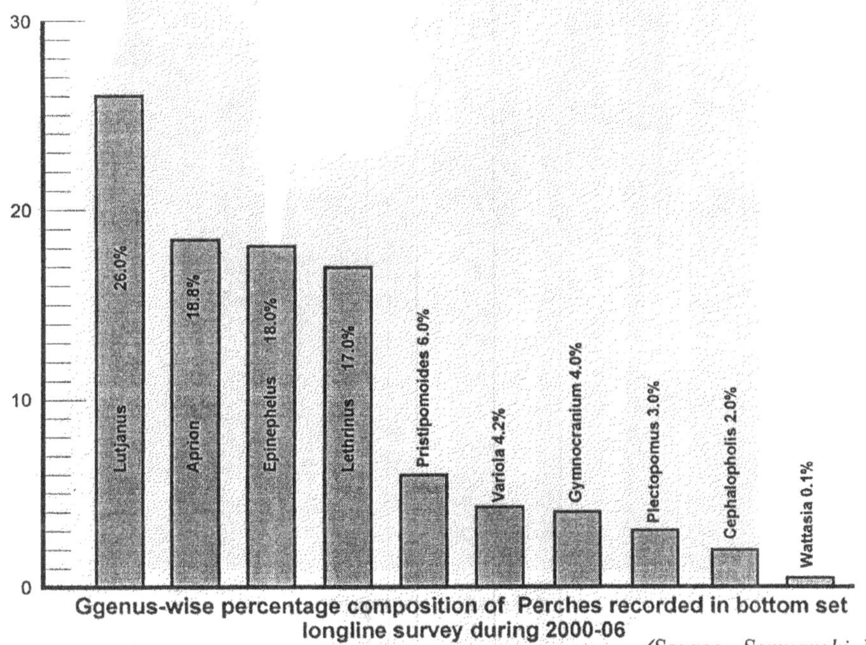

Ggenus-wise percentage composition of Perches recorded in bottom set longline survey during 2000-06

(Source - Somvanshi, V.S.)

"The aggregate hooking rate of perch and allied resources, registered from Andaman waters is 1.91%, whereas, aggregate hooking rate is 4.81% from the Nicobar waters.

The rich fishing grounds for snappers are in Lat. 7°, 8°, 10° and 13° N in Long 91° - 93°E. The groupers were found to be abundant in Lat. 7°, 8°, 10° and 13°N in Long 91° and 93°E only, whereas emperors were recorded in hooking rate above 3% in areas in Lat. 7°, 8°, 10° and 13°N in Long. 91° to 93°E.

Based on the knowledge gained through the demersal trawl surveys and bottom set long line surveys of limited nature the standing stock of perch resources in the continental shelf area around Andaman & Nicobar Islands is estimated. The standing estimated stock thus be 16,100 tonnes. Considering the current production of perches as 9,000 tonnes (2003 - 2004) and considering that the natural mortality rate of perches as 0.61 based on the earlier studies, it is estimated that the annual potential yield of the perches from Andaman & Nicobar waters is 9,500 tonnes ± 20%, which falls in the range of **7,600 tonnes to 11,400 tonnes."**

(Somvanshi, V. S. 2006)

Giant Grouper
(Epinephelus lanceolatus)

3.PERCH FISHERIES :-

Today there is a mad rush to export Snappers, Groupers and Emperors, all associates of coral eco-system from A & N Islands in general, and South Andaman district in particular. Every day hundreds of kilograms are being airlifted from Port Blair; export houses are in full swing. A glance through the available statistical data for the last sixty years manifest 8% to 17% Perch contribution towards total fish landings.

These fishes have great traditional valve in many neighbouring countries but unfortunately their own stock is reported to have depleted and the coral eco-system degraded, hence the importers have turned towards this territory. Now the time has come for us to monitor this precious stock in line with the standard and established scientific norms. The resource potential propaganda sound lucrative and is indeed so, but for a very short span of time. Once over exploited commercial extinction will set-in, ultimately depleting the whole stock and eroding the fabric of coral reef bio-niche.

During a pre-Tsunami - 2004 study made to evaluate the extent and magnitude of Snapper, Grouper and Emperor export fishery in South Andamans. It has been observed that the export fishing pressure is upon 0.2% EEZ and 4% of continental shelf, at Western Fishing Zone of South Andaman (WFZ-SA) which contributes 90% of total export. Four, core offshore fishing grounds, extending from little Andaman upto Baratang island have been identified as shown in the map, covering a cumulative area of 1,441 sq.km. (1,44,100 Hect.) as follows :

Sl. No.	Name of the core fishing ground		Area	
			Sq.km.	Hect.
1	South Coral Bank	F 1	599	59,900
2	North Sentinel Is.	F 2	131	13,100
3	South Sentinel Is.	F 3	169	16,900
4	North of Little Andaman	F 4	542	54,200
	TOTAL		1,441	1,44,100

Wherein, CPUE is 565.5 kg/unit/month through mono or branched vertical line fishing and rate of harvest per sq.km per annum from the identified grounds is 0.15 t. per fishing unit (Mustafa, et. al. 2001).

Fortunately the export fish fishing fleet of traditional crafts over the Western Fishing Zone of South Andaman (WFZ-SA) could operate only for seven months due to inclement sea, for rest of the five months there is no fishing beginning mid May to mid October coinciding with the onset of South-west monsoon. There are two major fish landing centers namely Wandoor and Guptapara with a total fleet strength of 92 & 70 respectively. These fishing units operates at 50% to 70% level. Rest of the 50% to 30% remains idle at landing center. Duration of each fishing trip varies from 15 to 30 hrs. depending on the fishing ground. The fishing fleet operate line trolling on their way to fishing ground and while returning which gives an additional pelagic yield of 3%. While at fishing ground, vertical mono hook line or vertical branch line with 4 to 6 hooks are used to harvest export variety fishes. Month of February is found to be the most productive month for western Fishing Zone. Snapper is the largest group of fishes contributing 78.5% followed by Grouper 13% and Emperor 5.5% in the total landing. Yield per trip of 15 to 30 hr. is 95 kg. whereas 24 kg. is the average catch rate at 4 hrs. of actual time of Fishing (ATF) per trip.

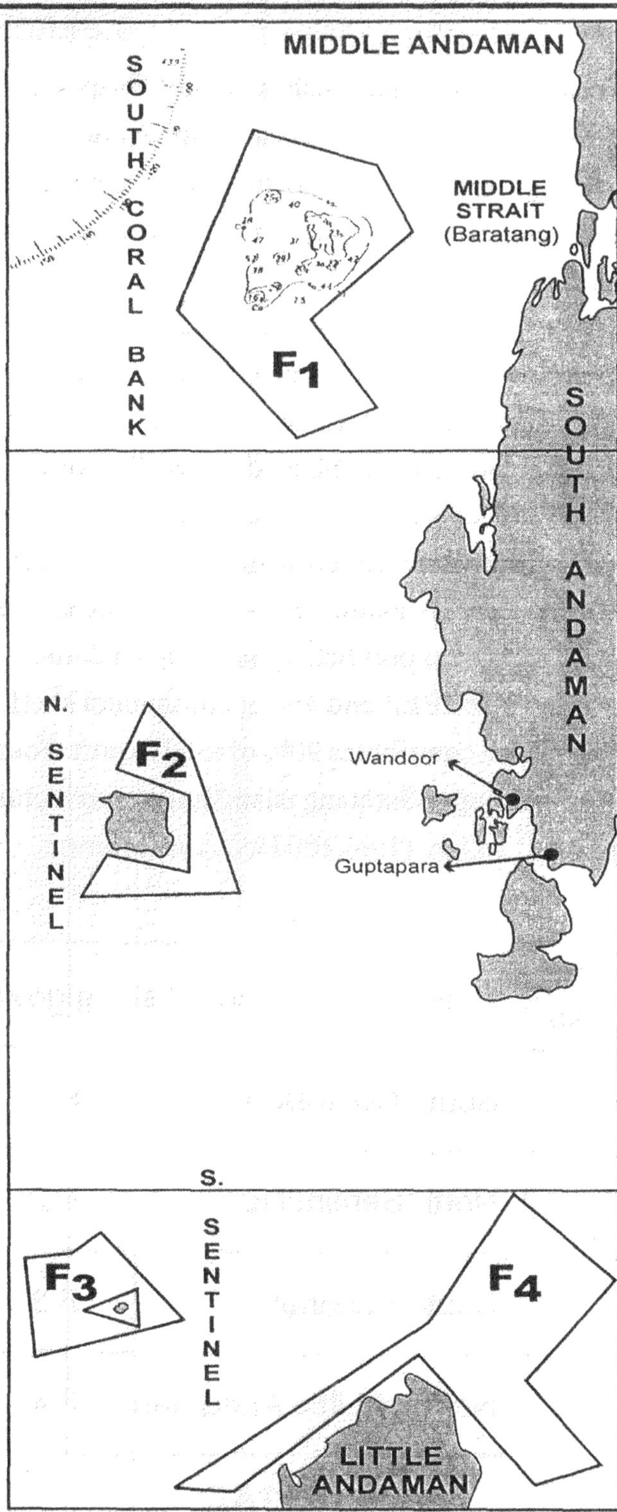

Perch fishing ground WFZ-SA.

4. PERCH THE FISH

Following 16 species of fishes belonging to 3 taxonomical families under 7 genera are found being exported to foreign countries and mainland India from Port Blair. Lutjanidae (Snapper) is found to be the largest family followed by Serranidae (Grouper) and Lethrinidae (Emperor) as follows :

		Lutjanidae		
Sl. No.	***Scientific name***	***Local name***	***Trade name***	***% comp. in export***
1	*Lutjanus malabaricus*	Lall	Red snapper	
2	*L.gibbus*	Kankata lall	Red snapper	
3	*L. bohar*	Kutta bhetki	Red snapper	43.3
4	*L.argentimaculatus*	Lall	Red snapper	
5	*L.sebae*	Lall	Red snapper	
6	*Pristipomoides filamentosus*	Mrigal	White snapper	
7	*P.typus*	Mrigal	White snapper	23.3
8	*Aphareus rutilans*	Mrigal	White snapper	

		Serranidae		
1	*Epinephelus chlorostigma*	Gobra	Brown/Black grouper	
2	*E. diacanthus*	Gobra	Brown/Black grouper	
3	*Cephalopholis sonnerati*	Lall Gobra	Red grouper	16.6
4	*Plectropomus leopardus*	Lall Gobra	Coral trout.	
5	*P. levis*	Lall Gobra	Coral trout.	

		Lethrinidae		
1	*Lenthrinus elongatus*	Kushal	Emperor	
2	*L.miniatus*	Kushal	Emperor	16.8
3	*L.lentjan*	Kushal	Emperor	

I. Family - **Lutjanidae (Snappers).** They are moderately oblong and fairly compressed in shape having terminal moderate to large prostrusible mouth; two nostrils on each side of snout, anterior of head (area between eye and mouth) devoid of scales.

Colour variable, mainly from yellow through red to blue, often with blotches, lines or other patterns.

Tropical in distribution coinciding with the range of reef building corals, mostly demersal from coastal waters to continental shelves and slopes upto - 500 m depth generally forming shoals (William, D; Anderson, Jr. - 1987). Most of them are unspecialized nocturnal carnivores. They are gonochoristic i.e. sexes are distinct. They spawn in groups probably twice a year. Fecundity ranges from a lakh to about a million eggs (Thompson, R.H. : Munro, J.L. 1974).

39 species are known to occur in A & N Islands (Rao, D. V.; Kamla Devi - 1997) out of which 8 species are being exported contributing 66.6% of export.

1. Genus - Lutjanus, Five Species collectively contributing about 43.3% of export, all are commercially called 'Red snapper'.

(i) L. malabaricus (Bloch & Schneidr - 1801) (Lall) A red or red orange colour fish, colour light on lower part of side and on belly. Fins reddish. Found in and around coastal and offshore reefs within - 100 m depth. Could grow upto 13.6 kg. in weight. (Grant, E.M. - 1982). A shoaling variety. Size : 90cm.

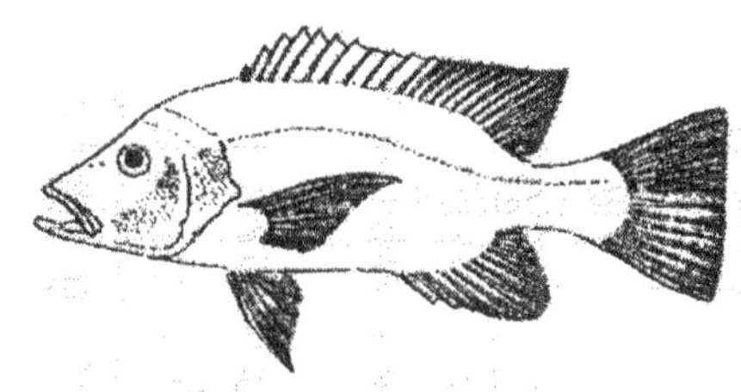

(ii) L.gibbus (Forsskal - 1775) Kankatta Lall) Generally deep red in colour, darker on back and upper portion of head, orange hue on lower part of opercle and in pectoral axil. Occurs mainly on coral reefs at a depth of around - 30m. often forming large static shoal during day time (Allen, G. R. & Talbot, F.H. 1985). Size : 50cm.

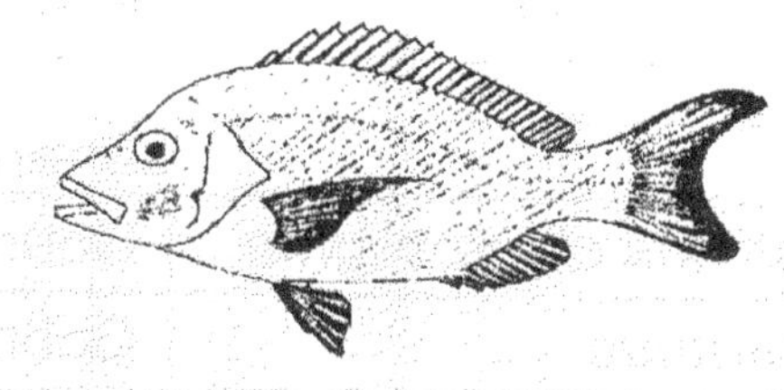

(iii) L.bohar (Forsskal - 1775) (Kutta bhetki). Red or purplish red in colour. Generally dark reddish brown on back, shading to pink or whitish ventrally, cheeks and lower half of sides often bright red in large adults. Usually inhabits coral reef areas but occassionally, solitary individuals occur down to - 70m in rocky areas. (Fischer, W. & Bianchi, G. - 1984). Size : 75cm. L. bohar is avoided by many buyers due to its suspected ciguatera fish poisoning acquired through feeding upon certain herbivorous fishes which feeds on a toxic dinoflagellate Gambierdiscus toxicus. This dinoflagellate lives on benthic algae or on dead corals. (Yasumoto, T. et.al. 1977).

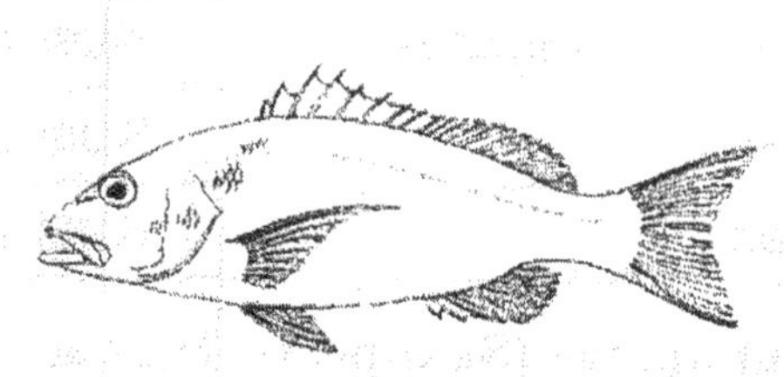

(iv) L.argentinaculatus (Frsskal - 1775) (Lall) Red brown, greenish brown on back, reddish on sides and ventral parts, deep water catch often overall reddish. Lives in sheltered coral reefs or in areas where siltation is heavy and coral growth is poor in the depth range of - 80 to 120m. Size : 120 cm.

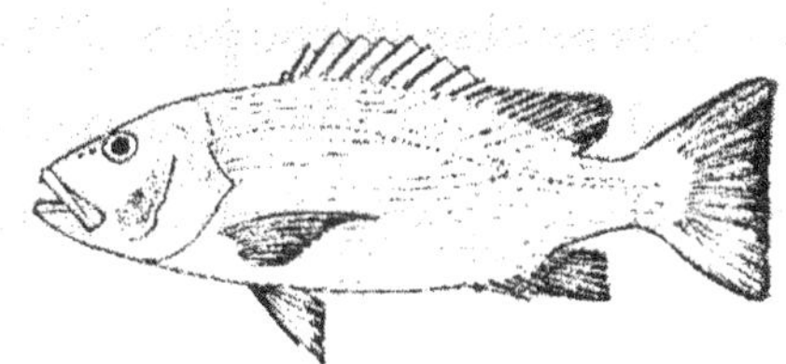

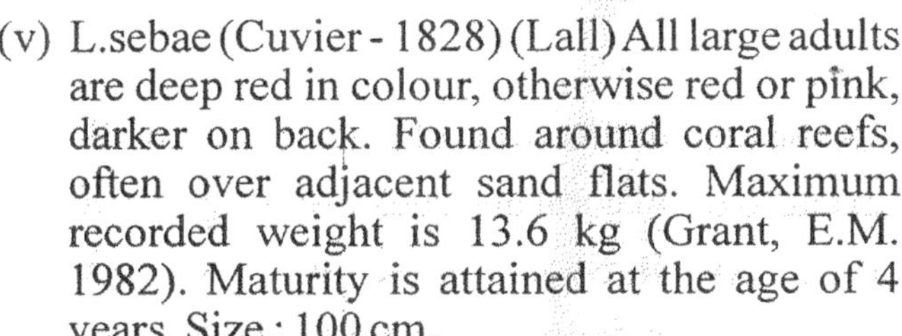

(v) L.sebae (Cuvier - 1828) (Lall) All large adults are deep red in colour, otherwise red or pink, darker on back. Found around coral reefs, often over adjacent sand flats. Maximum recorded weight is 13.6 kg (Grant, E.M. 1982). Maturity is attained at the age of 4 years. Size : 100 cm.

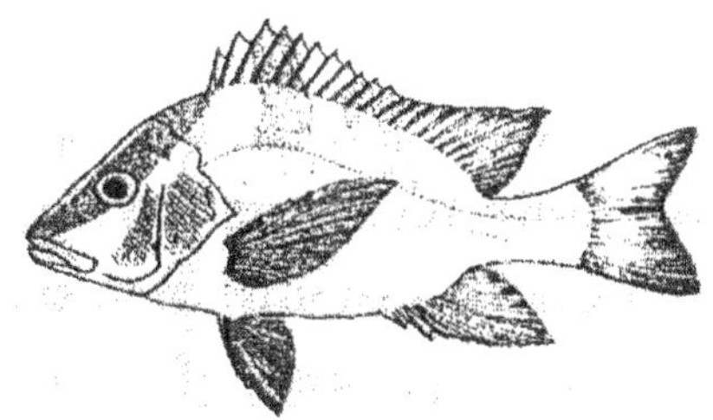

2. Genus - Pristipomoides - Two species contributing 16.6% of the export, collectively called 'White snapper'.

(i) P.typus (Bleeker - 1852) (Mrigal) Body rosy in colour, fins with yellow tinge, brownish yellow longitudinal vermiculation on top of head, dorsal fin with paler spots or rosy reticulate pattern. Generally caught from - 40 to -80m depth. Size : 70 cm.

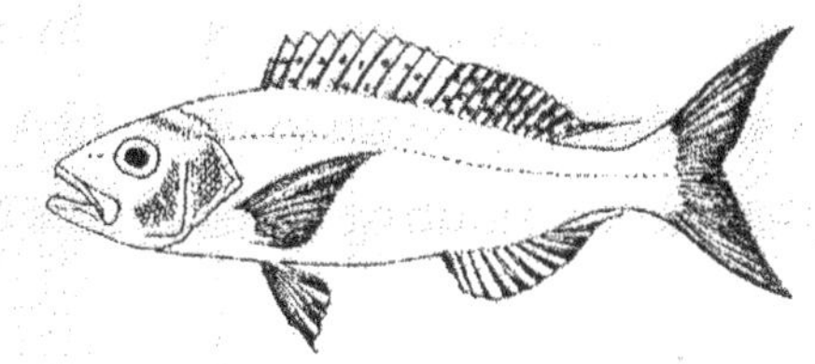

(ii) P. filamentosus (Valenciennes - 1830) (Mrigal) Colour varies with ground from reddish purple to lavender with blue tinge, small blue spots on top of head, dorsal fin with two yellowish longitudinal lines. Fished from deep waters mostly - 150 to -200 m depth. Size : 80cm.

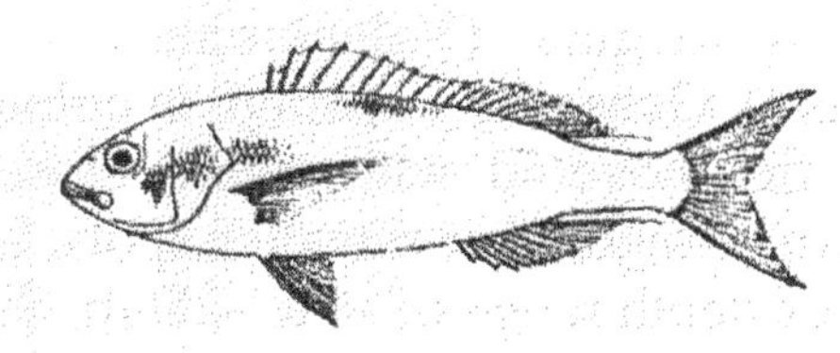

3. Genus - Apharens - This white snapper is represented in export only by one species.

(i) A. rutilans (Cuvier - 1830) (Mrigal) which constitute 6.6% of export. Colour is overall blue-grey, purplish - brown above with little yellow or pinkish fusion; edges of preopercle and opercle outlined with black. inhabit coral and rock reefs at depth between -60 to -70m. Size : 40cm.

II. Family - **Serranidae (Groupers).** They have robust or somewhat compressed, oblong-oval to rather elongated body. This is a group of closely related fishes (Smith, C.L. - 1972). Mouth large, with small slender, inwardly depressible teeth. A single dorsal fin with 7 to 12 strong spines; anal fin with 3 spines.

Colour changes instantaneously when under stress or due to a change in environmental conditions. Colour variable with patterns of light or dark stripes, spots, vertical or diagonal bars or nearly plain.

Distribution of groupers corresponds roughly to the distribution of reef building corals. They are demersal, carnivores and solitary in nature living in and around coral and rocky reefs upto a depth of -200m. Sexually, majority of them are typical to exhibit Synchronous and Protogynous hermaphroditism i.e. presence of ovotestis and sex change from female to male (Thompson, R & Munro, J. L. 1974; Shapiro, D.Y. 1987). All groupers for which there is evidence spawn in dense shoal for a restricted period of one to two weeks possibly once in a year. Fecundity ranges from 90,000 to about 6.5 lakhs. Intra generic hybridization is reported to occurs in this family.

43 species are known to occur in A & N islands (Rajan, P.T., 2001) out of which 5 species are being exploited contributing 16.6% of the total export.

1. Genus - Epinephelus - Two species collectively contributes about 11.6% of export, both are commercially called 'Brown/Black grouper'.

(i) E.chlorostigma (Valenciennes - 1828) (Gobra). Generally brownigh in colour body and fins covered with small, closely set, brown hexagonal or roundish spots. Lives in a wide depth range of -4 to -280m. Size : 75 cm.

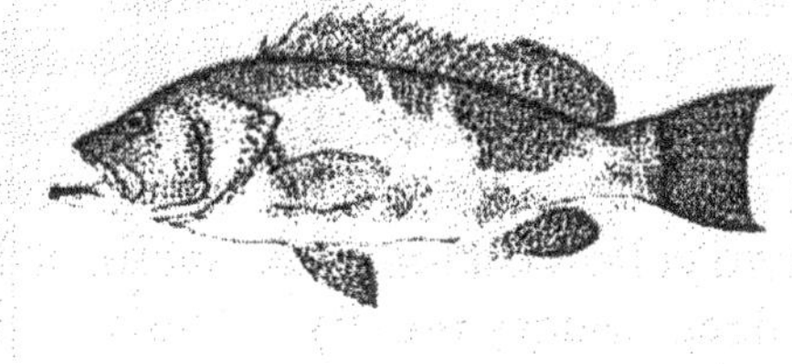

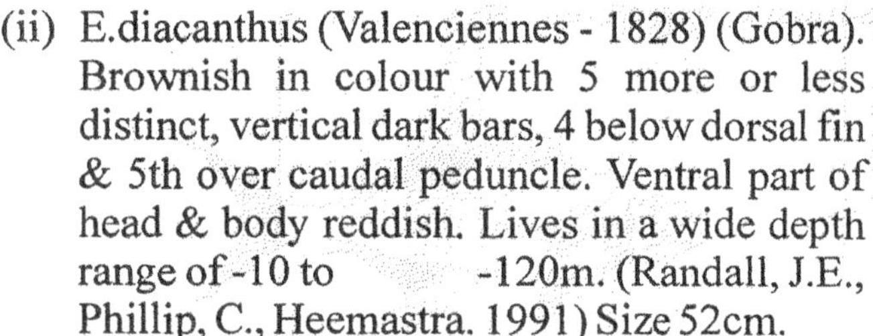

(ii) E.diacanthus (Valenciennes - 1828) (Gobra). Brownish in colour with 5 more or less distinct, vertical dark bars, 4 below dorsal fin & 5th over caudal peduncle. Ventral part of head & body reddish. Lives in a wide depth range of -10 to -120m. (Randall, J.E., Phillip, C., Heemastra. 1991) Size 52cm.

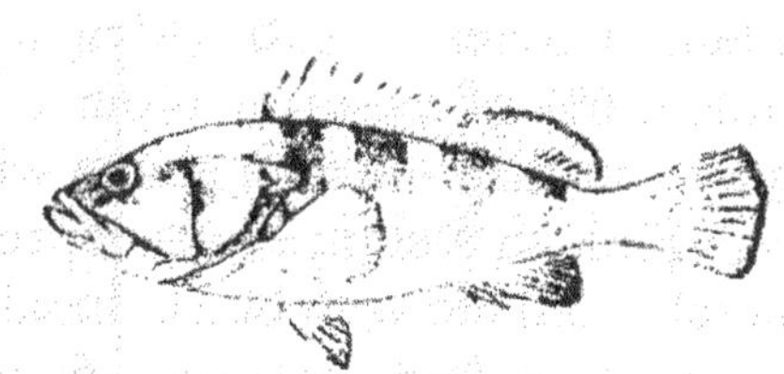

2. Genus - Plectropomus - Two species making thin marginal contribution of 3.3% towards export, both are commercially called 'coral trout'.

(i) P.leopardus (Lacepede - 1802) (Lall gobra). A reddish to olivaceous coloured fish impregnated with small dark-edged blue dots over head, body and medians fins. A blue ring on edge of orbit. Lives in coral reefs generally in the depth range of -10 to -30m. Size : 70 cm.

(i) P.levis (Lacepede - 1802) (Lall gobra). It shows two colour phases, one whitish or pale yellowish with five dark to black saddle like bars fainting ventrally. The second colour phase is brown, olivaceous, red or nearly black with or without five dark bars. Both colour phase will have a few widely scattered, small, dark edged blue spots on body; Lives in coral reefs in the depth range of -10 to -30 m. Size : 100 cm.

3. Genus - Cephalopholis - Only one species of Red grouper is being exported.

(i) C. sonnerati (Valenciennes - 1828) (Lall gobra) having least share of 1.6% to export. Bright orange-red in colour, usually with scattered faint bluish white spots, head purplish with numerous close-set orange red spots. Lives in coral reef areas upto - 100m depth. Size 57 cm.

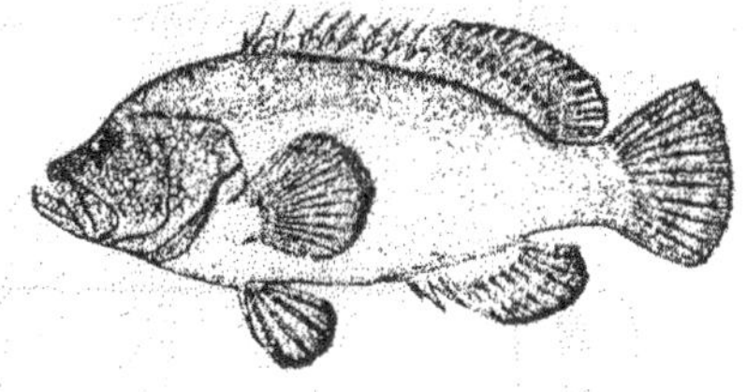

III. Family - **Lethrinidae (Emperors).** Large head fishes with wide sub-orbital space making the snout rather pointed, mouth moderate mostly with thick & fleshy lips. A single dorsal fin with 10 spines.

Colour pattern is usually brown, green or grey with tints of red, pink, yellow or blue, tinted areas brightens when excited or after death. Many species show scarlet coloration to inside of lips.

Mostly associated to coral reefs but also found over rocky reefs and soft bottom coastal areas upto - 100 m depth. All are demersal, carnivores mostly hunting at night. Sexually they mature as female progressively becoming hermaphrodite. Males are reported to be longer than females. They spawn in shoal throughout the year mostly in lagoons or near outer reef edge.

19 species are reported to occur in A & N Islands (Rao, D. V.; Kamla Devi - 1997) out of which possibly 3 species belonging to one genus Lethrinus are being exported, contributing about 16.8% of export. The species of this genus are taxonomically most difficult to distinguish.

(i) L.elongatus (Valenciennes - 1830) (Kushal) This species has not been reported earlier from A & N islands. This is a greenish - grey fish with brown patches above; colour fades towards ventral side. A prominent red line is always seen above and below the lips.

Lives upto a depth of -185 m. in the vicinity of coral reefs. Size : 100 cm (Sato, T. & Walker, M. - 1984).

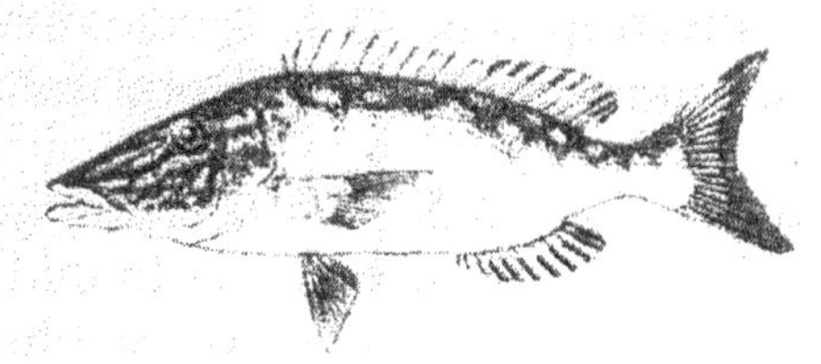

(ii) L.miniatus (Bloch & Schneider - 1801) (Kushal) Body olive - brown above, sometimes pink below, occasionally 2 or 3 blue streaks radiating from eye. Vertical fins pink to red, with brighter margins; paired fins yellow. Inhabits coastal waters Size : 90 cm.

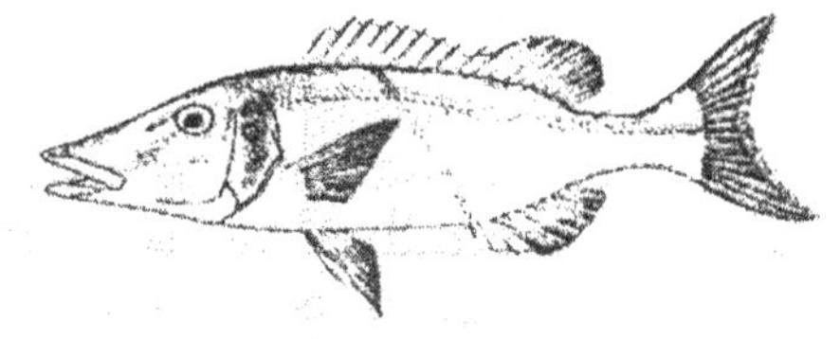

(iii) L.lentjan (Lacepede - 1802) (Kushal) Body is olive/green above, paler below. A bright red spot on posterior edge of operculum and often another on outer pectoral fin base. Each scale on back sometimes with a white center. Dorsal and caudal fins stripped with orange. Lives in coastal coral reefs down to a depth of -50 m. Size : 50cm.

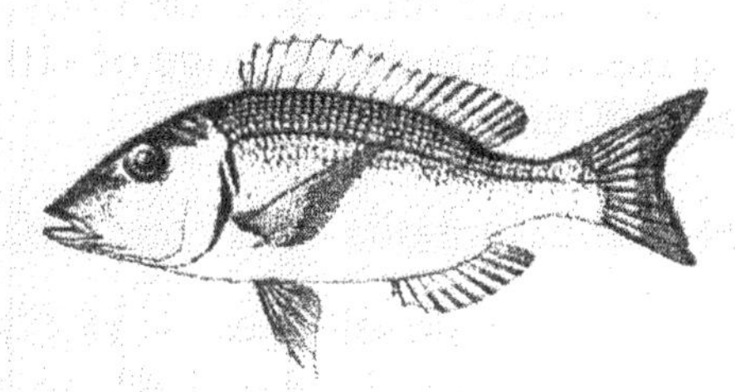

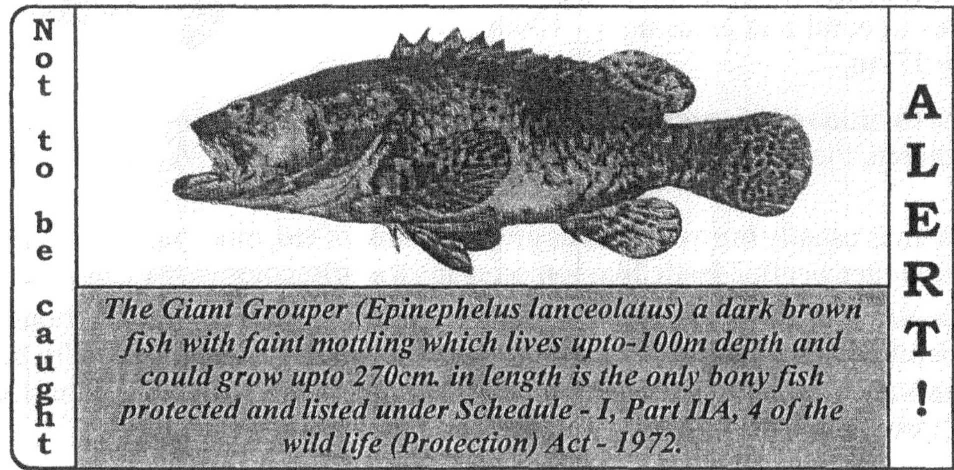

The Giant Grouper (Epinephelus lanceolatus) a dark brown fish with faint mottling which lives upto-100m depth and could grow upto 270cm. in length is the only bony fish protected and listed under Schedule - I, Part IIA, 4 of the wild life (Protection) Act - 1972.

5. MARKET

Export - Direct export of live Perches to South - East Asian countries from South Andaman Zone was made by M/s. Oya Exim (P) Ltd., during the period from 2005-2007 by engaging live fish carrier vessel from abroad. This has been discontinued due to various immigration hurdles with regard to direct export.

Shipment of fresh and frozen consignment in bulk, through refer containers to mainland India continues by six registered Fish Trading firms.

There are 144 registered Fish Traders at Port Blair, they keep despatching ice packed consignments to mainland India by air, depending on availability of cargo space.

At mainland India the consignment is dressed and re-packed for export.

EXPORT IN TONES

Year	Fresh and Frozen							Total	Live	Total
	AFL (1)	IMPL (2)	AB (3)	NR (4)	TSG (5)	AA (6)	Others		OE (7)	
1997	155	-	-	-	-	-	32	187	-	187
1998	111	39	-	-	-	-	21	171	-	171
1999	60	-	-	-	-	-	05	65	-	65
2000	54	-	-	-	-	-	06	60	-	60
2001	50	17	-	-	-	-	18	85	-	85
2002	81	92	-	-	-	-	78	251	-	251
2003	38	58	-	-	-	-	51	147	-	147
2004	40	23	-	-	-	-	64	127	-	127
2005	26	23	-	-	-	-	88	137	08	145
2006	57	31	-	-	-	-	135	223	03	226
2007	-	189	-	-	-	-	279	468	04	472
2008	-	86	-	-	-	-	456	542	-	542
2009	-	158	21	05	-	14	452	650	-	650
2010	-	36	07	41	02	-	491	577	-	577

(1) Andaman Fisheries Ltd (2) M/s Island Marine Products Ltd (3) M/s Andabar Cold Storage (4) M/s New Rich Marine (P) Ltd (5) M/s TSG Aqua (6) M/s Alberta Agro (P) Ltd (7) M/s Oya Exim (P) Ltd. Source: Dept. of Fisheries

Local Consumption : Annually about 400 tones of exportable varieties is consumed locally in South Andaman, out of which 50 tones goes to Shipping Services and 30 tones to Defence establishments.

The traders and middle men are becoming inclined towards export outlets due to better renumeration.

INFERENCE

The prevailing facts demand that time has now come for the Fisheries Managers to put in their knowledge and come up with prudent, management guidelines for the industry such as dispersal of effort, opening new grounds, cold chain, zonal quantum of stock that prevails. The level to which it could be harvested ? and the suitable fishing technique that guarantee least eco-degradation.

Popularisation of less expensive, low volume, high return, live demersal fish export, technique. The available 'strata resource potential figure' are collective, encompassing a huge area indicating probable level of fish mass presence and density. Such hypothetical projections are a mere indicator of richness and not an absolute tool to manage this dynamic resource. An appropriate way to achieve a realistic management policy is to have a heterogeneous quantitative assessment of resources rather than a homogenous projection.

Current export industry is totally dependent on raw material from coral reef and adjoining areas thriving upon the narrow coral infested continental shelf. Scientific community the World over agrees that coral reefs have low standing crop of exploitable species, say to a tune of only 10% of the standing stock. Thus, the presently targeted fishes are susceptible to depletion, if exploited, a little beyond limit, as has widely occurred in many tropical nations. Secondly, the ecosystem on which they depend should stay intact. The major threat to that ecosystem is bottom trawling, which unlike other fishing methods, is extremely destructive to corals and the coral reefs when deployed upon it. It mercilessly sweeps the sea floor and sucks-in whatever comes through. Hence in the interest of present day export industry, there shall not be any bottom trawling over coral infested areas. Even the bottom-set-long-lining in the coral area does not appear sensible.

Although A & N Islands have 5,96,554 sq.km. of EEZ the present day fish export industry thrives upon a narrow, fragmented area of 1,441 sq.km which is only 0.2% of the EEZ and 4% of the restricted (34,965 sq.km) narrow and elongated continental shelf. The fishing pressure on targeted varieties is being augmented through the induction of more and more traders and entrepreneurs to operate from Port Blair. The Western Fishing Zone of South Andaman (WFZ-SA) is known to hold rich coral bio-diversity and that is one of the prime reason of its being the abode of Groupers, Snappers and Emperors.

Recent past has witnessed extinction of many fish stocks including Groupers, Snappers and Emperors elsewhere in the World due to unmanaged and uncontrolled fishing. The wild stock of larger Groupers and Snappers are found to be the first to show instant depletion through a shift in catch composition when overexploited (Clark & Brown, 1977; Munro & Smith, 1985; Munro & Williams, 1985 and Russ, 1985).

Prevailing fishing intensity of one fishing unit per 57sq.km. of shelf area studied, is yielding between 100 to 200 kg of targeted fish per annum. This rate of exploitation appears quite conservative, but since we do not know the maximum or permissible limit and extent of this harvest and there is no policy to regulate fishing and export in time and space; an unckecked trade expansion may lead to catastrophic devastation of the whole marine eco-system over WFZ-SA in particular and elsewhere.

RECOMMENDATION

In view of the prevailing facts presented above, we are required to -

1. *Evolve a realistic population dynamic model for sustainable utilization of demersal export fish resources through the involvement of professional, research and development organisations.*

2. *A bio-mathematical population model is indispensably needed as an essential input in shaping a prudent demersal fisheries management plan.*

8. REFERENCES

1. **Allen G. R. & Talbot. F. H. 1985**. *Indo-Pacific Fishes no. 11 pp. 88. Bernice Pauahi Bishop Museum, Honolulu, Hawali.*
2. **Annon. 1980.** *The Andaman sea Tech. report 02/80. 1-208., NIO. Goa. Annon 1991. Plan for the development of deep sea fishing in A & N Islands, Report of the working group.*
3. **Antony raja, B. T. 1980.** *Current knowledge of fisheries resources in the shelf area of the Bay of Bengal. BOBP/WP/8.*
4. **Clark, S. H. and B. E. Brown. 1977**. *Changes in biomass of fin fishes and squids from the Gulf of Marine to Cape Hatteras, 1963-74, as determined from research vessel survey data. Fish. Bull. U.S. 75 : 1-21.*
5. **Devaraj, M. 1979.** *Fish population dynamics (course manual). CIFE 3 (10) 83: 1-98.*
6. **Devaraj, M. 1979.** *Studies on fish population dynamics - V. Chapter - III : 1-15, CIFE lecture note.*
7. **Fischer, W. and Bianchi, G. - 1984**, *FAO species identification sheets, (Fishing area - 51). Vol. III. LUT. Lut.2.*
8. **Gulland, J. A. 1962**.*Manual of sampling methods for fisheries biology. FAO. FB/T26.*
9. **Grant, E. M. 1982.** *Guide to Fishes. Pp. 896. 574 Col. Pls. Dept. Harbours and Marine, Brisbane.*
10. **John, M.E. et.al - 2005**, *Bull Fish Surv. India. 28, 1-38.*
11. **Mustafa, A.M. et.al. 2001** *Fish and Fisheries of exportable snappers (Ltujanidae) Groupers (Serranidae) and Emperors (Lethrinidae) from the Western Fishing Zone of South Andaman. Proc. Nat. Symp. on Biodiversity. ASA. 236-248.*
12. **Mustafa, A.M. et.al. 1987.** *Endangered coral reefs of bay islands and their ornamental fishes. Proc. Symp. Mang. of Coastal ecosystem and oceanic resources. ASA : 60-65.*
13. **Munro, J.L. 1984.** *Coral reef fisheries and world fish production. ICLARM, News letter 7 (4) : 3-4.*
14. **Munro, J. L., and I. R. Smith. 1985.** *Management strategies in multi species complexes in articanal fisheries. Proc. 36th Gulf. Cariab. Fish. Inst.*
15. **Munro, J. L., and D. M. Williams. 1985** *Assessment and management of coral reef fisheries. Biological environmental and socio-economic aspects. Proc. 5th Int. Coral Reef. Cong. Tahiti. 4 : 543-565.*
16. **Rao, G. C. 1999.** *Marine ecosystem around A & N islands. UNDP, Project report, ZSI, Port Blair. 1-33.*
17. **Rao, D. V. & Kamla Devi - 1997.** *Snapers (Family - Lutjanidae) of Andaman & Nicobar Islands. Environment and Ecology 15(4) : 924-931.*
18. **Rajan, P. T. 2001** - *A field guide to marine food fishes of Andaman islands. 1-260. ZSI.*
19. **Rao, D. V. 2003** - *Guide to reef fishes of A & N Islands. 1-555, ZSI.*
20. **Randall, J. E., Phillip, C. & Heemastra. - 1991.** *Indo-pacific Fishes no. 20. Berrnice Pauahi Bishop Museum, Honolulu, Hawaii.*
21. **Rao, D. V. & Kamla Devi 1997.** *Emperor fishes (family Lethrinidae) of Andaman & Nicobar Islands. Environment & Ecology 15(4) : 899-903.*
22. **Rao, D. V. 2003** - *Guide to reef fishes of A & N Islands. ZSI. : 1-555.*
23. **Russ, G. 1985** - *Effects of protective management on coral reef fisheries in the Central Phillippines Proc. 5th Int. Coral Reef. Congr, Tahiti 4: 219-224.*
24. **Sudershan, D.et. al. 1989.** *Assessment of oceanic tunas & allied fish resources in Indian EEZ. CMFRI. Nat. Cong. on Tunas, Cochin, 21-22 April.*

Reference

25. **Somvanshi, V. S. 2006** - *Letter No. 35-85/2006-F1 dt. 6.11.2006.*

26. **Smith, C. L. - 1971.** *A revision of the American groupers : Epinehelus and allied genera. Bull. Am. Mus. Nat. Hist. 146 : 67 - 242.*

27. **Shapino, Y. D. - 1987**. *Reproduction in Groupers. Pp.295-327. Tropical snappers and Groupers : Biology and Fisheries Management. Pub : West view Press / Boulder and Lonation.*

28. **Sato, T & Walker, M.** *(reviewed by Randall, J.E.) FAO species identification sheets (Fishing area 51). Vol.II. LETH. Leth. 5.*

29. **Thompson, R. & Munro, J. L. - 1974.** *The Biology, ecology and bionomics of the snappers, Lutjanidae. pp. 94-109. Caribbean Coral Reef Fishery Resources. Pub : ICLARM, Phillipines.*

30. **Thompson, R. & Munro, J. L. 1974.** *The biology, ecology and bionomics of the Hinds & Groupers, Serranidae. Pp. 59 - 81. Caribean Coral reef Fishery Resources Pub : ICLARM, Philippines.*

31. **White, A.T. 1987.** *Coral reefs, ICLARM No. 386 : 1-36.*

32. **William, D. & Anderson, Jr. - 1987.** *Systematics of the fishes of the family Lutjanidae (Perciformes : Percoidei). the snapper. Pp. 1-32. Tropical snappers and Groupers : Biology & Fisheries Management. Pub : Westview press/Boulder and London.*

33. **Yasumoto, T. et.al. 1977.** *Finding of a dinoflagellate as a likely culprit of ciguatera. Bull. Japan. Soc. Sci. Fish. 42(8) : 1021-1026.*

Author with a Perch

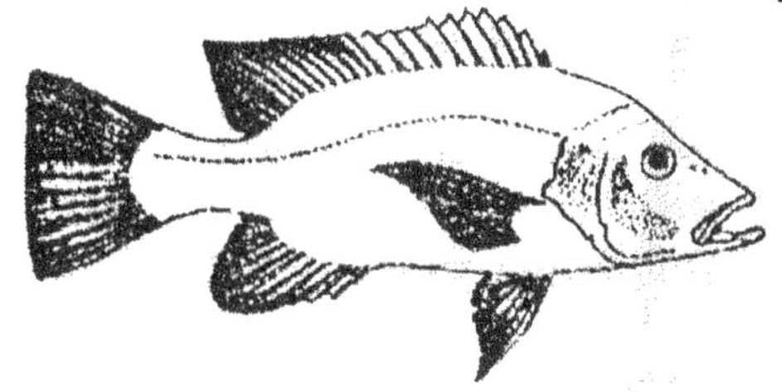

Lutjanus

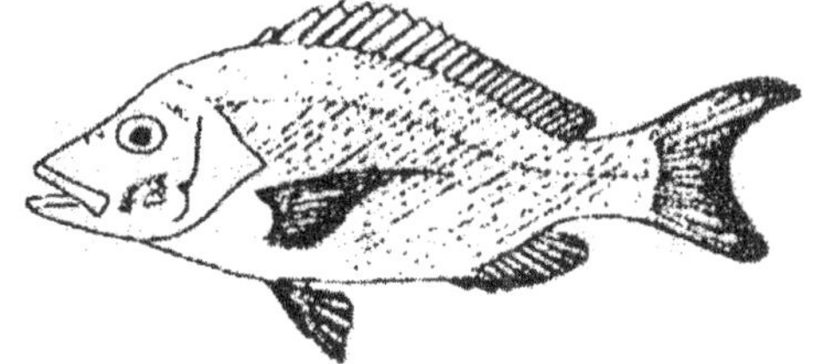

L.malabaricus

L. gibbus

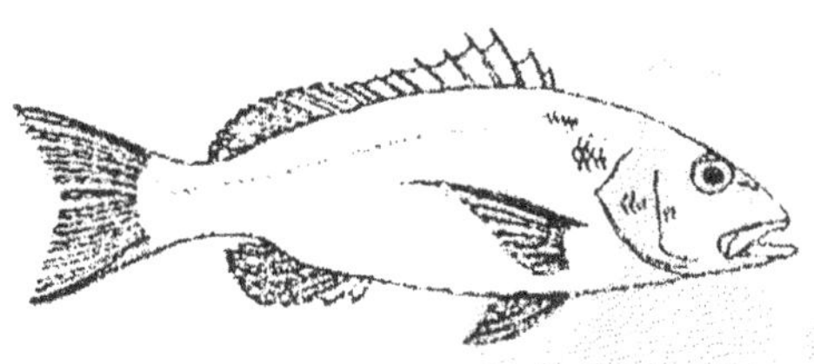

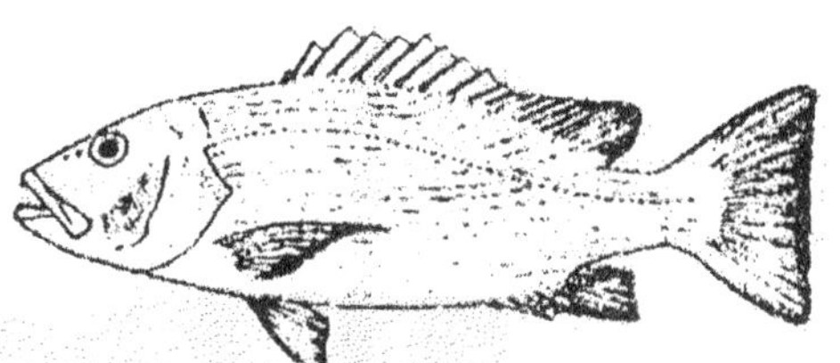

L.bohar

L.argentimaculatus

Pristipomoides

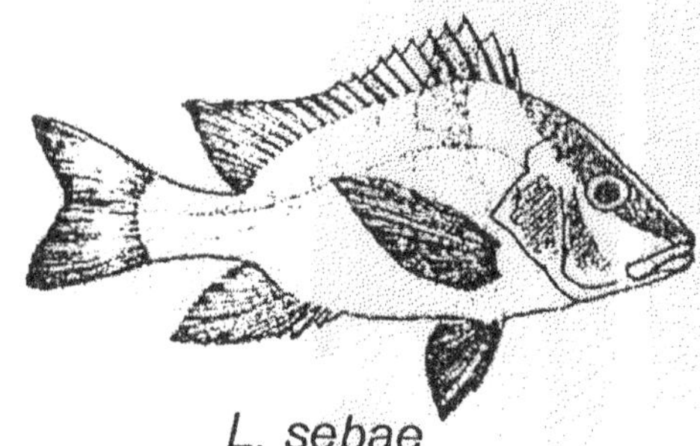

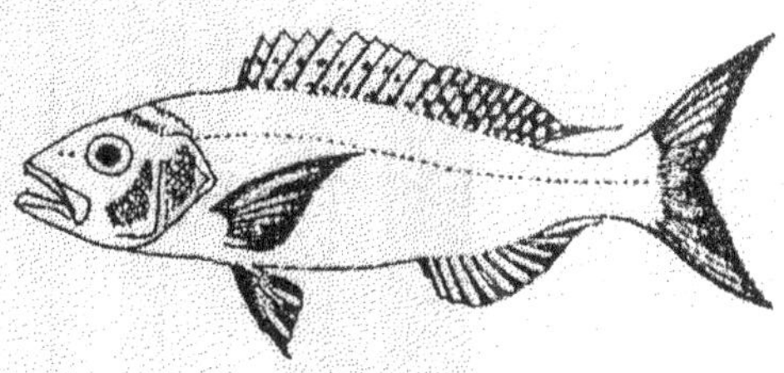

L. sebae

P.typus

Aphareus

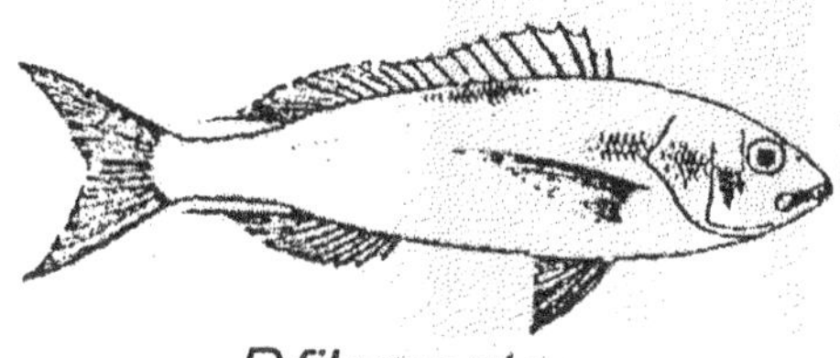

P.filamentosus

A.rutilans

Serranidae

Epinephelus
E.chlorostigma
E.diacanthus
Plectropomus
P.leopardus
P.levis
Cephalopholis
C.sonnerati

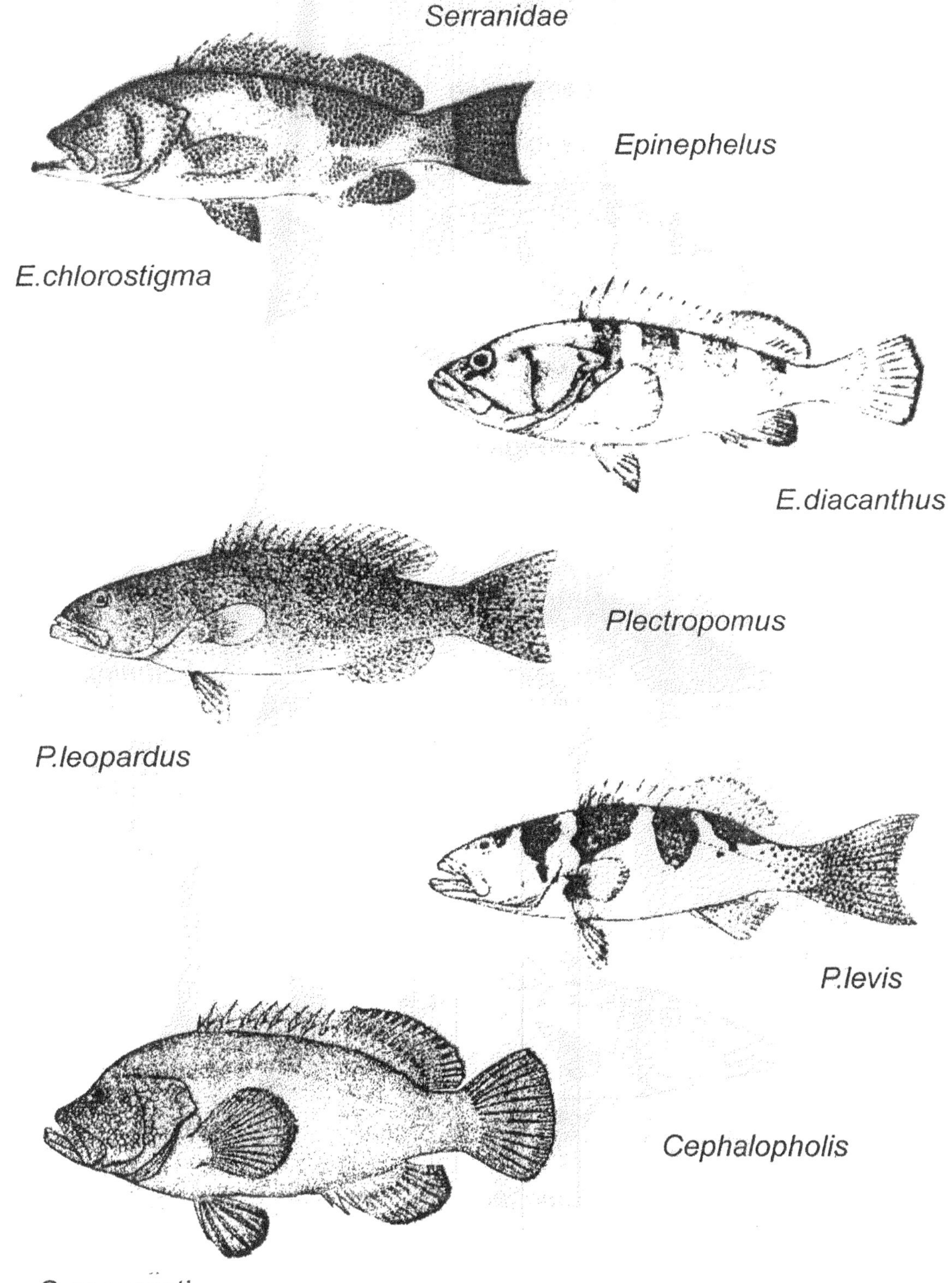

Lethrinidae

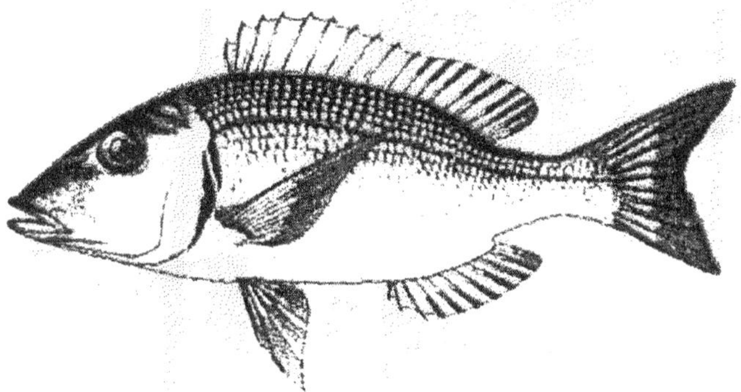

L.elongates

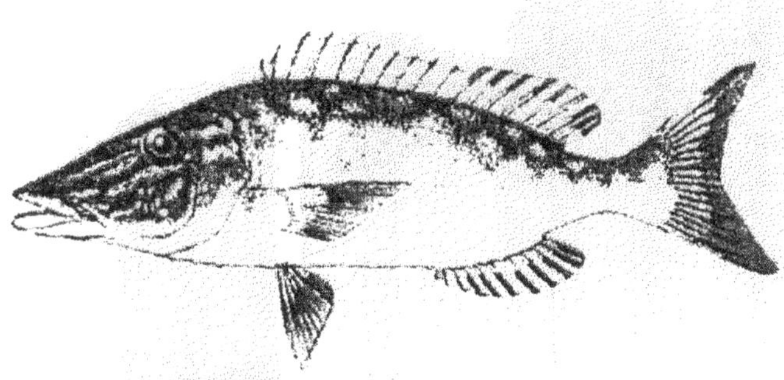

Lethrinus

L.miniatus

L.lentjan

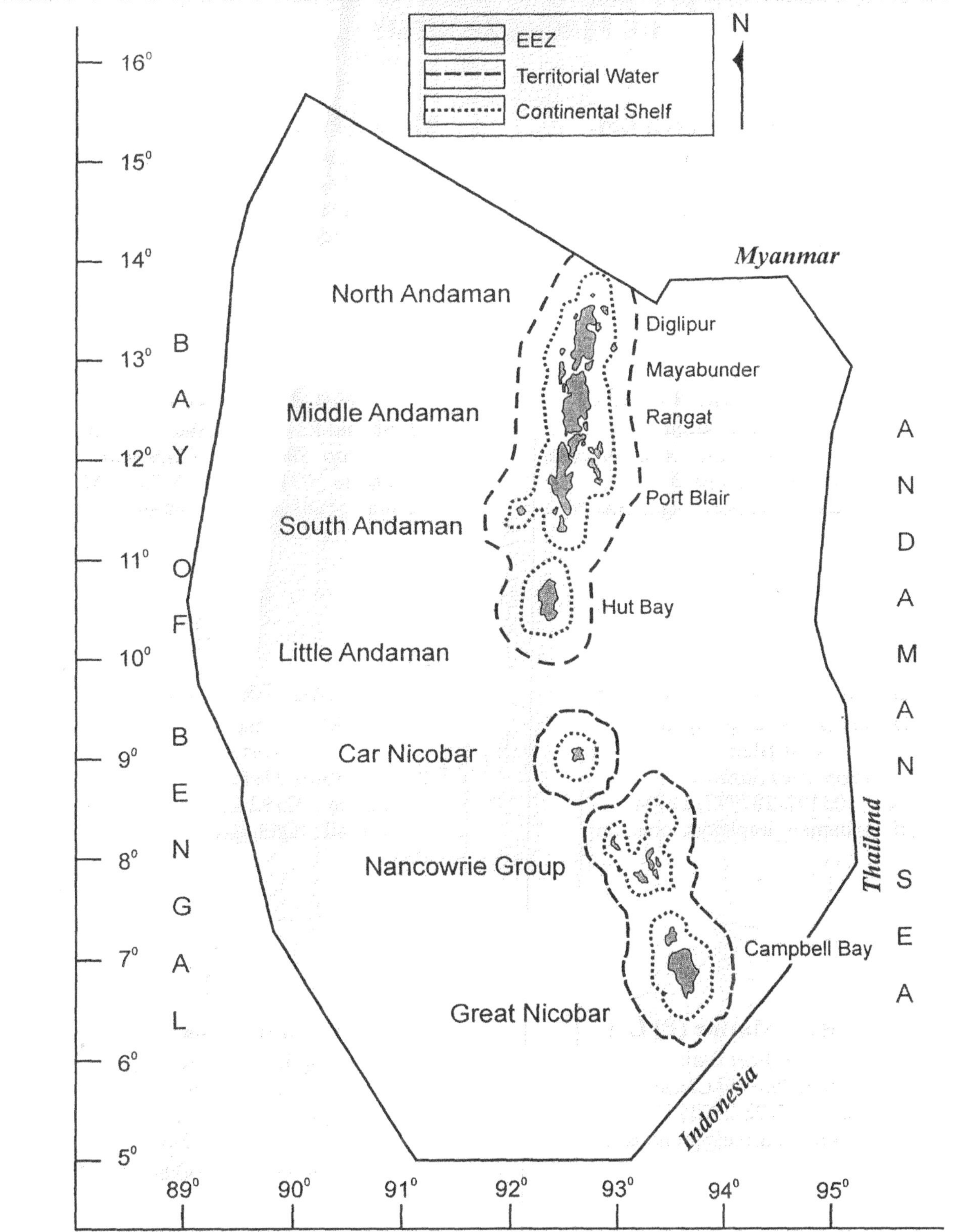
EEZ
Territorial Water
Continental Shelf
N
16°
15°
14°
13°
12°
11°
10°
9°
8°
7°
6°
5°
89°
90°
91°
92°
93°
94°
95°
B A Y O F B E N G A L
A N D A M A N S E A
Myanmar
Thailand
Indonesia
North Andaman
Middle Andaman
South Andaman
Little Andaman
Car Nicobar
Nancowrie Group
Great Nicobar
Diglipur
Mayabunder
Rangat
Port Blair
Hut Bay
Campbell Bay

10. PERCH EXPORTERS

M/s. Aqua World Export (P.) Ltd.
Jungliighat, Port Blair
Prop. Shri Haridas
Phone : 03192-235506
e-mail : aquaworld@sify.com

M/s. Andabar Cold Stores (P) Ltd.
Lamba line, Port Blair
Prop. Shri V. A. Kurian
Phone : 03192-233457
e-mail : pdsudarshan@rediffmail.com

M/s. Britto Seafoods Export Pvt. Ltd.
Phoenix Bay, Port Blair
Prop. Smti. Susila Christian / Shri Albin Michael
Phone : 09933292075
e-mail : brittoseafoodsexports@gmail.com

M/s. Rubin Sea Food
3, Shahid Road, Dignabad, Port Blair
Prop. Shri Rubin Fernando
Phone : 03192-240833 / 251752
e-mail : rubinseafoods1@gmail.com

M/s. Island Marine Products Ltd.
Dhanikhari, Humphreygunj
Port Blair
Prop. Shri Sasidhar
Phone : 03192-287777, 287243
e-mail : andaman_impl@yahoo.co.in

M/s. T.S.G. Aqua
25, M.A. Road, Phoenix Bay
Port Blair
Prop. Shri G. Bhasker
Phone : 03192-231894, 232894
e-mail : tsgbh@rediffmail.com

M/s. New Rich Marine (P) Ltd.
Wandoor, Port Blair
Prop. Shri Vincent Cletus
Phone : 03192-280121
e-mail : newrichmarine@gmail.com

M/s. Indian Sea Foods
Fishing Harbour, Junglighat
Port Blair
Prop. Shri U. Nasar Khan
Phone : 03192-237037, 9434280894, 9933222686
e-mail : nasarkhan19@gmail.com

Exported Groupers of Andaman Islands.

Epinephelus chlorostigma
(Gobra)

E.diacanthus
(Gobra)

Cephalopholis sonnerati
(Lall Gobra)

Plectropomus leopardus
(Lall Gobra)

P.levis
(Lall Gobra)

Courtesy - FAO / GBRMP / RFFG / Queensland.

Exported Emperors of Andaman Islands.

Andaman & Nicobar Islands Marine Fishing Rules - 2004

- Catching of any grouper below 30 cm in size
- Export of Brood stock & Juvenile

Courtesy - FAO / GBRMP / RFFG / Queensland.

Vol.24 No.10 FISHING CHIMES JANUARY 2005

A Glimpse of Icthyo-Beauties & Their Habitat Around Andamans

S.A.H. Abidi
Member, ASRB, ICAR
New Delhi

V. Krishnamurthy
Director of Fisheries
A & N Islands

Arif M. Mustafa
Asst. Fish. Dev. Officer
A & N Island

Geetanjali V Deshmukhe
CIFE
Mumbai

"Our arrival is theatrical. Misty-green and turquoise bursts of ocean clouds, ocean jungle, ocean sand and an indigo swell of sea. We fly low over waters that are gold in their shallow, sandy banks; over cliff-bound aquamarine bays, insanely careening gulls and mangrove forests in a jade sea; over Grubb, a knobby green island patch girdled by coral beds where the bay's waters glisten dark and blue", said a novelist while flying over Andaman & Nicobar Islands on his maiden voyage.

The Andaman & Nicobar Islands sprawl in the Bay of Bengal forming a crescent chain of 585 islands, islets and rocks extending for about 850 km between 6° 45'N & 13° 45' N latitude and 92° 15'E and 94° 00'E longitude separating Bay of Bengal from Andaman sea. It covers a land area of about 8,249 sq.km. and a coastline of nearly 1,962 km. which is about one fourth of the total coastline of India and which enjoys the status of Union Territory. It is rimmed discontinuously with saline and marshy lands which generally extend 1-2 km inland and at places for 4-5 km. Saddle peak, in the North Andaman, at the height of 732 m above sea level is the highest point on these islands.

From geological point of view, the Andaman & Nicobar group of islands form one of the most interesting regions of India. They constitute an Island arc, that can be separated into two concentric arcs, an outer (western) sedimentary comprising major islands of the Andaman & Nicobar extending to southeast forming Indonesian orogenic belt; and the inner (eastern) volcanic that are showing up above sea level as the conical volcanoes of Narcondum and Barren islands, (Srinivasan, 1986).

The coastline of these islands is variably dotted with mangrove vegetation. Most of the seabed near the coast is rocky with coral growth. Within few kilometers from the shore, the sea very deep, thus limiting the continental shelf area to about 34,965 sq.km. The shelf area of the western coast is much wider than the east coast. The Exclusive Economic Zone (EES) of the Andaman & Nicobar Islands covers an area of about 6 lakh sq.km. which is about 30% of the total for the whole country.

In Andaman sea there are three channels (a) the Preparis channel, divided into north and south portions by the islands of the same name, (b) the Ten degree channel, between the Andaman & Nicobar group of islands, and (c) the Great Channel between Great Nicobar islands and Sumatra.

Coral Reefs

Coral reefs constitute one of the most potent and productive marine ecosystem of A & N Islands. Theoretically in this part of Indian Ocean the coral reef area generates about 24,600t of biota per annum. These tropical reefs are the natural abode for almost all the marine ornamental fishes. Several authors have made valuable contributions on the related aspects of coral reefs of these islands (Seawell, 1922, 1935; Nair & Pillai, 1972; Scheer, 1966; Mahadevan & Eastersib, 1983; Nair & Gopinathan, 1983; Marichamy, 1983; Dorairaj, Soundararajan and Singh, 1987; Mustafa A,M. *et.al.* 1987, 1990, 1991). Among all the marine ecosystems, coral reef niche is the most fragile, delicate, complex, beautiful and productive ecosystem, sustaining about 3,000 living micro and macro biota. Islanders depend much on reef resources like fish and shell fish to meet their protein requirement, recreation, sport and for tourists etc.

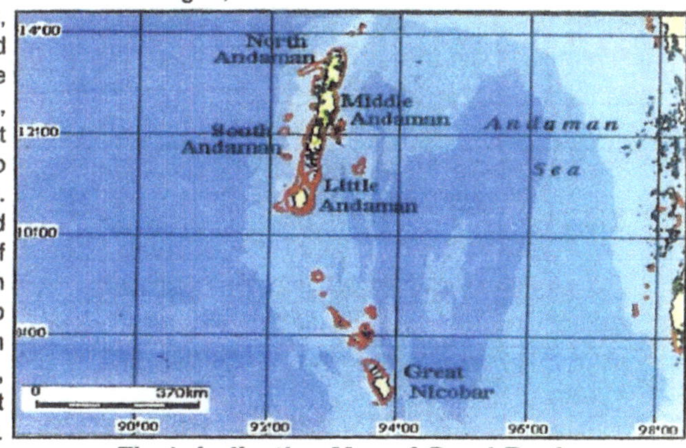

Fig.1: Indicative Map of Coral Reefs around A & N Islands *(Source: Reef Base 2000)*

Small marine animals that possess a stony skeleton and live in colonies are named corals. They belong to animal group Anthozoa under phylum Cnidaria. The corals are of two type viz., Soft corals (Alcyonarians or Octocorallians) and true or Stony corals (Solearctinian or Madroporarian corals). In Alcyonarians the skeleton is inside the body whereas in true stony corals it is completely outside and is known as theca or cup. These animals are capable of secreting a massive calcareous skeleton and of collectively depositing calcium carbonate to build large ornate colonies. Concerted growth of a variety of corals in a localized habitat gives rise to a coral reef. An account of coral reefs of the island and ornamental fish distribution has been given by Tikadar *et al.*, 1986; Mustafa *et al.*, 1987).

Because of the faunistic wealth these islands possess and the diversity and uniqueness they exhibit, these islands have often been referred to as 'Zoologists' Paradise'. There are about 2,500 known species of corals distributed in the world seas, of which 250 species of corals have been reported from these islands. In Andaman & Nicobar Islands, most of the coastal stretch on either side, eastern as well as western are dotted with fringing type of coral formation. The

Butterfly

1	*Chaetodon trifasciatus*	Over butterfly fish
2	*C.vagabundus*	Vagabond buttefly
3	*C.triangulum*	Triangle butterfly
4	*C.meyeri*	Meyers's butterfly
5	*C.octofasiatus*	Eight stripped butterfly
6	*C.descussatus*	Black finned vagalond
7	*C.auriga*	Threadfin butterfly
8	*C.melannotus*	Blackback butterfly
9	*Heniochus acuminatus*	Longfin banner fish
10	*H.diphreutes*	Shoaling banner fish

Angel

1	*Centropyge bicolor*	Bicolor angel
2	*C.vrolikii*	Pearl scaled angel
3	*Pomacanthus xanthometapon*	Yellow mask angel
4	*P.semicirculatus*	Semicircular angel
5	*P.imperator*	Emperor angel

Surgeon

1	*Acanthurus leucosternon*	Powder blue surgeon
2	*A.lineatus*	Blue striped surgeon
3	*A.tristegus*	Convict surgeon
4	*Naso lituratus*	Orangespine Unicorn
5	*Zebrasoma scopes*	Brush tail tang

Trigger

1	*Balistapus undulates*	Orange striped trigger
2	*Balistoids conspicillum*	Clown trigger
3	*Odonus niger*	Red tooth trigger
4	*Rhinecanthus aculeatus*	Black bar trigger
5	*R.rectangulus*	Wedgetailed trigger

File

1	*Oxymonacanthus longirostris*	Beaked file fish
2	*Pervagor melanocephalus*	Black headed file fish
3	*Aluterus scriptus*	Scribbled fish fish
4	*Monocanthus chinensis*	Fan bellied file fish

Wrasse

1	*Gomphosus caeruleus*	Bird wrasse
2	*Thalassoma quinquevittatum*	Five stripe wrasse
3	*T.purpureum*	Surge wrasse
4	*T.herbaricum*	Gold bar wrasse
5	*Labroidse dimidiatus*	Cleaner wrasse

Damsel

1	*Chromis atripectoralis*	Black-axil chromis
2	*Chrysiptera leveopoma*	Blue-ribbon damsel
3	*Dascyllus aruanus*	Zebra humbug
4	*D.trimaculatus*	Three-spot dascyllus
5	*Pomacentrus moluccensis*	Lemon damsel

Clown

1	*Amphiprion ocelaris*	False Clown
2	*A.clarkii*	Yellow tail clown
3	*A.frenatus*	Tomato clown
4	*A.bicinctus*	Two band clown
5	*A.percula*	Common clown

Associated Fauna

1	*Heteractis magnifica*	Common anemone
2	*Cryptodendrum adhaesivum*	Sticking anemone
3	*Ophiocoma* spp	Brittle star fish
4	*Linckia* spp	Reef star fish
5	*Stenopus* spp	Banded shrimp
6	*Echinothrix* spp	Banded urchin
7	*Panulrisu* spp	Rock lobster
8	*Pinctada* spp	Pearl oyster
9	*Tridacna* spp	Giant clam
10	*Holothua* spp	Sea cucumber

Eastern side has less coral formation than the Western side. About 21 km away from the Andaman coast is located a pseudo barrier reef in three patches, namely North coral bank, Middle coral bank and South coral bank. The so called barrier reef extends from latitude 10° 26' N to 13° 40'N covering a longitudinal distance of about 320 km. and separated from the coast by a trench of about 80 m deep. The Ritchie's archipelago comprising islands of tourist interest like Havelock and Neil have some of the best patches of shallow water fringing reefs and offshore vertically raised reefs of great SCUBA interest respectively; raised coral benches with knolls in their channel are seen from Outram to Lawrence island. Path reefs are in abundance in Nancowry. A flat reef about 500 m wide is reported from East Bay in Katchal island (Pillai, 1983). In these islands three distinct types of reefs are reported: (1) a Wind-ward reef at North Bay (Port Blair), Cinque islands and Joly Buoy island; (2) Channel reef/ Leeward reef at Camorta-Nancowry area and (3) Bay reef at Havelock, Neil and Nancowry Islands. Extent of coral reef formation around A&N island is shown in Fig.1.

Prime survey undertaken during 1986-87 to evaluate the extent of damage to the coral reefs of A&N islands, their distribution and icthyo-beauties which thrive upon it, revealed that out of an estimated area of 2000 sq.km. of coral coverage, an area of about 360 sq.km. was found intensively damaged mostly due to unwise human intervention (Mustafa, A.M.1987). The said survey report timely alerted all concerned round the globe and resulted in timely imposition of suitable conservation and management legislation and widescale awareness campaign among stakeholders, thus minimizing the threat of human induced degradation.

Icthyo-Beauties

The most spectacular and dazzling beauties in a coral reef niche are the ornamental fishes. While snorkeling along the shallow coral reefs, one comes across darting and dancing flamboyant fishes. They are difficult to catch as they swim around at a very fast speed. These fishes stay below the tossing waves amid corals. Many of them are well camouflaged to match their surroundings. Some even change their sex, colour and pattern in accordance with environmental need.

These fishes live in crevices, fissures and pits of branching and massive corals (Fig.2&3). An encounter with them is common throughout the island coast while snorkeling or diving. A few very popular ones among these are many varieties of Butterfly fishes, Angels, Surgeons, Triggers, File fishes, Wrasses, Damsels, Clowns and Sea horses. Many types of Shells, Star fishes, Sea lilly, Sea urchin, Sea anemone and Reptiles are also in the area of coral infested shallow bays and lagoons around Andaman & Nicobar Islands. Commonly occurring icthyo-beauties and some associated fauna list is given here under.

Prospects of Marine Ornamental Fishes

Since this renowned coral ecosystem of Andaman & Nicobar Islands is rich in ornamental fishes, it possesses great scope for the development of market for ornamental fishes in domestic as well as foreign markets. However, in order to protect this beautiful resource from depletion it is neessary to mount R&D efforts towards captive breeding and natural propagation through artifical reefs.

Development of such innovative R&D programmes on breeding and propagation will help the islanders in adopting technology for the development of export oriented cottage industries, for the unemployed tribal and local youths of this Union Territory.

It has been estimated that USA imports ornamental fishes worth one billion dollars, whereas Netherlands spends US dollars 37,08,486 on import of these tropical fishes for about five lakh fish hobbyists (Golden, 1991).

In order to have a long term sustainable supply of such exportable value products for development of domestic market, it will be essential that proper research on sexual and asexual propagation of ornamental fishes is mounted in an organized manner supported with comprehensive regulatory measures to conserve these resources and to avoid dwindling.

Fig.2: Lion/Fire Fish near soft coral Gorgonia

Fig.3: Clown trigger

US Trade Body Invites Comments on Tsunami Impact on Seafood Sector

The US International Trade Commission (ITC) has invited comments from the public and interested parties on the impact of tsunami on seafood industries in India and Thailand to determine whether the present situation warrants a review off the anit-dumping duty imposed on shrimp imports from these countries.

The ITC said in a statement that written comments must be filed with its Secretary by March 25 this year. "Information submitted to the Commission should address the impact of the tsunami on the ability of the shrimp industries in India and Thailand to produce and export shrimp to the US," the statement said.

The information may include an analysis of the condition of shrimp hatcheries, ponds, fishing fleet and processing and storage facilities after the tsunami disaster struck in December 2004. Estimates of the share of the countries' re-stock, or rebuilding of any damaged or destroyed production, storage, or transportation infrastructure may also be provided to the ITC.

Besides, any data on current inventories of shrimp in these countries, which may be exported to the United States, can also be presented. The Commission will make its determinations regarding institution of review investigations within 30 days of the close of the comment period. All submissions, except for business proprietary information, will be available for public inspection, the statement said.

While upholding anit-dumping duties imposed by the US Department of Commerce on shrimp imports, the ITC had stated that it may review the case of Thailand and India in view of the impact of tsunami on the countries' seafood sectors.

A decision whether in the changed circumstances reviews will be instituted in the case of India and Thailand is to made after the collection and analysis of the information submitted, ITC said.

Correction

At Page 48 of December 2004 issue (Vol 24 No 9) of *Fishing Chimes*, the name of the Union Joint Secretary (Fisheries) was given as 'Ajit Bhattacharya'. This may be read as 'Ajay Bhattacharya'. The error is regretted.

Icthyo-beauties

Few Marine Ornamental Fishes

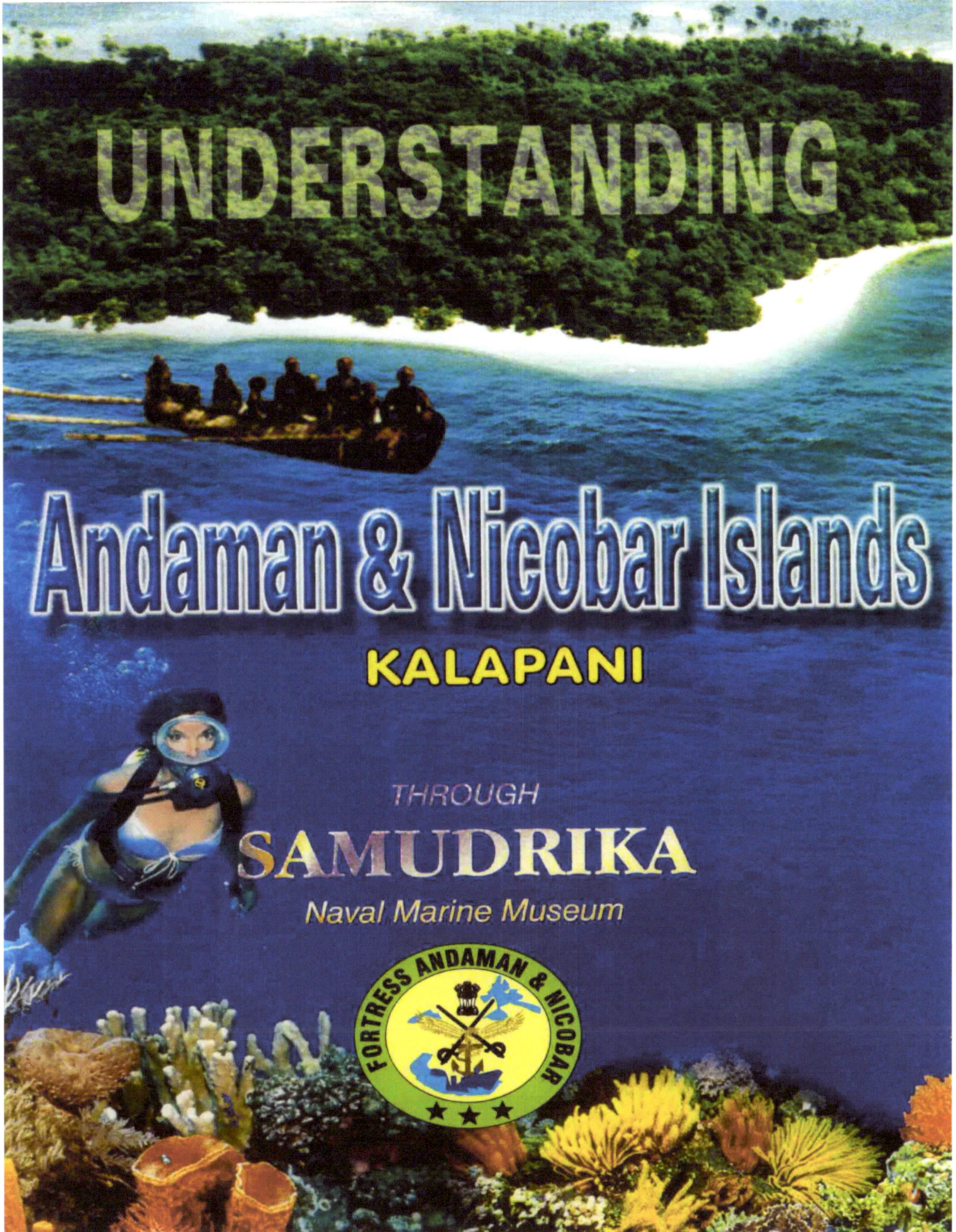
UNDERSTANDING
Andaman & Nicobar Islands
KALAPANI
THROUGH
SAMUDRIKA
Naval Marine Museum
FORTRESS ANDAMAN & NICOBAR

VIEW OF CELLULAR JAIL PRIOR TO EARTHQUAKE OF 1941

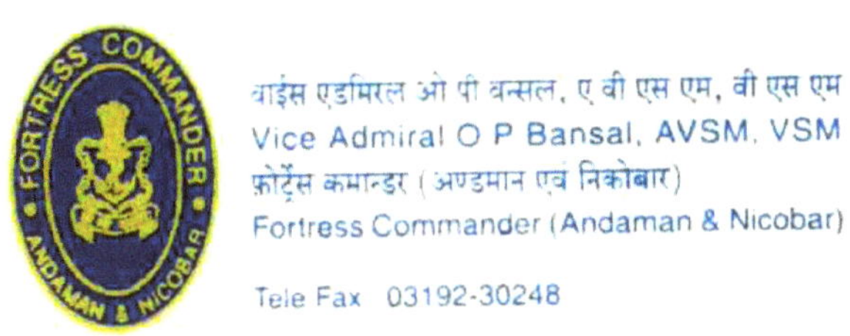

फोर्ट्रेस मुख्यालय
पोर्ट ब्लेयर - ७४४ १०२
Fortress Headquarters
Port Blair - 744 102

25 January, 2001

FOREWORD

The Andaman & Nicobar Islands are unique due to diverse ecosystem and abundance of flora and fauna. The evergreen tropical forest, the little known aborigines, colourful butterflies, moths, ornamental fishes, larger number of endemic species such as Robber Crab, Dugong, Hornbill, Megapode etc., make this region a paradise for nature lovers.

In an effort to spread awareness of this vast marine ecosystem and the bio wealth of these Islands, Indian Navy had set up and dedicated 'Samudrika', the Naval Marine Museum, to the nation in 1992.

This ecological guide 'Understanding Andaman & Nicobar Islands (Kalapani) through Samudrika' is a milestone in Navy's effort in this direction. The guide covers all aspects of the ecosystem of A&N in a comprehensive manner with emphasis on marine aspects. The section on Archaeology gives an insight into the ancient civilisations which existed in the Indian Subcontinent.

Sd/-
(OP Bansal)
Vice Admiral

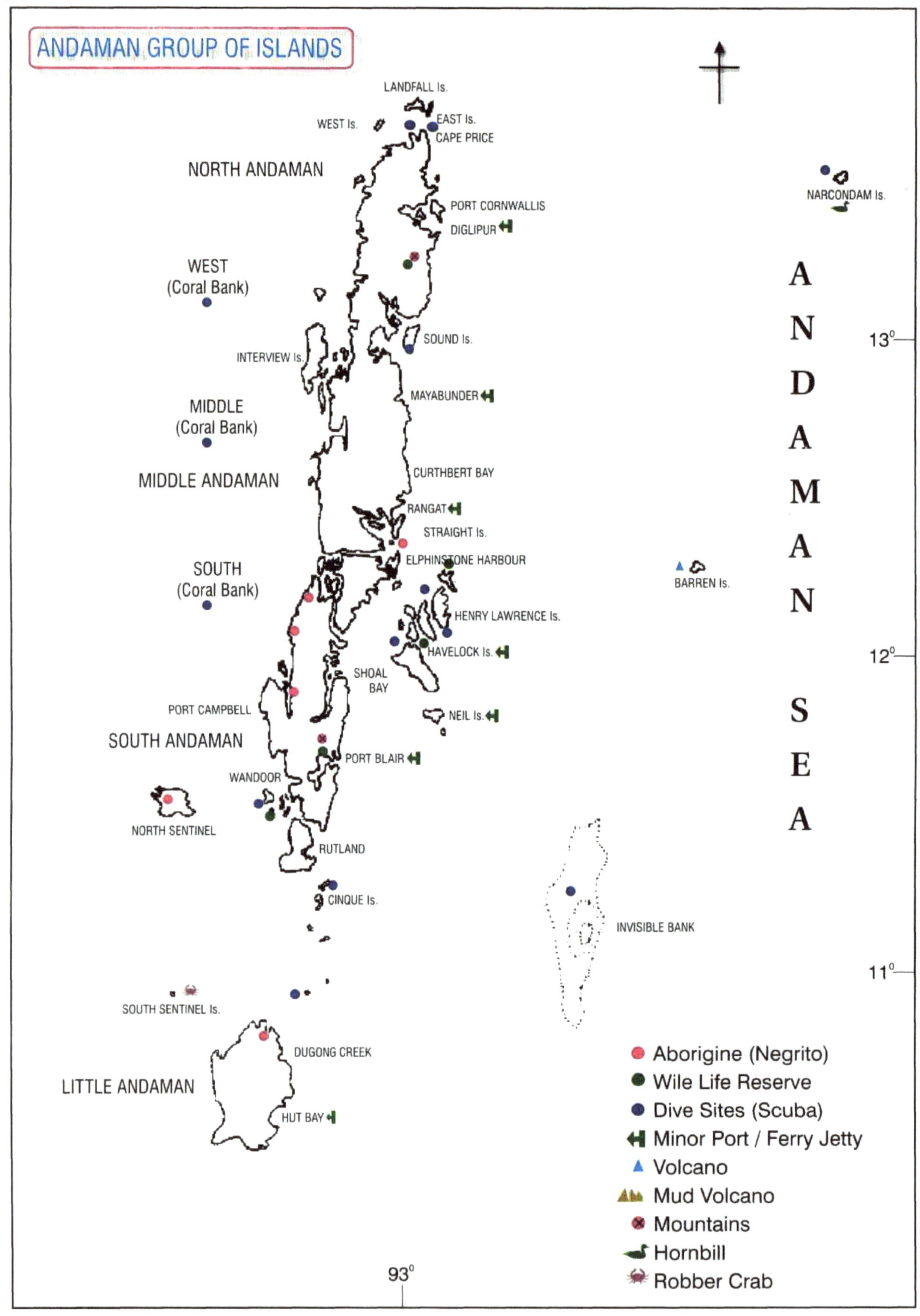
ANDAMAN GROUP OF ISLANDS
LANDFALL Is.
WEST Is.
EAST Is.
CAPE PRICE
NORTH ANDAMAN
NARCONDAM Is.
PORT CORNWALLIS
DIGLIPUR
WEST
(Coral Bank)
A
N
D
A
M
A
N
S
E
A
13°
SOUND Is.
INTERVIEW Is.
MAYABUNDER
MIDDLE
(Coral Bank)
MIDDLE ANDAMAN
CURTHBERT BAY
RANGAT
STRAIGHT Is.
ELPHINSTONE HARBOUR
BARREN Is.
SOUTH
(Coral Bank)
HENRY LAWRENCE Is.
HAVELOCK Is.
12°
SHOAL
BAY
PORT CAMPBELL
NEIL Is.
SOUTH ANDAMAN
PORT BLAIR
WANDOOR
NORTH SENTINEL
RUTLAND
CINQUE Is.
INVISIBLE BANK
11°
SOUTH SENTINEL Is.
DUGONG CREEK
LITTLE ANDAMAN
HUT BAY
93°
Aborigine (Negrito)
Wile Life Reserve
Dive Sites (Scuba)
Minor Port / Ferry Jetty
Volcano
Mud Volcano
Mountains
Hornbill
Robber Crab

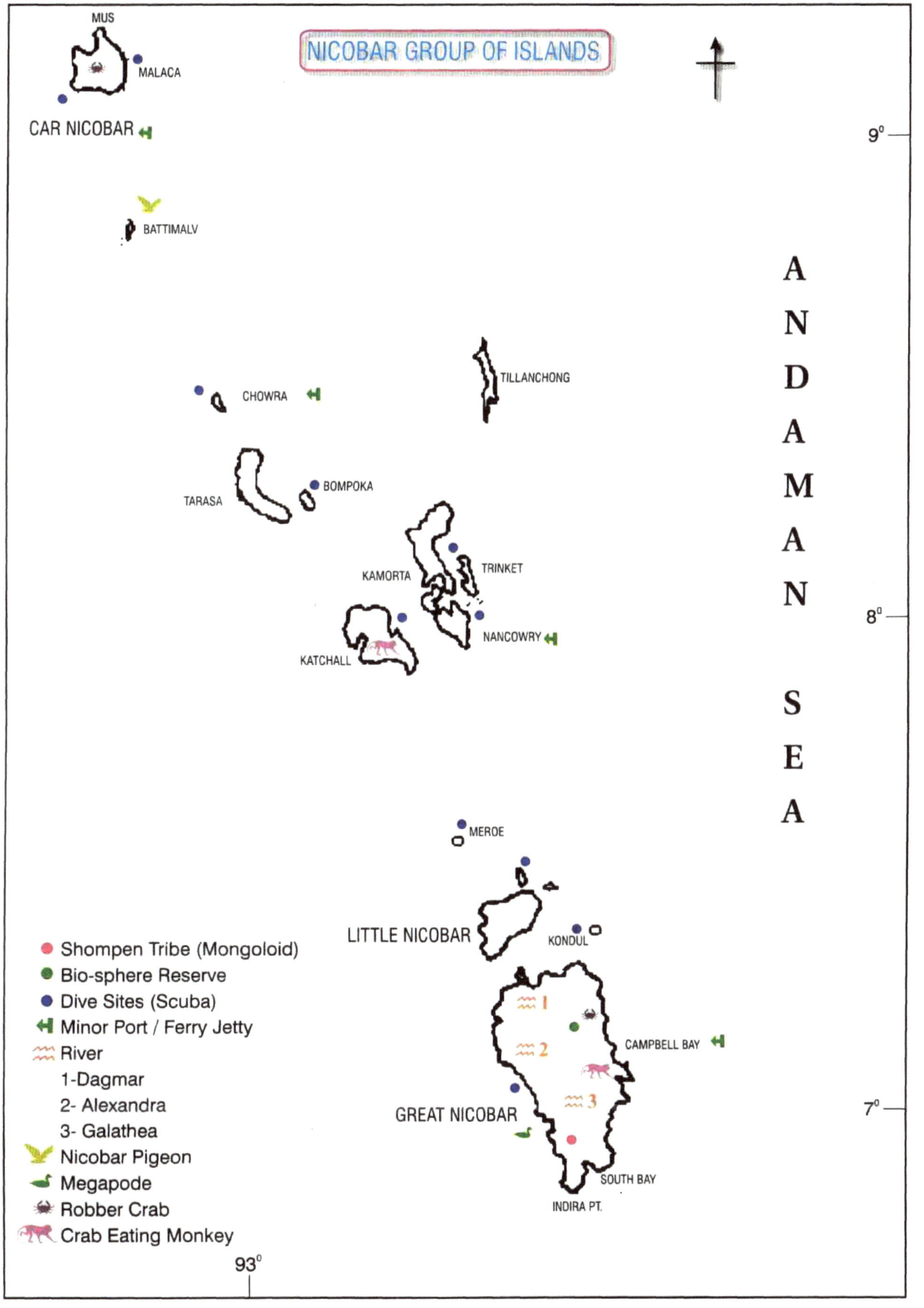
NICOBAR GROUP OF ISLANDS
MUS
MALACA
CAR NICOBAR
9°
BATTIMALV
A N D A M A N S E A
TILLANCHONG
CHOWRA
TARASA
BOMPOKA
TRINKET
KAMORTA
8°
NANCOWRY
KATCHALL
MEROE
LITTLE NICOBAR
KONDUL
CAMPBELL BAY
GREAT NICOBAR
7°
SOUTH BAY
INDIRA PT.
Shompen Tribe (Mongoloid)
Bio-sphere Reserve
Dive Sites (Scuba)
Minor Port / Ferry Jetty
River
1-Dagmar
2- Alexandra
3- Galathea
Nicobar Pigeon
Megapode
Robber Crab
Crab Eating Monkey
93°

CONTENTS

1. **Concept** 1
2. **Setup** 2
3. **Kalapani** 3
   - Genesis 4
   - Discovery 6
   - Geography 6
   - Volcanic Activity 8
4. **Advent of Modern Man** 9
   - British Rule 9
   - Japanese Rule 10
5. **Aborigines & Tribes** 13
   - Sentinelese 14
   - Jarawa 14
   - Onge 14
   - Great Andamanese 15
   - Shompen 15
6. **Forest** 16
   - Timber 17
   - Fauna 17
   - Snakes 18
   - Birds 18
   - Butterflies and Moths 20
7. **Shells** 21
   - Shell Middens 24
8. **Corals** 25
9. **Ocean** 33
   - Fishes 34
   - Associated Fauna 35
10. **Marine Aquarium** 37
11. **Marine Archaeology** 39
    - Submerged Dwaraka 39
    - Lothal Tidal Dock 40

CONCEPT

1

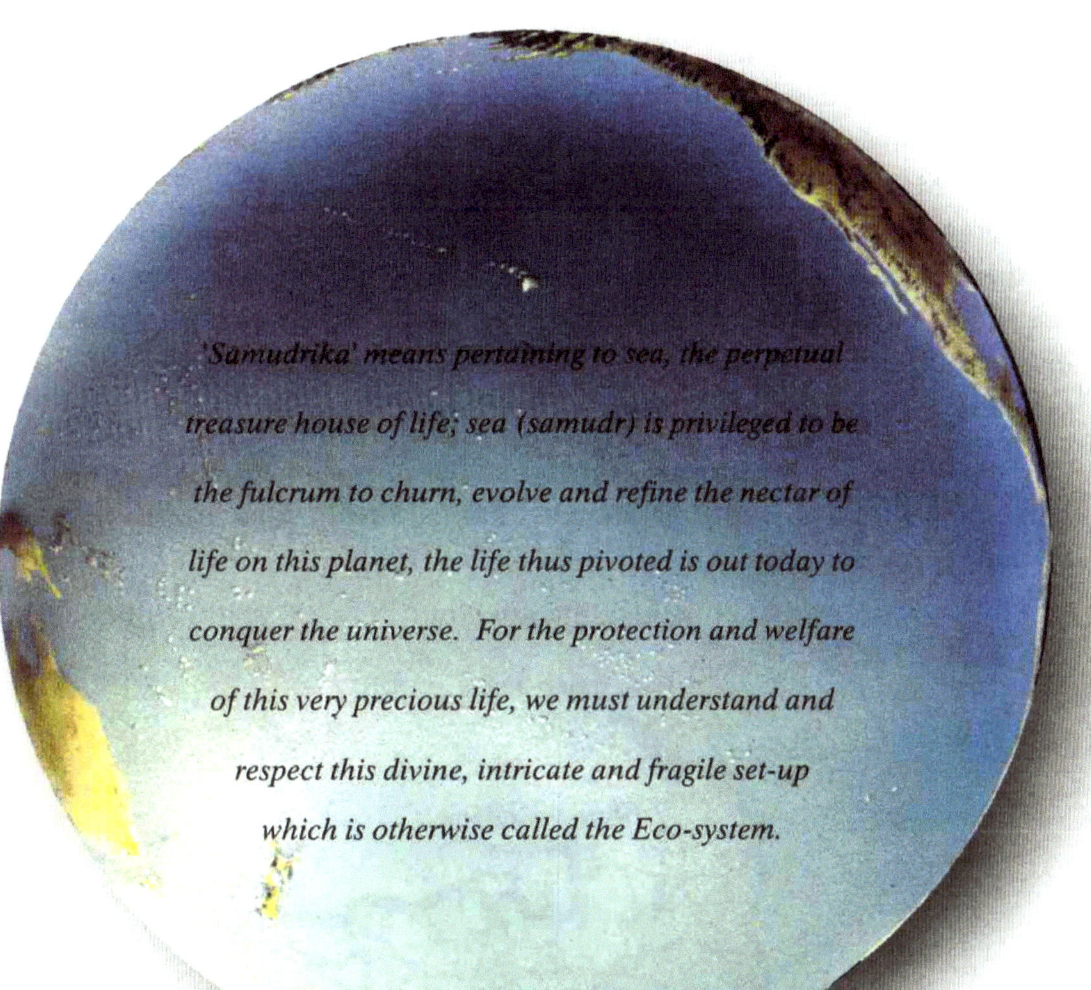

The naval marine museum 'Samudrika' was thus established in the year 1992 to give the masses an insight to the vast marine eco-system. The museum is unique due to the integrated depiction of life forms. It exhibits and explains various strata of oceanic life through Geology, Geography, Habitation and the History of settlement. This wide spectrum covered makes 'Samudrika' the star attraction for tourists in Port Blair.

The museum is broadly divided into five display rooms comprising: (1) Introductory section on A&N Islands (2) Shells (3) Corals (4) Oceanic Regime and (5) Marine Archeology. The introductory section provides basic information related to A&N Islands through write-ups, charts and models. The shells section exhibits about 200 varieties of decorative and commercial shell wealth of A&N Islands. Many rare and uncommon shells are also on display. The coral room presents a cross sectional view of stony and soft corals, which form the most complex, sensitive and fragile marine eco-system around these islands. A few significant deep sea and coloured soft corals are also on display. The section on oceanic regime displays magnificent art on evolution, preserved and stuffed specimens and four state of the art marine aquariums containing live fishes and sea animals.

The coral and the oceanic regime sections also have many real underwater photographs from around A&N islands. The last room is on marine archeology related to mainland-India.

2.1 LOCATION

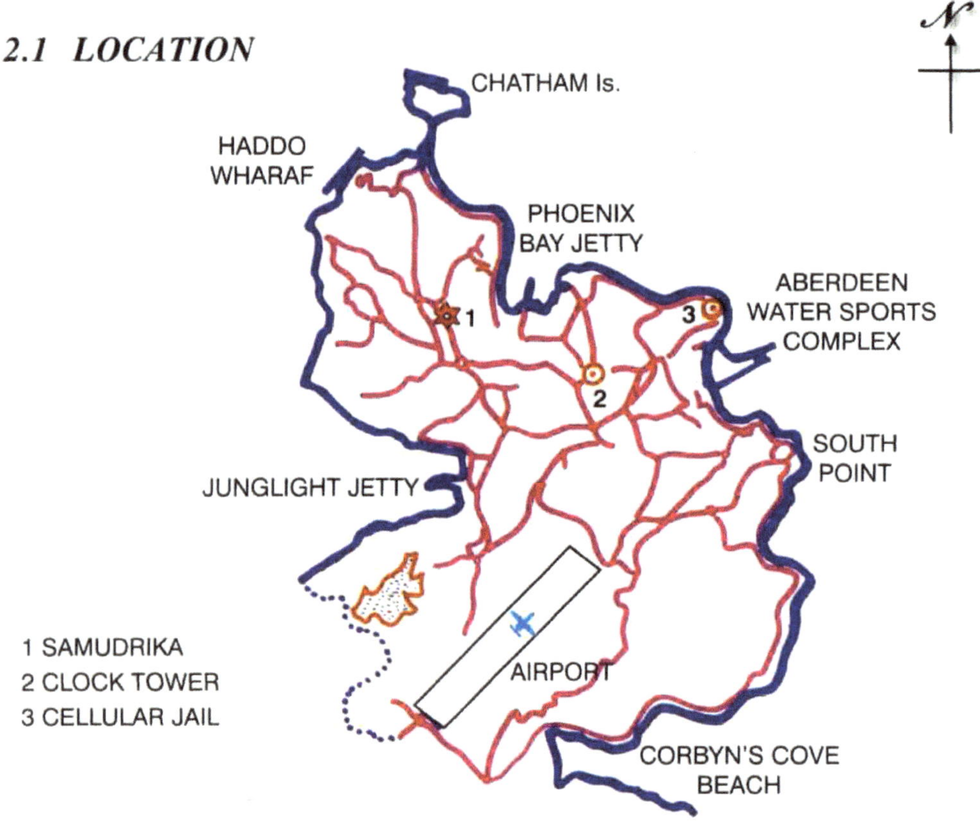

PORT BLAIR (TOWN MAP)

POCOCK Is
(NORTH ANDAMAN)

AN ISLET
(KATCHAL)

Courtesy: Cdr. Paul James

3.1 GENESIS

Far back in time some 150 million years ago, molten magma might have oozed from the earth's crust along the dilated ocean floor at the foot of a submarine ridge. The phenomenon appears to have taken a long long time thus creating mid-ocean rises known to many Geologists as 'whale back tumours'. These islands with a few exceptions are volcanic in origin. We are infact at the tip of a risen undulating lava plateau, the core of which goes deep down beyond the ocean floor. This hypothesis is in conformity with the expanding earth theory.*[1]

SMITH AND ROSS ISLANDS
(NORTH ANDAMAN)

JANSIN BAY
(KATCHAL)

Courtesy: Cdr. Paul James

3.2 DISCOVERY

The Chinese knew of these islands for more than a thousand years ago and called it *'yeng-t' omag'*in the first millenium. Ptolemy, the renowned Roman geographer during the second century called it ***'Angdaman islands'*** (Islands of good fortune). I' T Sing, a Buddhist monk, named it ***'Lo-jen-kuo'*** (Land of Naked) during the sixth century. Two Arab travellers during the eighth century referred to these islands as ***'Lakhabalus or Najabulus'*** (Land of Naked). The great traveller Marco Polo called it ***'Angamanain'***. However, when the current name 'Andaman' crept in is shrouded.[*2&3] The name Nicobar seems to be a corruption of the South Indian term 'Nakkavaram' (Land of the Naked) as indicated in the great Tanjore inscription of A.D.1050.

3.3 GEOGRAPHY

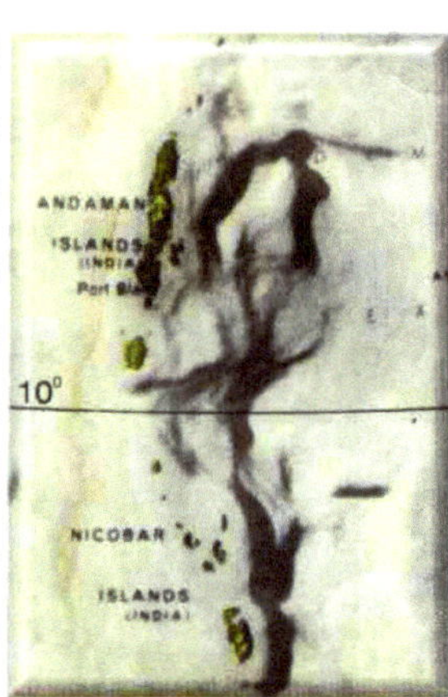

The Andaman and Nicobar archipelago comprising of 572 islands, islets & rocks is situated 1200 km off the southeastern coast of India in the Bay of Bengal. They are also called 'Kalapani'. Together they constitute one of the Union Territories of India, and is divided into two districts. Andaman to the north of the 10^0 channel and Nicobar to its south. The two are separated by about 160 km of sea. Being close to the equator and surrounded by the sea, these islands have a tropical climate. Precipitation is heavy with both North-east and South-west monsoons being received. It rains for about eight months a year. Cyclones do occur, temperature is moderate, and relative humidity is high. These Islands once believed to be a continuation of the Arakan Yoma mountain range of Burma upto Achin head of Indonesia, have undulating terrains with main ridges running north to south. There are a few hills running east west. In between the main ridges, deep inlets and creeks exist. The average width is about 20 km. There are a few flatlands and perennial streams. Ground water reserve is limited; and soil is mostly acidic and poor in nutrients. Soil types vary, from heavy clay to loamy sand. The cumulative land area is 8,24,900 ha. (8249 sq.km), out of which 7,09,400 ha. (7094 sq.km) including 1,07,046 ha. (1070 sq.km) of tribal reserve area is claimed to be under forest cover.

VOLCANIC ACTIVITY

THERMAL FURROW AT THE FOOT OF VOLCANO
(A YEAR AFTER ERUPTION)

SOLIDIFIED EDGE OF THE LAVA RIVER
(TWO YEARS AFTER ERUPTION)

Courtesy: Arif M. Mustafa

3.4 VOLCANIC ACTIVITY

The only live volcano in the Indian peninsula, Barren lies around 120 km north east of Port Blair in the Andaman sea (12" 17' N Lat. and 93" 50' E Long). This tiny, circular Island covers an area of 8 sq.km. It belongs to general Sunda group and is believed to have been born out of an eruption, which occurred during the late Post Pleistocene period. Later in the course of geological evolution, the prime giant cone got transformed into the present day Barren Volcano which in fact, is the central part of the blown off cauldron.

The first recorded eruption was in 1795 as observed by Lieutenant A. Blair. The Volcano emitted enormous smoke and ejected red-hot stones. Some were of giant size weighing 3 to 4 tons and were thrown hundreds of yards from the cone. Thereafter, feeble activity was noticed during 1857, 1866 and 1890. After about a century it roared and erupted again in March 1991. This violent volcanism continued upto January 1992.

Contrary to its name Barren, the island is covered with lush green forest and is inhabited by 13 species of birds, 10 species of butterflies, 9 species of insects, 7 species of mammals, 6 species of flies and 2 species each of centipedes and spiders. The mammals include a stock of feral goats, which depend on seawater to survive in the absence of a perennial fresh water source on the Island.

BARREN ISLAND VOLCANO

Courtesy: Surg. Cdr. Ravi Verma

4.1 BRITISH RULE

Modern history of Andaman islands can be traced back to 1789 when the Governor General of British India commissioned a survey of these islands by Lt. Archibald Blair, who conducted the first ever topo-cum-hydrographical survey and reported suitability for human settlement. Immediately thereafter, in 1790, the first settlement was established at Port Blair (then Port Cornwallis) in the present day Chatham Island by bringing in hard core criminals from undivided India. However, high mortality due to malaria and frequent attacks by aborigines forced the settlement to be shifted to a new port at North Andaman during 1792 wherein again similar problems cropped up and that settlement was also abandoned in 1796.

Penal settlement - It was in 1857, after India's First War of Independence - the *Sepoy Mutiny,* that a penal colony was attempted at Port Blair with the first lot of 200 'rebels'/mutineers of the Indian army, who, for the first time attempted to overthrow British rule in India. The convicts count increased to 773 in three months. Dr.J.P.Walker was the first Administrator of these convicts. The famous *Battle of Aberdeen* between civilized man and the Stone Age aborigines of Andamans was fought on 14th May 1859 at Aberdeen Bazaar. During 1861 the administrative control of A&N Islands was transferred to the Chief Commissioner of Burma. During 1869-70 many *Wahabi Movement* activists were deported to Andaman, one among them was Mohd. Sher Ali Khan (a Pathan), who assassinated Lord Mayo on 8th Feb.1872 at Hope Town jetty (now called Panighat). Later, in the same year, Sher Ali Khan was executed in Viper Island by the British. On 13th Sept. 1893 the British Government of India, vide settlement Order no.423, ordered the construction of a Cellular Jail to accommodate 600 prisoners at a cash expenditure of Rs.95,881/-. Mr. Mc Quilen, Sub-Engineer was made in-charge of the project. Prior to construction of the Cellular Jail, male convicts were held in a jail on Viper island and women convicts in South Point barracks (near the present day hotel Sinclair). Then occurred the great uprising of moplahs, the *Moplah Rebellion* during 1921. About 1400 moplahs mostly from Ernad, Walluvanad and Calicut were sent to Andamans with their families. Then came the *Rampa Revolution* during 1922-24. As a result many Rampa revolutionaries were also sent to Andamans.

The missionaries entered Nicobar group of islands in 17th Century and in 1756 the Dutch colonised Nancowry group of islands. The Dutch stayed there upto 1787. After several unsuccessful attempts to build up a colony there, the Dutch Government ultimately handed over Nicobar group of islands to the British, who took possession in 1869.

4.2 JAPANESE RULE

During the IInd World War the British hastily evacuated and abandoned these islands in the face of advancing Japanese forces, allowing Japanese occupation of Andaman and Nicobar Islands. The Japanese brutally ruled the territory for four years from 1942 to 1945. During this period, Japanese took-up massive fortification of these islands through construction of airfields (Port Blair, Rutland, Car Nicobar), installation of radars and guns for air defence network, chain of foreshore concrete pill boxes, underground ammunition dumps with trolley lines connecting some of them to Port Blair airport and elaborate underground bunkers for the troops. The remains of pillboxes, radars and anti aircraft guns are still visible at some places. Port Blair harbour was extensively used as a forward surveillance base for seaplanes of the advancing Japanese forces. A few months after the Japanese occupation, allied forces succeeded in blocking sealanes threatening island population to the brink of starvation. Japanese successfully averted the disaster through enforced intensive community farming of tubers like tapioca and sweet potato. Extensive road network expansion was also undertaken at that time for connecting Port Blair with outlying villages and cultivable land.

Local Borns - Much before the Japanese invasion, very many rebels or convicts surviving under constrained, horrifying and hypnotic atmosphere of imperialism were hypocritically encouraged to accept the conceptual halo of penal settlement in the Andamans. The basic objective behind this exercise was their disintegration from the Indian mainstream. Ultimately this socially distracted, politically isolated, penniless human bunches, blessed with an intense sense of patriotism and enormous will to continue their struggle for a free and independent India, laid the foundation of penal settlement in and around Port Blair. The spectrum of Indian culture spun around the fulcrum of hostility, hardship, barbarism and extradition spontaneously shaped itself into a unique unified Indian culture and rose into a synergetic stream of classically blended culture free from social discrimination. Prolific genetic infusion enriched the progeny further. This class of humans was later termed as 'Local Born' by the British and subsequently categorised by Govt. of India as 'Pre-42 settlers'. Perhaps their enduring struggle and sacrifice brought them the privilege of being the first group of Indians on Indian soil to unfurl the national flag with full honour at Gymkhana ground by the head of state of Azad Hind Government, Netaji Subash Chandra Bose on 30th Dec.1943. Indian Independence League surfaced silently but mightily to welcome their Commander in Chief on his arrival at Port Blair airport. However, after Netaji's departure hell broke loose on the local born community. The hypocrisy of Japanese diplomacy vanished like summer dew and the Japanese occupational forces went hysteric, embossing barbaric atrocity upon local population. The educated and intellectual were openly shot dead and homes raided. The fabric of settlement was shredded and humanity subjugated for two more long years.

Surrender - In the morning of 7th October 1945 a divine help ascended from the eastern horizon, when the armada carrying 116 Indian infantry brigade of South East Asian allied land force under the command of Brigadier A.J.Solomon surrounded Port Blair compelling about 20,000 armed Japanese force to surrender on 9th October, 1945. It was an occasion of rejoices for the islanders.

Mr. Arfat Ahmed Khan, a prominent civilian gives an account of the Japanese surrender ceremony, which was held at Gymkhana Ground thus; "One table was laid just facing the Andaman club. All officers were gathered there. The military was in charge. A car drove up. It carried the Commanding Officer of the Japanese Naval Force. He was in full uniform and was carrying his sword. Two aides accompanied him. Brigadier Solomon was seated on the other side of the table. The Japanese Admiral saluted. Brigadier Solomon stood up and shook hands with him. Then the Japanese Admiral took his sword out of the scabbard, surrendered it to Brigadier Solomon. Brigadier Solomon unbuckled his belt and badges of honour. A prepared document was read out. Brigadier Solomon signed the document on behalf of the British Government and the Japanese Admiral on behalf of the Japanese. Contrary to the instructions, the crowd that had gathered burst into cheers. The whole ceremony did not last for more than two or three minutes. When the ceremony was over, the Japanese Admiral walked quietly with his bowed head to the car. His two aides were walking behind him. They got in and drove away".

On 28th Nov.1945, Lord Louis Mount Batten, the Supreme Allied Commander of South East Asia Command visited these islands. About Port Blair his diary reads : "At 14:30 we landed at Port Blair, the capital of Andaman Islands. Here I met M.Patterson, High Commissioner (SIC), Brigadier Solomon Commanding the 116 Indian Infantry Brigade, Group Captain Pope and Captain Blair, RNR, the great nephew of the original Lt. Blair who founded the port. After inspecting the Naval Guard of Honour, I visited and addressed practically the whole of 116 Brigade in groups, and later went round the magnificent harbour with Captain Blair and went on board HMS Kistna. The population consists of some 12,000 natives and 6,000 convicts who have been released and pardoned. Wherever we drove, every single native, without exception, stood up and saluted as we passed, a relic from the Japanese who beat them if they did not salute. However, I think they were pleased to see us. They were being starved to death by the Japanese".

With the advent of Indian Independence on 15th Aug.1947 these islands were merged with the Indian main stream.

NETAJI SUBASH CHANDRA BOSE INSPECTING THE GUARD OF HONOUR PRESENTED BY INDIAN NATIONAL ARMY ON HIS ARRIVAL AT PORT BLAIR ON 29TH DECEMBER, 1943.

Courtesy: Gouri Shankar Pandey

5  ABORIGINES & TRIBES 13

The original inhabitants of Andaman & Nicobar Islands, account for only 12% of the total population. The Andaman group of islands presently have six living aborigine groups namely Sentinelese, Jarawas, Onges, and Great Andamanese of Negrito origin; whereas the Nicobarese and Shompens are offshoot of Mongoloid stock and live in the Nicobar group of islands.

It is believed that the Andaman aborigines might have reached these islands very early in time, possibly by boat from South East Asia since they show a strong affinity to the Semangs of Malaysia and the Aetas of Philippines. However, a recent DNA match is indicative of direct links with the Pygmies of Southern Africa.[4] The Nicobarese must have migrated sometime before the Christian era. The origin of Shompens is not known but it appears that the Shompens have a Malayan strain. The Andaman aborigines are primarily hunter-gatherers, whereas the tribes of Nicobar are mainly horticulturists and herders. Sentinalese continue to be hostile to outsiders whereas the once ferocious Jarawas are gradually becoming friendly to civilized population. Onges and the Great Andamanese have accepted the presence of outsiders. The Nicobarese have integrated well with outsiders and have joined the Indian main stream. Shompens still avoid contact but are non-hostile.

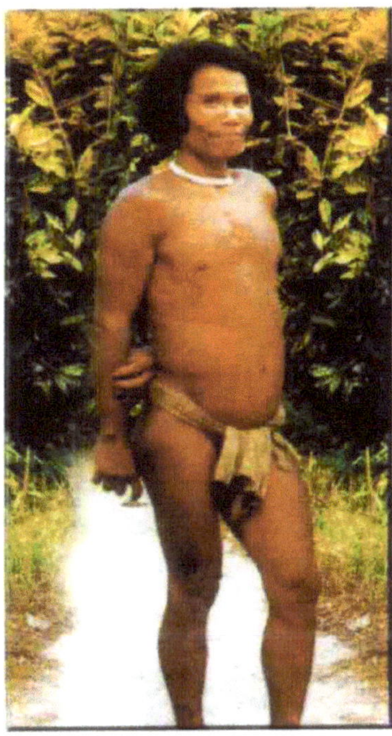

SHOMPEN

JARAWA

Courtesy: C.P. Operai

5.1 SENTINELESE

Sentinelese are the sole inhabitants of the North Sentinel Island, which is 60 sq.km. in area and located 34 km west of South Andaman. They live in complete isolation. They are hostile, semi-nomadic and their settlement consists of about 20 single huts close to one another in a cluster. In February 1982 a team from the local Administration contacted a small batch of this tribal community who responded with friendly gestures for the first time. Since then persistent efforts are being made to befriend them. Their present population is estimated at 250.

5.2 JARAWA

The Jarawas live in the reserve forest belt of about 639 sq.km area in the western coast of South and Middle Andaman. The members of this tribe are now becoming friendly. However, they continue with their primitive way of life. The Jarawa live in communal huts, which lack a sleeping platform. They utilise naturally available materials such as palm leaves, barks, canes, seashells, corals etc. for their ornaments as well as for their dress. Since 1974 continuous efforts are being made to develop friendly relations with them. Their population is around 350.

5.3 ONGE

The Onges live in Little Andaman, an island 130 km south of Port Blair. They inhabit in two settlements, one at Dugong Creek and the other at South Bay over an area of 25 sq.km. The Onge differ from the Great Andamanese in language and in some aspects of their culture. Their settlement pattern and subsistence activities are however similar to those of the Great Andamanese. They have their local groups with well defined territories for hunting and collecting forest products. Each member of the clan lives in his/her respective communal hut. This tribe still exists as hunters. Monogamy is their traditional social norm. Onges are friendly with neighbours and visitors. There are about 98 Onges alive today.

5.4 GREAT ANDAMANESE

The vanishing Great Andamanese tribe are settled at Strait Island, which is about 46 nm. away from Port Blair and has an area of 60 ha. Present population of 39 are the remnants of ten clans of the once flourishing aborigines who inhabited the South & Middle Andaman and were first to be befriended. Mass mortality occurred among them soon on contact with civilised man through influenza, small pox, measles and sexually transmitted infectious diseases. Today they are considered perhaps the weakest ethnic group.

5.5 SHOMPEN

The Shompens inhabit 119 sq.km. interior forest area of Great Nicobar. They are shy by nature and avoid contact with outsiders. Shompens live invariably along or around the perennial fresh water rivers, streams or rivulets in the forest. Shompen villages generally comprise 2 to 10 huts at a place. Their huts are constructed on piles, the height of which varies form 1.5 m to 3 m. They use palm leaves and padanus leaves to thatch their huts. The floor space on the piers is used for living while the space below is utilised for their domestic animals. Monogamy as well as polygamy is a social norm. In each village the oldest male is the head of the tribe. Burial is their traditional custom for disposal of the dead. Shompens population is estimated to be between 250-300.

A TYPICAL NICOBARI HUT

Courtesy: Directory, IIPA

6 FOREST 16

These Islands are blessed with a unique luxuriant evergreen tropical rainforest canopy, sheltering a mixed germ plasm bank, comprising of Indian, Myanmarese, Malaysian and endemic floral strain. So far, about 2200 varieties of plants have been recorded out of which 200 are endemic and 1300 do not occur in mainland India.

"The South Andaman forests have a profuse growth of epiphytic vegetation, mostly ferns and orchids. The Middle Andamans harbours mostly moist deciduous forests. North Andamans is characterised by the wet evergreen type, with plenty of woody climbers. The north Nicobar Islands (including Car Nicobar and Battimalv) are marked by the complete absence of evergreen forests, while such forests form the dominant vegetation in the central and southern islands of the Nicobar group. Grasslands occur only in the Nicobars, and while deciduous forests are common in the Andamans, they are almost absent in the Nicobars". (Shekhar Singh et.al.1991, Directory of national parks and sanctuaries in A&N islands. IIPA). This atypical forest coverage is made-up of twelve types namely : (1) Giant evergreen forest (2) Andamans tropical evergreen forest (3) Southern hilltop tropical evergreen forest (4) Cane brakes (5) Wet bamboo brakes (6) Andamans semi-evergreen forest (7) Andamans moist deciduous forest (8) Andamans secondary moist deciduous forest (9) Littoral forest (10) Mangrove forest (11) Brackish water mixed forest (12) Submontane hill valley swamp forest. The present forest coverage is claimed to be 86.2% of the total land area.

EMERGENT

CANOPY

UNDERSTORY

SHRUB

HERB

6.1 TIMBER

WET BAMBOO BRAKES

Andaman Forest is abound in plethora of timber species numbering 200 or more, out of which about 30 varieties are considered to be commercial. Major commercial timber species are Gurjan (Dipterocarpus spp.) and Padauk (Pterocarpus dalbergioides). Ornamental wood such as (1) Marble Wood (Diospyros marmorata) (2) Padauk (Pterocarpus dalbergioides), (3) Silver Grey (a special formation of wood in white chuglam) (4) Chooi (Sageraea elliptica) and (5) Kokko (Albizzia lebbeck) are noted for their pronounced grain formation. Padauk being steadier than teak is widely used for furniture making.

Burr and the Buttress formation in Andaman Padauk are World famous for their exceptionally unique charm and figuring. Largest piece of Buttress known from Andaman was a dining table of 13' x 7'. The largest piece of Burr was again a dining table to seat eight persons at a time.

The holy Rudraksha (Elaeocarps sphaericus) and aromatic Dhoop/Resin trees also occur here.

6.2 FAUNA

This tropical rain forest despite its isolation from adjacent land masses is surprisingly enriched with many animals.

Mammals - About 50 varieties of forest mammals are found to occur in A&N Islands, most of them are understood to be brought in from outside and are now considered endemic due to their prolonged insular adaptation. Rat is the largest group having 26 species followed by 14 species of bat. Among the larger mammals there are two endemic varieties of wild pig namely Sus Scrofa andamanensis from Andaman and S.S.nicobaricus from Nicobar. The spotted deer Axis axis, Barking deer and Sambar are found in Andaman District. Interview island in Middle Andaman holds a fairly good stock of feral elephants. These elephants were brought in for forest work by a private contractor who subsequently left them loose.

(Macaca fascicularis umbrosa)

They are coastal inhabitants of Great Nicobar Island; dusky brown in colour and superficially resemble the bonnet monkey. Omnivorous by nature marine crabs appear to be their favorite food.

Snakes - The Andaman & Nicobar Islands hold 46 species of snakes out of which 13 are endemic. India's largest snake, the Regal Python (Python recticulatus) belongs to Nicobar group. The Snake Island, an islet opposite Corbyn's cove is said to be the favourite mating and nesting ground of the only egg laying Indian sea snake (Laticuda).

Birds - The Andaman & Nicobar Islands also hold a unique avi-fauna exhibiting a high degree of endemism with 270 species, of which 39% are endemic to these islands. Endemism among birds is significant when compared to the total faunal endemism of 13%. Birds of the Andamans include Streaked Grasshopper, Warbler, Yellow Wag tail, Olive Backed Sunbird, Red Whiskered Bulbul, Red Rumped Swallow, Three Toed King Fisher, Chestnut Headed Bee Eater, Shikra, Moorhen, Crested Serpent Eagle, White Bellied Sea Eagle, Drongo, Sparrow Hawk, Teal, Grey Heron, Cattle Egret, Chestnut Bittern, Eastern Golden Plora, Curlew, Common Sand Piper, Wimbrel, Rosy Tern, Black Naped Tern, Little Stint, Koel, Brown Hawk Owl, Barn Owl, Glossy Stare, Hill Myna, Oriole, Paradise Fly Catcher, Ground Thrush, Magpie Robin, Fairy Blue Bird, Green Breasted Pitta, Dark Thrush and the Siberian Blue Chat.

A few endemic birds are:-

(1) Nicobar Megapode
(Megapodius freycinet nicobariensis)

They frequent the open jungle area near the shore where the soil is so light to build its mould easily. Each egg weighs around 150 g., which is approx. 1/6th of the bird's own weight. The Megapode does not take part in the incubation of its eggs. Decaying leaves of its mould give enough humidity and heat for incubation.

(2) Swiftlets *(Collocalia fuciphaga inexpectata)*

Locally known as Hawabill. This bird is capable of flying upto a speed of 150 kmph. This very versatile flyer is capable of spending an entire night in flight and can even have catnaps while flying. They build nests using their saliva, which is said to have medicinal and aphrodisiac values. They are fast becoming extinct.

(3) Narcondum Hornbill
(Rhyticeros narcondami)

It is a pariah kite size bird having a conspicuous yellow beak with light brown casque and is seen only in the Narcondum Island.

(4) Andaman Teal
(Anas gibberifrons arbgularis)

This duck like migratory bird measuring upto 43 cm lives in fresh water ponds, tidal creeks and swampy grasslands abundant with weeds and vegetation.

ANDAMAN MORMON
(Papilo mayo)
FEMALE

Butterflies and Moths - With about 225 species, the A&N Islands house some of the larger and most spectacular butterflies of the world. Ten species are endemic to these Islands. Mount Harriet National Park is one of the richest areas of butterfly and moth diversity on these Islands.

The Andaman Mormon was the first endemic butterfly to be described from these Islands. This large butterfly was christened Papilo mayo in honour of Lord Mayo, the Viceroy of India who was killed in harness at the foot of Mount Harriet on 08 Feb 1872 by an Afghan convict.

ANDAMAN MORMON
(MALE)

Other endemic species are Andaman Flat, Andaman Swordtail, Andaman Clubtail, Andaman Palmking, Andaman Visconut, Andaman Oak Leaf and Tailless oak blue. The Nicobar Yeoman and the Nicobar Map are butterflies endemic to the Nicobar Islands.

TREE NYMPH
(Idea agamarschana Cadelli)

Butterflies pick and choose the environment they would like to live in. They are equally particular about the plants they live on. Some like the Andaman Blue - Nawab are so fastidious that they will feed on only one species of plant. In the event of non-availability of their chosen plant species, they would rather starve and die than choose an alternate.

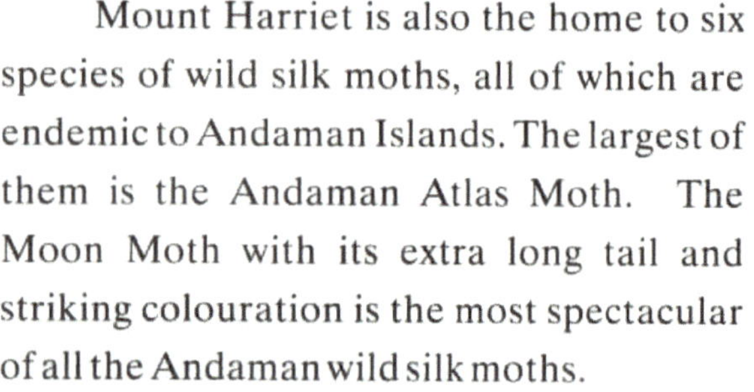

Mount Harriet is also the home to six species of wild silk moths, all of which are endemic to Andaman Islands. The largest of them is the Andaman Atlas Moth. The Moon Moth with its extra long tail and striking colouration is the most spectacular of all the Andaman wild silk moths.

Atlas Moth

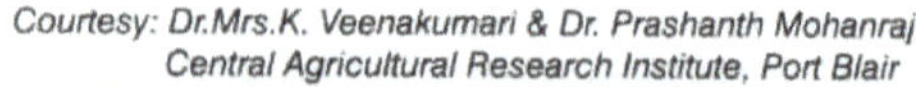

Courtesy: Dr.Mrs.K. Veenakumari & Dr. Prashanth Mohanraj, Central Agricultural Research Institute, Port Blair

Shells are perhaps the most colourful and fascinating objects known to man other than Gems since time immemorial. They served as money, ornaments, musical instruments, drinking cups, in magic and in the making of fine porcelains. They were also the symbols in rituals and religious observances, and the returning pilgrims wore them as a token of divine pardon.

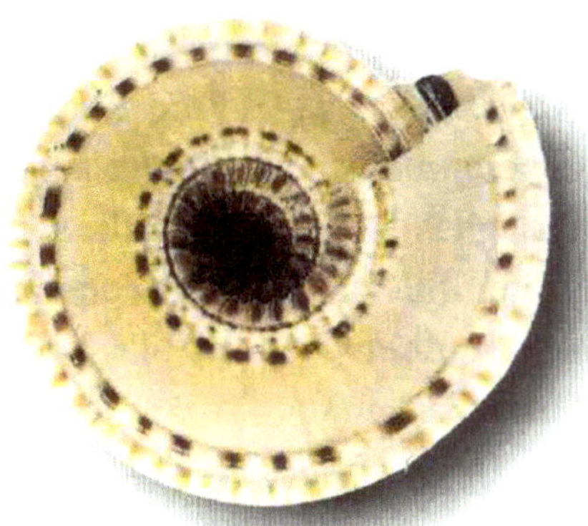

SUN DIAL SHELL

Shells are formed, on principles that are still quite mysterious, by a soft bodied creature known as 'Mollusca'. A 'shell' is loosely defined as an external skeleton made up of calcareous matter exhibiting a wide variety of shapes and colours specific to the animal that creates it through complex physiological process. This group of animals includes a seemingly infinite variety of forms in their evolution. Many authorities believe this group to be the second largest in the animal kingdom. Shells are commonly divided as Univalve and Bivalve depending on the number of calcareous valves or shells they possess. Further, they are divided into many taxonomical families. These mobile home dwellers were probably born some 600 million years ago. Today they inhabit all the oceans, lakes, rivers and ponds. Their blood is usually blue due to the presence of a copper-based blood pigment called haemocyanin. They are generally hermaphrodite or bisexual. There are more than one lakh living and thirty five thousand extinct varieties documented in the world so far. In A&N Islands about a thousand varieties are known to occur. Samudrika displays 1129 shells of the range.

These islands are traditionally known for their shell wealth specially Turbo, Trochus, Murex and Nautilus. Earliest recorded commercial exploitation began during 1929. Shells are important to these islands because some like Turbo, Trochus & Nautilus etc. are being used as novelties supporting many cottage industries producing a wide range of decorative items & ornaments. Shells such as Giant clam, Green mussel and Oyster support edible shellfishery, a few like Scallop, Clam and Cockle are burnt in kiln to produce edible lime.

SUN DIAL SHELL

The Univalve or one shell group belongs to the class Gastropoda having more than 80,000 species. Sacred Chank belongs to this group. Their body, in the course of development, go through a complicated process, 'torsion' i.e. the visceral mass is twisted though 90^0 together with the shell that covers it. Under mysterious circumstances many a time this process proceeds in the reverse direction thus creating an abnormal shell which otherwise lives like a normal shell. A classic example is the most wanted left-handed chank.

The Bivalve or Pelecypoda has about 20,000 living species. Majority of them burrows in sand or mud such as Pearl Oyster, Wing oyster, Giant clam etc.

A third group, which is comparatively smaller, is called Cephalopoda, which includes Octopus, Squid, Nautilus etc.

The soft body animal, which lives inside the shell, is covered with a thick layer of specialised epithelium cells known as mantle, which in turn secretes a two tier shell material making the shell. The outer layer having a different colour pattern is organic in constitution, technically called 'periostracum'. Calcium ions from the environment are absorbed into the blood and deposited evenly under this layer. The next inner layer is called 'nacre' or 'mother of pearl' responsible for the pearly lustre common to many shells.

Few varieties

1. Turbo

Turban shell is considered to be the island's pride and is the most sought after commercial shell. It has a mobile door like appendage called Cat's eye to protect itself from predators by closing the large underside opening. They live from the lowest water line upto the reef edge. We have many varieties and some among them grow quite large. This shell possesses high quality mother of pearl.

2. Trochus

The Top shell is the second most important shell. They are robust and thick having regular conical high spire. Coloration consists of irregular reddish or brown flaming on a dirty white background. They are found on rocky shores grazing upon sedentary vegetation.

3. Cyprea

Cowries are wide spread throughout the islands. They are attractively coloured, smooth and gleam like marble. They vary in size, shape and colour. Many can be found in unpolluted inter tidal areas.

4. Murex

The Nancowy shell belongs to this group. They are characterised by spiny ornamentation. They inhabit open coastal as well as closed muddy lagoons. Best of the murex variety comes from Nancowry group of islands. They feed on other shells.

5. Conus

The cone shells come in a remarkable array of shape, colour and size. A few of them are extremely poisonous. Some rarest of the rare and highly priced shells belong to this group.

6. Xancus

The sacred chank, which once abounded the sea around A&N island, has depleted. This thick, spindle shaped shell could grow upto 14 inches in length. It is generally white in colour. However, a yellowish specimen having chestnut patches is also found. The shell's apex through spiral cord is cut open for puja purposes. Left-handed chanks are considered sacred and thus valued.

7. Cassis

Two well-known varieties of helmet shells are locally known as King and Queen shells. King is invariably handsome and robust as compared to Queen. They could grow upto 14 inches. They possess appealing orange colour over a creamy background, or pale pink contrasting with white or rich brown on beige.

8. Strombs

Lambis or Punja shells are common in occurrence. They walk differently. They eat green or red filamentous algae.

9. Bivalve

This is a large group of shells, which include pearl oyster, mussels, oysters, scallops, cockles, clam etc. They live in a vast array of habitats. Their body is enclosed in a pair of calcareous plates, which open and close as per their requirement. Majority of them is sedentary and are filter feeders.

10. Nautilus

A smooth creamy white shell marked with rich brown on the outside and brilliant mother-of-pearl within. A marvel of nature's engineering, the shell is divided into 30 compartments. The animal lives in the innermost compartment. Other chambers are used to regulate buoyancy. It can dive upto a depth of 600 m.

7.1 SHELL MIDDENS

Ancient coastal aborigine encampments in Andamans are marked with 'shell mound' generally covering basal area of 200 to 300 sq. m. and height of 4 to 5 m. These mounds are meanly made up of discarded shells known as 'Shell middens' or 'kitchen middens' as kitchen waste was dumped at these sites. The shells recovered from one of these middens located in the Chouldari village of South Andaman has revealed an age of about 2200 years through radiocarbon (C^{14}) dating.

STONY CORAL (GONIOPORA SP.) GIVING OUT POLYPS
(PORT BLAIR - AQUARIUM)

SOFT CORAL PATCH REEF
(PORT BLAIR - NORTH BAY)

Courtesy: Arif Mustafa

CORALS

The noun 'coral' is believed to have derived its origin from an Arabic word 'garal', which means small stone, or Hebrew 'goral', which means pebble. Later, the Greeks adopted it as '*korallion*' and in Latin it appeared as '*Coralium*'. The present day English version means 'the hard stony skeleton secreted by certain marine polyps'. The animal, which secretes and builds this skeleton originated some 570 million years ago. This tiny, boneless, fragile creature is genetically endowed with exceptionally high architectural skill. Polyps create multispectral & multidimensional skeletons known to us as corals. A close look at a dead bleached coral piece will reveal its porousness. Millions of pores are found in a small piece of coral. Each 'pore' was the home for a 'polyp', and the piece of this calcareous '*garal*' held for observation is the outcome of cumulative effort put-in by billions of polyps for a long time.

Under specialised environmental and geographical conditions very many coral colonies flourish at selected sites, thus creating a coral reef which under normal ecological condition keeps growing like the Great Barrier reef of Australia, which is the greatest structure made by animal life on the earth and is visible from the Moon. A coral reef is an assemblage of more than 3,000 living organisms in perfect harmony, a magnificent manifestation of nature's ability to create, thread and balance various life forms in space and time. Coral reef ecosystem is the most intricate, diversified and aesthetically appealing ecosystems of this planet.

Courtesy(Upper): Veron, JEN.

The International Union for Conservation of Nature (IUCN) in world conservation strategy has identified coral reefs as an *essential ecological life support system necessary for human survival and sustainable development*. The coral reefs benefit mankind by :-

- Providing excellent nursery grounds for the growth and development of young ones of innumerable commercial marine fishes.
- They yield many varieties of shallow water fishes, shells, lobsters, sea urchin, sea cucumber, sea grasses both edible and ornamental. They provide nourishment and play a significant role in many maritime nation's economy. In A&N islands about 17% of annual fish catch is from coral reef areas providing livelihood to about 2000 fishermen.
- Coral reef is the first underwater guardian of the coastline, preventing seaward coastal erosion.
- Coral reef glamour supports tourism.
- Coral reefs hold complex biochemicals for the future. Anti-tumour, anti-leukemia, anti-microbial & ultraviolet blocker chemicals of high medicinal value are being extracted from reef inhabitants.
- Coral reefs provide support and sustenance to adjoining coastal and terrestrial ecosystems on which people depend.
- Coral colonies are excellent record keepers of the past meteorological events.

A traditional survey of A&N Islands conducted in 1986-87 estimated an area of 2000 sq.km. under coral coverage which amounts to 6% of the 34965 sq.km. of total continental shelf area. The human induced degradation was estimated to be about 360 sq.km. However, a recent survey using remote sensing technique indicates 1000 sq.km of coral coverage. Are the corals disappearing or is there an error in the surveys? The question is yet to be answered. So far about 200 varieties of corals have been documented from A&N islands, which is the highest in this part of the world. There are fringing reefs scattered throughout the territory, and also have many luxuriant coral banks in the adjoining open sea.

The renowned marine naturalist Cap. Jacques Cousteau with his associates on board Cousteau Foundations research vessel 'CALYPSO' made 70 hours of underwater observations over 12 dive sites in the Andamans in the year 1989 and recorded "INDIAN WATERS ARE PRESERVED, RICH AND VIRGIN".

BOULDER CORAL & BUTTERFLY FISHES
(NARCONDUM ISLAND)

BRANCHING CORAL WITH ANTHIAS FISHES
(NARCONDUM ISLAND)

Courtesy: Cousteau Foundation

8 CORALS 29

8.1 Corals belong to a large group of animals known as *Coelenterata* (stinging animals) or *Cnidaria* (thread animals). Corals grow slow, they have type wise site specific growth rates. The massive forms may grow upto 2 cm. in diameter and upto 1 cm in height a year, whereas, delicate branching forms grow between 5 to 10 cm. per annum. A true reef building stony coral may be unisexual or bisexual. They breed together once in a year at a pre-determined time after dusk. This process, at places is so intense that the water stays pinkish till next morning. A large number of baby corals are released in the open ocean this way. After sometime these baby corals settle over a suitable substratum and start forming new colonies through asexual reproduction. Their morphological features change with the environment in which they settle. Due to this peculiar character they are often called 'Plastic animals'.

Stony corals could be broadly divided into reef builders and non-reef builders, the reef builders are called *hermatypic* whereas others are known as *ahermatypic* corals. The reef builders possess hard calcareous skeleton and need sunlight like plants to survive. On the other hand, the non-reef builders are devoid of a true stony framework and can live well without sunlight. A few among them are capable of making protein based solidified skeleton.

Few varieties of Cinidarians

1. Hermatypic

Coral reefs all over the world are the handy work of hermatypic reef builders. In these islands there are two chief reef builders namely *Porites* and *Favia*. The other forms such as *Favites*, *Platygyra*, *Symphyllia*, *Ganiastrea* and *Diploastrea (Brain coral)* present here are the associated reef builders. All of them create massive dome shape colonies creating sheltered areas for other branching and delicate forms such as *Acropora (Staghorn coral)*, *Pocillopora (Thorny coral)*, *Stylophora*, *Seriatopora (Birds nest)* and *Montipora* to establish. All the massive and branching forms named above are sedentary. However, there is also a group of roaming corals, which belong to family *Fungiidae* popularly known as Mushroom corals.

CORAL REEF ~ RED SEAFAN UPON CHIEF REEF BUILDING CORAL
(INVISIBLE BANK)

SHARK ENCOUNTER ~ Dr. F. SARANO OF COUSTEAU FOUNDATION
(INVISIBLE BANK)

Courtesy: Cousteau Foundation

2. Ahermatypic

Since their survival does not depend on the availability of sunlight, they could grow at any depth. Majority of them is colonial. Some of them such as *Balanophyllia*, *Dendrophyllia* and *Tubastraea* occur in dark and shaded crevices of coral reefs. Very little is known about them, following varieties are common around A&N Islands.

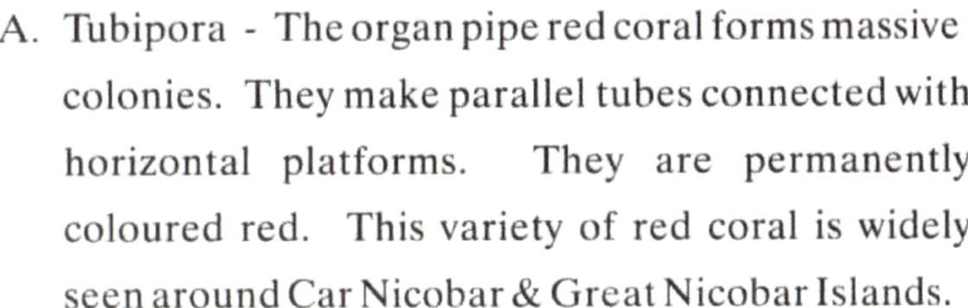

A. Tubipora - The organ pipe red coral forms massive colonies. They make parallel tubes connected with horizontal platforms. They are permanently coloured red. This variety of red coral is widely seen around Car Nicobar & Great Nicobar Islands.

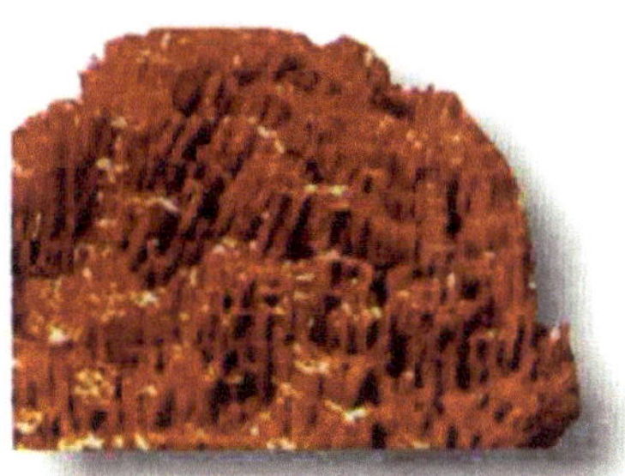

B. Heliopora - The Blue coral or the Fire coral, as commonly known, is permanently coloured blue. It's stings produce rashes with burning sensation. This variety is commonly seen around Port Blair, Diglipur and Rangat.

C. Isis - The marble coral is a shrub like branching form having a smooth, ringed black and white skeletal frame. This variety is common around Ross, Cinque, Brother & Nacowry group of Islands.

Samudrika has a fairly good cross sectional representation of corals found around A&N islands. Displayed underwater photographs depict many factual life events of corals and reflect their present status in the oceanic water around us.

ECO DIVERSITY DEPICTING CORAL, MANGROVE AND RAIN FOREST
(ATLANTO BAY - NORTH ANDAMAN)

LUXURIANT FRINGING REEF
(DELGARNO Is. - NORTH ANDAMAN)

Courtesy: Cdr. Paul James

9 OCEAN 33

Indian Ocean is a vast stretch of water comprising of many geographically demarcated seas and bays. The A&N Islands are surrounded by the Andaman Sea to the east and the Bay of Bengal on the west. Andaman Sea is the abode of volcanic activity, the sub crustal zone under the floor of Andaman Sea continues to be extremely active and volatile. Barren island volcano is the living testimony. The Andaman Sea, by virtue of its volcanic history, holds huge stocks of minerals, metals and hydrocarbon reserves. However, it is comparatively less productive in terms of biomass production.

Sea, the cradle of life abounds with life forms. Every drop of it is living and all animal groups known to science are present in the ocean. It holds infinite number of animal forms and plants ranging in size from nano-microns to gigantic whales. In the sea, as on land, life depends largely on plants. The plants in the ocean are almost as productive as the plants on land. The pastures of the sea and the basis of life cycles are myriads of free floating, microscopic plants known as *phytoplankton*. These are the food for minute animals called *zooplankton*. Zooplankton are preyed upon by larger animal species, which themselves provide food for still bigger creatures. Thus continues a ruthless and never ending cycle. Minerals derived in part from the decay of marine organisms nourish the plants, in their turn. The movement of currents, by which the oceans 'plough' themselves, is caused by three main forces, the wind, the earth's rotation and differences in sea's density.

By the provisions of various international conventions, we have sovereign right over about six-lakh sq.km. of sea area encompassing both eastern and western water masses. This part is fairly rich in living as well as non-living resources. Living resources are typically tropical in nature. Theoretically it could produce about 4,70,000 tonnes of biomass per annum. A rich variety of fishes exceeding 1200 in number are known to occur here. Important surface fishes are Sardine, Mackerel, Caranx, Tuna & Tuna like fishes, Shark, Ray etc. Among bottom dwellers we have Perche, Snapper, Grouper, Croaker, Silver belly, Deep sea Lobster and Shark.

9.1 FISHES

Each life form in the sea is confined to its own particular zone, where pressure, light, temperature and salinity are more or less constant. In this stable environment some creatures have remained unchanged throughout their entire history. The now famous *Coelacanth*, one of the groups of fishes thought to have been extinct for 60 million years, has remained essentially like its relatives as they appear in fossils. Fishes are the masters of water world. For more than 360 million years they have inhabited it. Today we have about 40,000 varieties of fishes known to science. They range in size from 10 mm (Philippine Gobie) to 21m. (whale shark). Some are flattened, others inflated, many spindle shaped, a few snakelike, still others are compressed depending on the environment in which they live or particular way of life. Fishes are broadly classified as : -

Jawless fishes - *Agnatha* are snake like primitive fishes, which include hagfishes and lampreys. Hagfish lives imbedded in deep waters and are excellent scavengers of the ocean floor. Lampreys stay near to land but could migrate to very deep water and feed mostly on crustaceans (prawn, lobster etc.). They enter into the body of large fishes and consume them from inside out.

Cartilaginous fishes - *Chondrichthyes.* They are devoid of true skeleton; this group includes Sharks, Skates and Rays. Shark, the most fearsome sea fish keeps replacing its lost teeth. Its females conceive through coitus. It appears that they don't perceive pain. They have a built-in immunity to cancer. The largest living fish belongs to this group.

Bony fishes - *Teleostomi/Osteichthyes.* The true bony fishes are the most modern and largest group of fish. The ray finned fishes, where fin rays or spines support the body of each fin, constitute the vast majority of bony fishes. Fertilization is generally external. They possess a specialised buoyancy regulation device called Gas bladder. Many of them are capable of sex reversion as per demand. Some have bio-luminous organs to glow the deepest ocean floor. With the exception of sharks all the edible and ornamental fishes belong to this group.

9.2 ASSOCIATED FAUNA

(1) Marine mammals

Whale, Dolphin and Dugong are the three marine mammals known to occur in these islands. Unlike fishes they are warm-blooded animals like us. They breast feed their young ones. The common Dolphin (*Delphinus delphis*) and the Dugong or Sea Cow (*Dugong dugon*) are residents of this area. The *whales (Balenoptera musculus and Physeter catodon*) are transitional. Whales regularly visit our sea for delivery of their young ones. The whole family enters these waters and stays here till the babies are born. The spent whale release a sort of faecal matter which is known as Ambergris, a fixative used in perfume manufacturing. It is said to have medicinal and aphrodisiac values. The present market value of Ambergris is about two lakh Rupees per kg.

Once upon a time Dugongs were common in these islands. The Dugong creek at Little Andaman has since been so named. Our aborigine and tribal population traditionally hunt dugongs. Today they are one among the many rare and depleting marine animals.

(2) Turtles

Five types of turtles live in the Andaman waters; Loggerhead turtle, Green sea turtle, Olive ridley turtle, Hawksbill turtle, and the Andaman giant leather back turtle. Giant leather back turtle is massive in size and need a thick layer of sand for nesting. Galathia, South Bay of Great Nicobar Island and Twin Is. are its favourite nesting grounds. They lay upto 100 eggs in a clutch, generally during November-February.

3. Salt water crocodile (***Crocodilus porosus***)

They are seen mainly in Mayabunder and Diglipur area of North Andaman. They are fast becoming extinct due to the myth associated with the medicinal properties of their fat and oil, and of course, its hide.

4. Robber Crab (***Birgus latro***)

The term 'robber' came to existence because of the belief that they climb coconut trees to rob the nuts. However, experts believe that this crab climbs the coconut trees to nibble the tender shoots and also sip the water trapped in between the leaves. They have been found to occur in Great Nicobar, Car Nicobar and South Sentinel Island. An adult crab grows to 30-80 cm weighing about 2-3 kg. With the long walking legs they climb coconut, arecanut and sago trees. The young hatchlings are born in the sea.

10 MARINE AQUARIUM 37

Marine fish and animal keeping still has a certain mystique attached to it. This is one of the most complicated aspects of live stock management. The animal husbandry involved in it is mainly nurtured through water chemistry and microbiology. The tropical coral reef inhabitants are generally maintained in glass boxes known to us as marine aquariums. These animals turn 'fragile' under captive atmosphere because the natural system to which they belong is so heterogeneous, complex and dynamic with every tide bringing in a different condition that is so difficult to create artificially. However, since May 1853 when the first tropical marine aquarium was made public in London, much has been understood and we are now able to practice a system where these animals are acclimatized and taught to be happy in their new environs.

Few varieties

The oceanic water surrounding us is the abode teeming with the most beautiful *icthyo* (fish). We have a large variety of brightly coloured dazzling fishes in many shapes and forms along with a vast number of amazing marine creatures. Even with all the advancement made in this regard, we are still not able to hold them all alive. However, a fraction of which could be displayed is briefly described hereunder :-

(1) Butterfly fishes

The *chaetodontidae* are among the most gracefully agile and attractive ones of reef fishes. They are laterally compressed with a vertical strip across the eye. Juveniles differ from adults in colour and shape. Their population is directly proportional to the health of coral reefs.

(2) Clown/Damsel/Anemone fishes

They belong to the family of *Pomacentridae* and have wide distribution. Generally on a coral reef they are the first to be noticed. As a group they are great opportunists for colonising and exploiting the reef. The anemone fish belongs to this group. They live in between the poisonous tentacles of sea anemone *Stoichactis / Radianthus / Metridium.* There are two sub groups of anemone fishes known from these islands, the Amphiprion and *Premnas.* The former having more than a dozen varieties out numbers the latter with only two known varieties. They are territorial and strongly defend their anemone from any intruder.

(3) Scorpion fishes

Scorpaenidae - They are easily noticed under natural environs due to their slow majestic movement, brilliant colour contrast and enlarged filamentous fins. They herd small prey using enlarged fins. They carry painful venom in their long dorsal, anal and pelvic fin spines and so are dangerous to touch. They are shy to sunlight.

(4) Wrasses

The Labridae are generally coloured in brilliant green. They have an elongated body with small fins. They keep moving fast during the day and dig into sand at night to sleep.

(5) Parrot fish

Scaridae. They are closely related to the Labridae. They have robust bodies with large scales. Teeth are fused into a beak. They produce sand by excreting crushed coral, which is their favourite food. They make noise while scraping the hard corals for polyps.

(6) Surgeon fish

A canthuridae derived their common name due to the presence of a pair of foldable scalpel on either side of the tail. They are herbivores.

(7) Sea horse

Syngnathidae is the family of sea horses and pipefishes. The common sea horse Hippocampus is well known and needs no introduction except for the fact that, it is the father sea horse that gives birth to babies.

11.1 SUBMERGED DWARAKA

The Marine Archaeology Unit of the National Institute of Oceanography, Goa found the submerged Dwarka of Lord Krishna in between Dwarka and Baith Dwarka. According to Mahabharata, Lord Krishna built Dwarka on the ruins of an earlier town known as Kusasthali founded by his Yadava ancestor Kakudim Raivata. In Harivamsa, an appendix to Mahabharata it is mentioned that the sea yielded 12 yojanas of land for its construction suggesting some kind of reclamation works. A recent investigation has unearthed eight distinguished settlements at Dwaraka. The first settlement made in the 15th Century BC was submerged or washed away and so also the second one made in the 10th Century BC. After a long gap, the third settlement was made in 1st Century BC/AD as suggested by the red polishedware and Copper coins known as 'karshapanas'. Temple-I was built during this time. The temple-II was built on the ruins of temple-I when the sea destroyed it. The tidal force of sea destroyed this temple also. The temple-III dedicated to Vishnu or Vasudeva was built in the 9th Century. It was perhaps a storm wave in the 12th century, which blew away the roof of this temple leaving only the walls. The Temple-IV came into existence soon thereafter. The present temple of Dwarkadhis is the fifth of the series. Temples - I to IV represent settlements III to VII respectively and the modern town is the eighth settlement in Dwarka.

Of the five temples were built from 2nd Century BC to 15th Century AD, except for the last two temples, the catastrophic effects of the sea destroyed them all. Dynamic tidal barrier of the Gulf of Kutchh is one of the main factors. A comparison of Dwarka's shorelines from the Admiralty chart of 1848 with the Indian Hydrographic chart of 1977 shows that during the intervening period of 130 years the coastline has retreated by about 550 m. Dwarka coast on the Gulf region has been eroding approximately 4-m. per year since the turn of the Century. It is the most active coastal erosion site along the Gujarat coast. Assuming this rate of erosion to have prevailed in the past, the causes of the destruction by the sea of the ancient Dwarka can easily be understood.

11.2 LOTHAL TIDAL DOCK

The architects of the Indus Valley civilization, the Harrappan, are found to have constructed the first tidal dock of the world at the port town of Lothal (Lat. 20^0 31' 25" N, Long. 70^0 14' 25" E) at the head of the Gulf of Cambay. This exciting discovery was made in 1954 by the underwater archaeologists of National Institute of Oceanography, Goa. The Saragwala village, which shelters Lothal (mound of dead) is now in the district of Ahmedabad, Gujarat. An artificial basin built in 2300 BC was discovered during excavation carried out to locate Harrappan settlements in India. The structure measured 210 x 35 m. and was lined with wall of kiln-baked bricks. The maximum drought is 3.5 m. The lock gate system in the outlet ensures automatic desiltation at high tide and allowed maneuverability at low tide with a drought of 1 m. There are other archaeological evidences suggesting that the Harrappans possessed a high degree of knowledge relating to ebb-and-flood of tides, hydrography and marine engineering. In this brick built structure, now accepted as dock, ships were sluiced through the river estuary flooded by an inlet channel at high tide from the Gulf of Cambay. Similarly, the ships had to leave the basin at high tide when the water level was maintained sufficiently high above the inlet channel. Precautionary measures were taken against erosion and scouring of the tidal waters. The port installation also includes 24 m. wide brick wharf for haulage of cargo and well-ventilated warehouse. Perforated stone anchors and Terracotta models of boats found during excavation substantiate the use of the basin as a dock. Lothal dockyard with high tidal range was probably cut off from marine environment due to shoaling of the Gulf of Cambay as a result of Holocene sea level rise.

* 1.Nunn, P.D., 1994 *Oceanic Islands.* Blackwell Publishers, Oxford. Pp 413.

2.Mouat,F.J., 1863 *The Andaman Islanders* (1979, reprint). Mittal Publications, Delhi 110 035., pp.viii + 1-367 (6-9)

3.Kloss.C.B., 1902 *Andamans and Nicobars* (1971, edition). Vivek Publishing House, 18-D, Kamla Nagar, Delhi 7., pp xiii + 1-373 (176-177, 208)

4.Mukherjee,M., 1999 *Out of Africa into Asia.* Scientific American. 20(1) p.14.

EPILOGUE 41

Publisher	-	Fortress Headquarters, IN. Port Blair *E-mail: samudrikaportblair@yahoo.com* *Fax 03192-32829*
Document No.	-	SAMUDRIKA / 2001
Year	-	2001

Edited

by

(1) **ARIF M. MUSTAFA**
General Secretary
Andaman Science Association

(2) **PAUL JAMES**
Commander
Indian Navy

Graphic Layout: ***CHS Srinivas***

First author, our associate consultant is a specialist on the Fish & Fisheries of A&N Islands. He is the pioneer in Marine Aquarium, Coral reef survey and conservation, artificial reef etc. in these Islands. He reported two economical varieties, a deep sea shark (*Centrophorus sp.*) & a lobster (*Linuparus andamanensis*) from Andaman waters.

He hails from a family of freedom fighters, his grand father, Pandit Ayodiya Rai Sharma from village Kothar, Shajahanpur was exiled to Andamans in 1871. His grandmother Chand Bibi was the daughter of first Mopilah lady convict to be exiled from Ponnani Tahasil of Malabar. His parents were jailed in Cellular jail and two of his uncles were shot dead by Japanese during the INA movement.

Cdr. P.James,
OIC, Samudrika

For bibliographic purposes this document may be quoted as :-

Mustafa, A.M., James, P., 2001 Understanding Andaman & Nicobar Islands (Kalapani) through Samudrika, Doc. No. SAMUDRIKA / 2001.

Printed By: FAST PRINTS Chennai-20. Tel : 4404806
Co-Ordination: By PRO WIDE Ch-20. Tel : 4464486

AN ECO GUIDE

Re-print

FISH AND FISHERIES OF EXPORTABLE SNAPPERS *(LUTJANIDAE)* GROUPERS *(SERRANIDAE)* AND EMPERORS *(LETHRINIDAE)* FROM THE WESTERN FISHING ZONE OF SOUTH ANDAMAN.

ARIF M.MUSTAFA, J.CHANDER SEKHER, S.DAM ROY *

Department of Fisheries, Andaman and Nicobar Islands
* Central Agricultural Research Institute, Port Blair

ABSTRACT

This paper presents prime findings based on ground realities related to Groupers, Snappers and Emperors exploitation from the Western Fishing Zone of South Andaman (WFZ-SA). The investigation unveils many shrouded facts related to fish export industry of Andamans through conclusive vital management inputs essential in understanding the magnitude, extent and intensity of harvest, fishing efforts and export. The industry thrives on 16 selected demersal species of fishes belonging to three families. These varieties are being harvested from 4 fragmented core fishing grounds (FG) covering an area of 1441 km^2. Identified FGs are predominantly coral covered areas or its vicinity. There are two major fish landing centers namely Guptapara and Wandoor with 56 nos. of moderate size (22') mechanized boats/fishing units (FU) functioning at an assessed operational level of 70% for a fishing season of 7 months. February is the most productive month. Composition-wise family Lutjanidae (Snapper) contributes 78.5% followed by Serranidae (Grouper) 13% and Lethrinidae (Emperor) 5.5% towards total landing. Fish catch per fishing trip of 15 to 30 hr. per FU is 95 kg. Actual time of fishing (ATF) is 4 hr. with FU catch rate of 24 kg. Catch per unit effort (Cpue) through sampling approach is found to be 565.5 kg./FU/month, whereas catch/FU derived through projected landing is 802 kg/FU/month. Thus the resultant yield/FU/month is 684 kg. Total exportable fish catch for a season from WFZ-SA is currently about 180.4 ton with an intensity of 37 km^2 per FU. Estimated rate of harvest per km^2 of fishing ground is 150 kg. The two export industries namely AFL & IMP are starving for want of raw material, their present export level is quite low compared to rated production capacity. There exist strong indications of superseding the industries by private cottage industries level traders specially when an International airport is established at Port Blair. Present level of exploitation, effort verses yield etc. appears to be normal, but since the fishing pressure is restricted to 0.2% and 4% area of EEZ and continental shelf respectively the fishing effort need to be restricted at present level till the formulation of a fisheries policy and its enactment.

The oceanic archipelago of Andaman & Nicobar Islands is experiencing rapid quantitative enhancement in export of fish through moderate industries & traders. The island's administration is being shadowed by fish export oriented entrepreneurs. The trend is indicative of a quantum jump in fish export in near future, opening new horizon of economic development for these islands thus realising the old dream of 'Blue revolution'. Apparently Port Blair is likely to emerge as a major fish export center during the current millennium, primarily due to obvious fact of owing the largest EEZ of 5,96,554 km^2 and longest coast line of 1962 km. which constitute 30% & 25% respectively of the national total.

Alarmingly, the export thrust and the race to open new doors in this direction is for a limited but presently flourishing groups of fishes called Snappers, Groupers and Emperors for which no stock assessment has been made and there is no Fisheries Policy in existence. We in fact at the moment have no fulcrum to hoist the prudent flag of management.

Vague estimates made earlier are suggestive of 20,000 to 22,500 tonnes of demersal fish potential per annum. (Kumaran, N. 1973; Jones, S. 1973; George, P.C. 1977; Sudersan, D.1978; Antony Raja, B.T. 1980; Joseph, K.M.1985; Sudersan, D. 1989). All the three targeted group of fishes are known to have affinity with corals and occurs only in that part of the world were coral grows. For A&N Islands we have about 1,000 to 2,000 km^2 of coral reef coverage estimated through remote sensing & traditional methods respectively (Mustafa, A.M. 1987, 2000 & 2001) the precise area is yet to be ascertained. However, out of available coral area it has been found that more than a decade ago about 360 km^2 was totally degraded mainly because of unwise human induced stress (Mustafa, A.M.1987). Hypothetically in theory one km^2 of coral reef could produce 15 ton. of biota per annum (Munro, 1984), this biota is an assemblage of more than 3,000 living species (White, A.T.1987). The coral reefs of A&N Islands are estimated to have 10% of fish component towards macro faunal niche (Rao, G.C.1999), which is a sub niche, this percentage constitution will drastically reduce when integrated with micro & nano niches. The above assumption is indicative of fractional fish production through coral eco. system whereas variety diversity is high to about more than 500 recorded so far from A&N islands. Proceeding further with the bio production model narrated above we have 15,000 to 24,600 ton. of annual total produce through coral reefs out of which 1,500 to 2,460 ton. or less are fishes.

Andamans export industry is targeting only 16 species which has immerse demand throughout S-E & Far-East Asia. Such species targeted, choice fishing elsewhere in World has led to commercial extinction of many fish stock, devastating the bio-fabrics of associated eco-system and eroding the livelihood of millions.

Through this systematic study an humble attempt is made to enlight all concerned of the factual ground reality.

MATERIAL AND METHOD

Samples of exportable fishes were collected from processing plants and landing centers, samples identified using standard procedure, intensive literature search was made to understand the bio-ecology of each species, photographs are incorporated mainly for the convenience of fishermen, their agents, traders and fisheries statistical personal. Information related to export was collected through appropriate govt. semigovt. and private agencies. Fishing grounds were identified through admiralty chart work using compass and GPS with the aid of experienced mariners.

Two major fish landing centers located over the Western coast of South Andaman in the vicinity of Port Blair namely Guptapara and Wandoor were selected for exportable fish catch characteristic evaluation through continuous and stratified random sampling. The evaluation was for the entire West Coast Fishing Zone of South Andaman, (WFZ-SA), which in addition to above two major fish landing centers also have two minor seasonal fish landing centers namely Chidiyatapu & Loha Barrick mainly harvesting pelagic varieties for local consumption. Ten identical fishing boats/units having LOA 22 ft, powered with 8.5 hp inboard engine from among 56 similar boats, 5 out of 30 at Guptapara and 5 out of 26 at Wandoor were randomly selected as sample units. Detailed data related to fishing frequency, total landing, catch composition, duration, direction, fishing ground and actual fishing time was closely monitored for a full export fish harvest season of seven months commencing from October 1999 upto April 2000 many times accompanying the fishermen in their fishing trip. Length – weight parameters were initially gathered for a short period, which later became unmanageable due to distance, weather and pre-occupation of investigators, hence it was scrapped.

Stratified cluster sampling method for catch evaluation (Gulland, J.A.1962; Devaraj.M.1983) was adopted with modification as 'continuous' for the entire fishing season. The randomly selected fishing units were identified as G1, G2, G3, G4, G5 ; W1, W2, W3, W4, W5 for Guptapara & Wandoor respectively and maintained till the end of season. 18% of the total fishing units in Western Fishing Zone were sampled against the 10% prescribed limit of sampling (Devaraj, M.1979). Actual composition wise landing and fishing days effort/trip for all the 10 sampled units were recorded for 7 months of fishing season. Total monthly yield, composition and fishing trips based on actual continuous sampled data was calculated for the season through simple arithmetical expansion to the order of operational units. Actual time of fishing was fixed and average monthly catch per trip and catch per hour was also calculated. Catch per unit effort (Cpue) was derived as follows :-

Landing center month 'n'	Oct. 99	Nov. 99	Dec. 99	Jan. 00	Feb. 00	Mar. 00	Apr. 00	= n=7
Fish catch (kg.) 'y'	a	b	c	d	e	f	g	= $^{Y}ab\ldots g$
No. of units/ efforts 'x'	x_1	x_2	x_3	x_4	x_5	x_6	x_7	= $\sum_{x_1}^{x_7} x$

$$\therefore \text{Yield (seasonal)} \quad \sum_{i=1}^{n} {}^{Y}ab\ldots g = \text{kg.}$$

$$\text{Yield (annual)} \quad \hat{Y} = \frac{{}^{Y}ab\ldots g}{\sum_{i=1}^{n}} \times N \quad \text{(where N=12)}$$

$$\text{Cpue} = \frac{\hat{Y}}{\sum_{1}^{N} x} = \text{kg./unit/month}$$

$$\text{Rate of harvest} = \frac{FG\ (km^2)}{X \,\&\, Y} = \text{ton/km}^2$$

RESULT

Fish - A total of 16 species of fishes belonging to 3 taxonomical families under 7 genera are found being exported to foreign countries and mainland India from Port Blair. Scientific names were matched with valid vernacular synonymous. *Lutjanidae* (Snapper) is found to be the largest family followed by *Serranidae* (Grouper) and *Lethrinidae* (Emperor) as follows :

Lutjanidae				
S.no.	Scientific name	Local name	Trade name	% comp. in export
1.	*Lutjanus malabaricus*	Lall	Red snapper	43.3
2.	*L.gibbus*	Kankata lall	Red snapper	
3.	*L.bohar*	Kutta bhetki	Red snapper	
4.	*L.argentimaculatus*	Lall	Red snapper	
5.	*L.sebae*	Lall	Red snapper	
6.	*Pristipomoides filamentosus*	Mrigal	White snapper	23.3
7.	*P.typus*	Mrigal	White snapper	
8.	*Aphareus rutilans*	Mrigal	White snapper	

Serranidae				
1.	*Epinephelus chlorostigma*	Gobra	Brown/Black grouper	16.6
2.	*E.diacanthus*	Gobra	Brown/Black grouper	
3.	*Cephalopholis sonnerati*	Lall gobra	Red grouper	
4.	*Plectropomus leopardus*	Lall gobra	Coral trout.	
5.	*P.levis*	Lall gobra	Coral trout.	

Lethrinidae				
1.	*Lethrinus elongatus*	Kushal	Emperor	16.8
2.	*L.miniatus*	Kushal	Emperor	
3.	*L.lentjan*	Kushan	Emperor	

Bio-ecology :

Family - *Lutjanidae* (Snappers). They are moderately elongated oblong and fairly compressed in shape having terminal moderate to large prostrusible mouth; two nostril on each side of snout, anterior of head (area between eye and mouth) devoid of scales.

Colour variable, mainly from yellow through red to blue, often with blotches, lines or other patterns.

Tropical in distribution coinciding with the range of reef building corals, mostly demersal from coastal waters to continental shelves and slopes upto – 500 m depth generally forming shoals (William, D ; Anderson, Jr. – 1987). Most of them are unspecialized nocturnal carnivores. They are ***gonochoristic*** i.e. sexes are distinct. They spawn in groups probably twice a year. Fecundity ranges from a lakh to about a million eggs (Thompson, R.H. : Munro, J.L. 1974).

39 species are known to occur in A&N Islands (Rao, D.V. ; Kamla Devi – 1997) out of which 8 species are being exported contributing 66.6 % of export.

II. Genus – *Lutjanus*, Five species collectively contributing about 43.3. % of export, all are commercially called 'Red snapper'.

(i) *L. malabaricus* (Bloch & Schneidr - 1801) (Lall) A red or red orange colour fish, colour light on lower part of side and on belly. Fins reddish. Found in and around coastal and offshore reefs within – 100 m depth. Could grow upto 13.6 kg. in weight. (Grant, E.M. – 1982). A shoaling variety. Size : 90 cm.

(ii) *L.gibbus* (Forsskal - 1775) (Kankatta lall) Generally deep red in colour, darker on back and upper portion of head, orange hue on lower part of opercle and in perctoral axil. Occurs mainly on coral reefs at a depth of around -30 m. often forming large static shoal during day time (Allen, G.R. & Talbot, F.H. 1985). Size : 50 cm.

(iii) *L.bohar* (Forsskal – 1775) (Katha bhetki). Red or purplish red in colour. Generally dark reddish brown on back, shading to pink or whitish ventrally, cheeks and lower half of sides often bright red in large adults. Usually inhabits coral reef areas but occasionally solitary individuals occur down to – 70 m in rocky areas. (Fischer, W. & Bianchi, G. – 1984). Size : 75 cm. *L.bohar* is avoided by many buyers due to its suspected ciguatera fish poisoning acquired through feeding upon certain herbivorous fishes which feeds on a toxic dinoflagellate *Gambierdiscus toxicus*. This dinoflagellate lives on benthic algae or on dead corals. (Yasumoto, T. *et. al.* 1977).

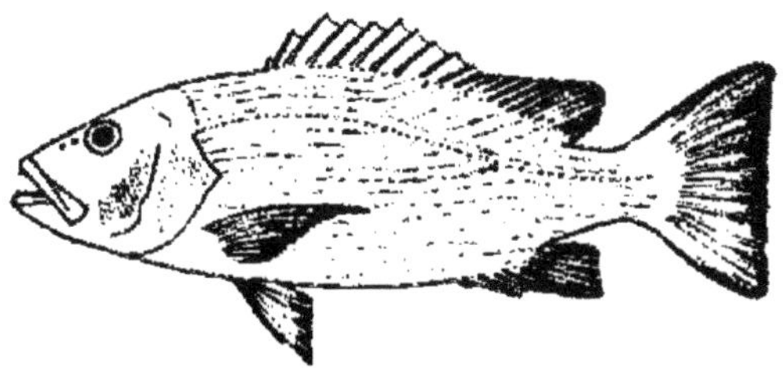

(iv) *L.argentimaculatus* (Forsskal – 1775) (Lall) Red brown, greenish brown on back, reddish on sides and ventral parts, deep water catch often overall reddish. Lives in sheltered coral reefs or in areas where siltation is heavy and coral growth is poor in the depth range of -80 to -120 m. Size : 120 cm

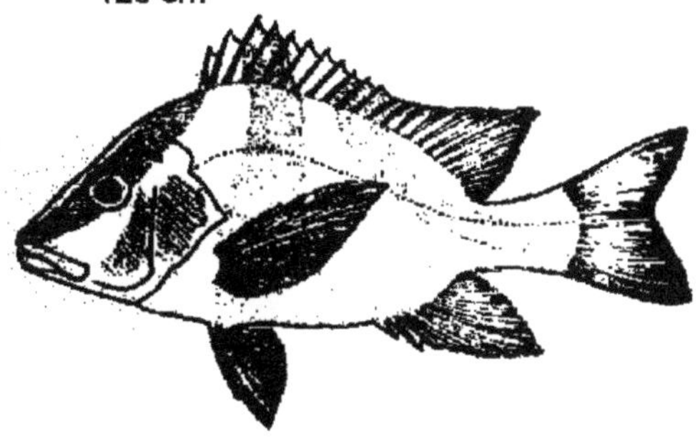

(v) *L.sebae* (Cuvier – 1828) (Lall) All large adults are deep red in colour, otherwise red or pink, darker on back. Found around coral reefs often over adjacent sand flats. Maximum recorded weight is 13.6 kg (Grant, E.M. 1982). Total maturity is attained at the age of 4 years. Size : 100 cm.

II. Genus – *Pristipomoides* - Two species contributing 16.6 % of the export, collectively called 'White snapper'.

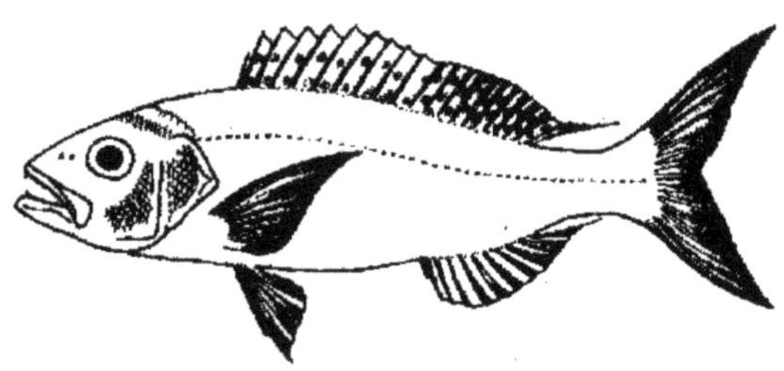

(i) *P.typus* (Bleeker – 1852) (Mrigal) Body rosy in colour, fins with yellow tinge, brownish yellow longitudinal vermiculation on top of head, dorsal fin with paler spots or rosy reticulate pattern. Generally caught from -40 to -80 m depth. Size : 70 cm.

(ii) *P.filamentosus* (Valenciennes – 1830) (Mrigal) Colour varies with ground from reddish purple to lavender with blue tinge, small blue spots on top of head, dorsal fin with two yellowish longitudinal lines. Fished from deep waters mostly -150 to -200 m depth. Size : 80 cm.

III. Genus – *Apharens* - This white snapper is represented in export only by one species .

(i) A.rutilans (Cuvier – 1830) (Mrigal) which constitute 6.6 % of export. Colour is overall blue-grey, purplish – brown above with little yellow or pinkish suffusion; edges of preopercle and opercle outlined with black. Inhabit coral and rock reefs at depth between -60 to -70 m. Size : 40 cm.

Family – *Serranidae* (Groupers) They have robust or somewhat compressed, oblong-oval to rather elongated body. This is a group of closely related fishes (Smith, C.L. – 1972). Mouth large, with small, slender, inwardly depressible teeth. A single dorsal fin with 7 to 12 strong spines; anal fin with 3 spines.

Colour changes instantaneously when under stress or due to a change in environmental conditions. Colour variable with patterns of light or dark stripes, spots, vertical or diagonal bars or nearly plain.

Distribution of groupers corresponds roughly to the distribution of reef building corals. They are demersal, carnivores and solitary in nature living in and around coral and rocky reefs upto a depth of -200 m. Sexually, majority of them are typical to exhibit ***Synchronous*** and ***Protogynous hermaphroditism*** i.e. presence of ovotestis and sex change from female to male (Thompson, R & Munro, J.L. 1974; Shapiro, D.Y.1987). All groupers for which there is evidence spawn in dense shoal for a restricted period of one to two weeks possibly once in a year. Fecundity ranges from 90,000 to about 6.5 lakhs. Intra generic hybridization is reported to occurs in this family.

43 species are known to occur in A&N islands (Rajan, P.T. (ZSI, Port Blair) (In Press) out of which 5 species are being exploited contributing 16.6 % of the total export.

I. Genus - *Epinephelus* - Two species collectively contributes about 11.6 % of export, both are commercially called 'Brown/Black grouper'.

(i) *E.chlorostigma* (Valenciennes – 1828) (Gobra). Generally brownish in colour, body and fins covered with small, closely set, brown hexagonal or roundish spots. Lives in a wide depth range of -4 to -280 m. Size : 75 cm.

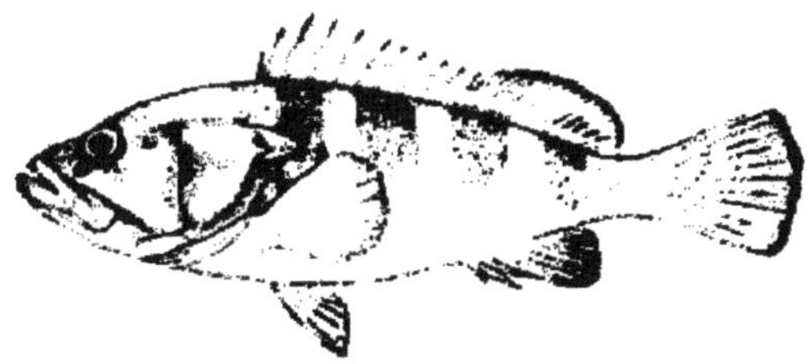

(ii) *E.diacanthus* (Valenciennes - 1828) (Gobra). Brownish in colour with 5 more or less distinct, vertical dark bars, 4 below dorsal fin & 5th over caudal peduncle. Ventral part of head & body reddish. Lives in a wide depth range of -10 to -120 m. (Randall, J.E., Phillip, C., Heemstra. 1991). Size 52 cm.

II. Genus – *Plectropomus* - Two species making thin marginal contribution of 3.3% towards export, both are commercially called 'coral trout'.

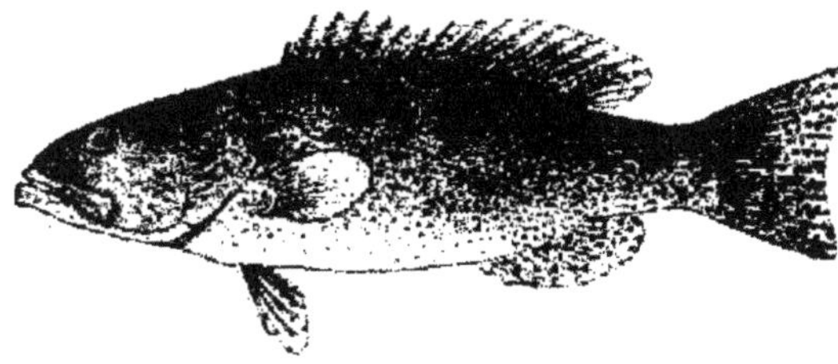

(i) *P.leopardus* (Lacepede – 1802) (Lall gobra). A reddish to olivaceous coloured fish impregnated with small dark-edged blue dots over head, body and median fins. A blue ring on edge of orbit. Lives in coral reefs generally in the depth range of -10 to -30 m. Size : 70 cm.

(i) *P.levis* (Lacepede – 1802) (Lall gobra). It shows two colour phases, one whitish or pale yellowish with five dark to black saddle like bars fainting ventrally. The second colour phase is brown, olivaceous, red on nearly black with or without five dark bars. Both colour phase will have a few widely scattered, small, dark edged blue spots on body. Lives in coral reefs in the depth range of -10 to -30 m. Size : 100 cm.

III. Genus – *Cephalopholis* - Only one species of Red grouper is being exported.

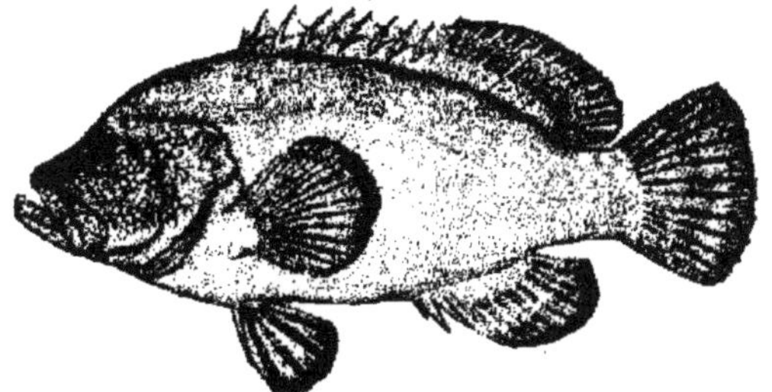

(i) *C.sonnerati* (Valenciennes –1828) (Lall gobra) having least share of 1.6 % to export. Bright orange-red to red in colour, usually with scattered faint bluish white spots, head purplish with numerous close-set orange red spots. Lives in coral reef areas upto -100 m depth. Size 57 cm.

Family – *Lethrinidae* (Emperors) Large head fishes with wide sub-orbital space making the snout rather pointed, mouth moderate mostly with thick & fleshly lips. A single dorsal fin with 10 spines.

Colour pattern is usually brown, green or grey with tints of red, pink, yellow or blue, tinted areas brightens when excited or after death. Many species show scarlet coloration to inside of lips.

Mostly associated to coral reefs but also found over rocky reefs and soft bottom coastal areas upto -100 m depth. All are demersal carnivores mostly hunting at nigh. Sexually they mature as female progressively becoming hermaphrodite. Males are reported to be longer than females. They spawn in shoal throughout the year mostly in lagoons or near outer reef edge.

19 species are reported to occur in A&N islands (Rao, D.V. ; Kamla Devi – 1997) out of which possibly 3 species belonging of one genus *Lethrinus* are being exported, contributing about 16.8% of export. The species of this genus are taxonomically most difficult to distinguish.

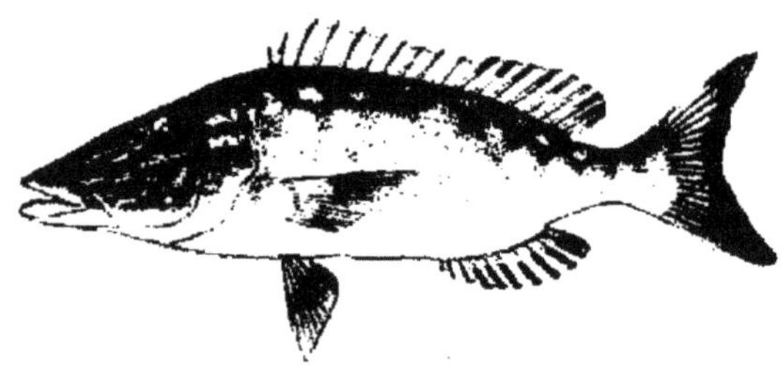

(i) *L.elongatus* (Valenciennes - 1830) (Kushal) This species has not been reported earlier from A&N islands. This is a greenish – grey fish with brown patches above; colour fades towards ventral side. A prominent red line in always seen above and below the lips. Lives upto a depth of -185 m. in the vicinity of coral reefs. Size : 100 cm (Sato, T. & Walker, M. – 1984).

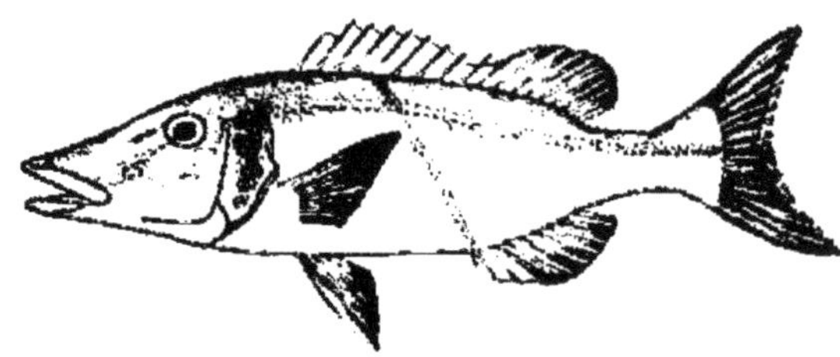

(ii) *L.miniatus* (Bloch & Schneider – 1801) (Kushal) Body olive – brown above, sometimes pink below; occasionally 2 or 3 blue streaks radiating from eye. Vertical fins pink to red, with brighter margins; paired fins yellow. Inhabits coastal waters. Size : 90 cm.

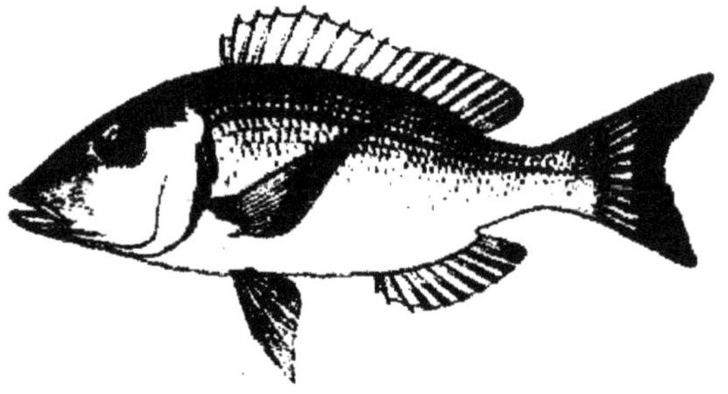

(iii) *L. lentjan* (Lacepede – 1802) (Kushal) Body is olive/green above, paler below. A bright red spot on posterior edge of operculum and often another on outer pectoral fin base. Each scale on back sometimes with a white center. Dorsal and caudal fins stripped with orange. Lives in coastal coral reefs down to a depth of -50 m. Size : 50 cm.

Fisheries - Western Fishing Zone of South Andaman (WFZ-SA) contributes 99% of total export. Four core offshore fishing grounds extending from Little Andaman upto Baratang island have been identified as shown in Fig.1 covering a cumulative area of 1441 km^2. (1,44,100 Hect.) as follows :

S.no.	Name of the core fishing ground	Area	
		Km^2	Hect.
1.	South Coral Bank (SCB)	599	59,900
2.	North Sentinel Is. (NSI)	131	13,100
3.	South Sentinel Is. (SSI)	169	16,900
4.	North of Little Andaman (NLA)	542	54,200
		1441	1,44,100

Export quality fish landing is made through two major fish landing centers namely Guptapara and Wandoor with 30 & 26 fishing boats respectively. Catch pattern, % composition, fishing trips and monthly/seasonal estimated fish landing etc. from Guptapara and Wandoor were recorded and tabulated. Fishing pattern revealed that about 50% to 70% units participate in each fishing trips, rest of the 30% to 50% remains idle at landing center. Duration of each fishing trip varies from 15 to 30 hrs. depending on the fishing ground. The fishing fleet operate line trolling on their way to fishing ground and while returning which gives an additional yield of 3% Seer. While at fishing ground, vertical monohook line or vertical brachline with 4 to 6 hooks are used to harvest export variety fishes. Month of February is found to be the most productive month for Western Fishing Zone. Snapper is the largest group of fishes constituting 78.5% followed by Grouper 13%, and Emperor 5.5% in the total landing. Unexportable variety Seer is negligible to the tune of 3%. Yield per trip of 15 to 30 hr. is 95 kg. whereas 24 kgs. is the average catch rate at 4 hrs. of Actual Time of Fishing (ATF) per trip.

Catch per unit effort
(Cpue)

Landing center month 'n'	Oct. 99	Nov. 99	Dec. 99	Jan. 00	Feb. 00	Mar. 00	Apr. 00	n=7
Fish catch (Kg.) 'y'	4044	10057	7094	6779	12635	8671	7271	y=56551
Units/Effort (no.) 'x'	10	10	10	10	10	10	10	$\sum_{x_1}^{n}$ x=70

$$\therefore \text{Yield (seasonal)} \quad y = \sum_{1}^{7} Y_{ab\ldots g} = 56551 \text{ kg.}$$

$$\text{Yield (annual)} \quad \hat{Y} = \frac{y}{\sum_{1}^{10} x} \times N = \frac{56551}{10} \times 12 = 67861 \text{ kg.}$$

Total effort for 7 months = Σx = 70

Total effort for 12 months = $\frac{70}{7}$ x 12 = 120 Units

∴ **Cpue**	=	$\frac{67861}{120}$	=	**565.5** kg/unit/month	A

No. of operational units at Guptapara (70% level)	=21	
Estd. catch for 7 months at Guptapara	= 98364 kg.	
No. of operational units at Wandoor (70% level)	=18	
Estd. catch for 7 months at Wandoor	= 120593 kg.	
∴ Total estd. catch for 7 months by 39 units in WFZ-SA	=218957 kg.	
∴ **Total estd. catch for 1 month by 1 unit in** WFZ-SA	= $\frac{218957}{7x39}$ = **802** kg/u/m	B
∴ Total estd. catch for 12 months by 39 unit in WFZ-SA	= 802x12x39 = 375336 kg	
Estd. yield per unit per month in WFZ-SA $\frac{A+B}{2}$	= $\frac{565.5+802}{2}$ = **684** kg.	X
Hence estd. landing for 7 months from WFZ-SA = (684x7x39)	= 186t (-3%) = **180.4** t	Y
Hence estd. landing for 12 months from WFZ-SA = (684x12x39)	= 320t (-3%) = **310.4 t**	Z
∴ *Current estimated rate of harvest per annum per km² of fishing ground*	= *0.1t to 0.2t* ≈ *0.15 t / Fu.*	

i.e., 216 t per annum from an area of 1441 km^2.

Market :

Export - There exist three potential buyers for Groupers, Snappers and Emperors namely; (1) M/s Andaman Fisheries Ltd. (AFL) a subsidiary company of A&N Islands Integrated Development Corporation (ANIIDCO), (2) M/s Islanders Marine Products Limited (IMP) and (3) Group of Private Traders (PT). Former two have the infrastructural facility for freezing and packing, whereas the traders do it at cottage industry level. The whole export is in the form of 'fresh & frozen'. AFL is the leading exporter followed by private traders.

Export in ton.

Year	AFL	PT	IMP	Total
1996-97	155	32	-	187
1997-98	111	21	39	171
1998-99	60	5	-	65
1999-00	54	6	-	60
2000-01	50	18	17	85

Local Consumption (LC) : Annually about 50 ton. of exportable varieties is consumed by Shipping Services and 20 ton. by Defence establishments. Rest is consumed fresh by the local population & hotels.

The traders and middle men are more incline towards PT & LC outlets due to better remuneration because AFL and IMP purchases at an average rate of Rs.20/kg. or so, which is essential to protect their own economy, moreover the export industries ensure quality of raw material and as such the fishermen community is not much interested in selling their catch to them.

DISCUSSION

Although A&N Islands have 5,96,554 km^2 of EEZ the present day fish export industry thrives upon a narrow, fragmented area of 1441 km^2 which is only 0.2% of the EEZ and 4% of the restricted (34965 km^2) narrow and elongated continental shelf. The fishing pressure on targeted varieties is being augmented through the induction of more and more traders and entrepreneurs to operate from Port Blair. The Western Fishing Zone of South Andaman (WFZ-SA) is known to hold rich coral biodiversity and that is one of the prime reason of its being the abode of Groupers, Snappers and Emperors.

Recent past has witnessed extinction of many fish stocks including Grouper, Snapper and Emperor elsewhere in the World due to unmanaged and uncontrolled fishing. The wild stock of larger Groupers and Snappers are found to be the first to show instant depletion through a shift in catch composition when overexploited (Clark & Brown,1977; Munro & Smith, 1985; Munro & Williams, 1985 and Russ, 1985).

Prevailing fishing intensity of one fishing unit per 37 km^2 of shelf area studied, is yielding between 100 to 200 kg of targeted fish per annum. This rate of exploitation appears quite conservative, but since we does not know the maximum or permissible limit and extent of this harvest and there is no policy to regulate fishing and export in time and space; an unchecked trade expansion may lead to catastrophic devastation of the whole marine eco-system over WFZ-SA.

RECOMMENDATION

In view of the prevailing facts presented above, the A&N Administration may consider to adopt following measure for sustainable utilization of export fish resource :-

Long-term measure.

(1) To workout a bio-mathematical population model preferably in consultation with renowned fish population dynamics experts to evaluate various essential parameters, indispensably needed as essential inputs in shaping a prudent fisheries policy for demersal stock management.

(2) Bottom trawling over WFZ-SA and elsewhere in the territory shall immediately be banned and made punishable.

(3) Areas upto – 250 m. depth shall be kept fishable only by traditional fisher community and no large (> 50') fishing vessel shall be allowed to fish within that depth zone.

Interim measure

(1) Export demand over Western Fishing Zone of South Andaman shall immediately be 'freeze' at present level till the finalisation of fisheries policy.

(2) The starving AFL shall be advised for partial trade diversification from fresh and frozen to much more lucrative live fish export preferably through the involvement of private entrepreneurs possessing in-line expertise by opening new fishing grounds.

(3) The IMP (P) Ltd. may expand fishing ground/collection centers to Eastern Fishing Zone of South & Middle Andaman.

ACKNOWLEDGEMENT

First author express his deep thanks & gratitude to Cdr. **P.James**, IN for navigational assets, my colleagues Mr.**E.K.Raveendran**, FDO for encouragement, Ms. **V.Sheeja**, AFDO for logistic, Mr. **Shaji Thomas**, SIF for admiralty chart work and Mr. **CHS Srinivas Rao** for graphics.

REFERENCES

Antony raja, B.T. 1980. *Current knowledge of fisheries resources in the shelf area of the Bay of Bengal.* BOBP/WP/8.

Joseph, K.M. 1985. *Marine fishery resources in India.* A system frame work of marine food industry in India, by Kulkarni, G.R. & Srivastava, G.K.

Sudersan, D.et. al. 1989. *Assessment of oceanic tunas & other fish resources of the Indian EEZ.* CMFRI.

Mustafa, A.M. et. al. 1987. *Endangered coral reefs of bay islands and their ornamental fishes.* Proc. Symp. Mang. of Coastal ecosystem and oceanic resources. ASA : 60-65.

Mustafa, A.M. et. al. 2000. *A comprehensive analysis of the coral ecosystem vis-a-vis resources exploitation around A&N islands.* UNDP, Project report, ZSI, Port Blair : 1-29.

Mustafa, A.M. et. al. 2001. *Understanding A&N Islands.* Document no. SAMUDRIKA/20001. Indian Navy, FHQ, Port Blair.

Munro, J.L. 1984. *Coral reef fisheries and world fish production.* ICLARM, News letter 7 (4) : 3-4.

White, A.T. 1987. *Coral reefs.* ICLARM No.386 : 1-36.

Rao, G.C. 1999. *Marine ecosystem around A&N islands.* UNDP, Project report, ZSI, Port Blair; 1-33.

Gulland, J.A. 1962. *Manual of sampling methods for fisheries biology.* FAO. FB/T26.

Devaraj, M. 1983. *Fish population dynamics (course manual).* CIFE 3 (10) 83 : 1-98.

Devaraj, M. 1979. *Studies on fish population dynamics – V.* Chapter-III : 1-15, CIFE lecture note.

William, D. & Anderson, Jr. – 1987. *Systematics of the fishes of the family Lutjanidae (Perciformes : Percoidei), the snapper.* Pp.1 – 32. Tropical snappers and Groupers : Biology & Fisheries Management. Pub : Westview press/Boulder and London.

Thompson, R. & Munro, J.L. – 1974. *The biology, ecology and bionomics of the snappers, Lutjanidae.* pp. 94-109. Caribbean Coral Reef Fishery Resources. Pub : ICLARM, Phillippines.

Rao, D.V. & Kamla Devi – 1997. *Snappers (Family – Lutjanidae) of Andaman & Nicobar Islands.* Environment and Ecology 15(4) : 924- 931.

Yasumoto, T. et.al. 1977. *Finding of a dinoflagellate as a likely culprit of ciguatera.* Bull. Japan. Soc. Sci. Fish. 43(8) : 1021-1026.

Grant, E.M. 1982. *Guide to Fishes.* Pp 896. 574 Col. Pls. Dept. Harbours and Marine, Brisbane.

Allen, G.R. & Talbot, F.H. 1985. *Indo-Pacific Fishes* no.11 pp.88. Bernice Pauahi Bishop Museum, Honolulu, Hawaii.

Fischer, W. and Bianchi, G. – 1984. *FAO species identification sheets*, (Fishing area – 51). Vol.III LUT Lut2.

Smith, C.L. – 1971. *A revision of the American groupers : Epinephelus and allied genera*. Bull. Amr. Mus. Nat. Hist. 146:67 – 242.

Thompson, R. & Munro, J.L. 1974. *The biology, ecology and bionomics of the Hinds & Groupers, Serranidae*. Pp.59 – 81. Caribbean Coral Reef Fishery Resources. Pub : ICLARM, Philippines.

Shapino, Y.D. – 1987. *Reproduction in Groupers*. Pp.295-327. Tropical snappers and Groupers : Biology and Fisheries Management. Pub : Westview Press/Boulder and Lonation.

Randall, J.E., Phillip, C. & Heemstra. – 1991. *Indo – Pacific Fishes* no.20. Bernice Pauahi Bishop Museum, Honolulu, Hawaii.

Rao, D.V. & Kamla Devi – 1997. *Emperor fishes (Family Lethrinidae) of Andaman & Nicobar Islands*. Environment is Ecology 15(4) : 899-903.

Sato, T & Walker, M. (reviewed by Randall, J.E.). *FAO species identification sheets* (Fishing area 51). Vol.II LETH Leth 5.

Clark, S.H. and B.E.Brown. 1977. *Changes in biomass of fish fin fishes and squids from the Gulf of Marine to Cape Hatteras*, 1963-74, as determined from research vessel survey data. Fish. Bull. U.S. 75: 1-21.

Munro, J.L., and I.R.Smith. 1985. *Management strategies in multispecies complexes in artisanal fisheries*. Proc. 36th Gulf. Cariabb. Fish. Inst.

Munro, J.L., and D.M.Williams. 1985 - *Assessment and management of coral reef fisheries*. Biological environmental and socio-economic aspects. Proc. 5th Int. Coral Reef. Congr. Tahiti 4 : 543-565.

Russ, G. 1985 – *Effects of protective management on coral reef fisheries in the Central Phillippines*. Proc. 5th Int. Coral Reef. Congr. Tahiti 4: 219 – 224.

Findings and views narrated in this paper are of the investigators and not of any Department, Company or Organization.

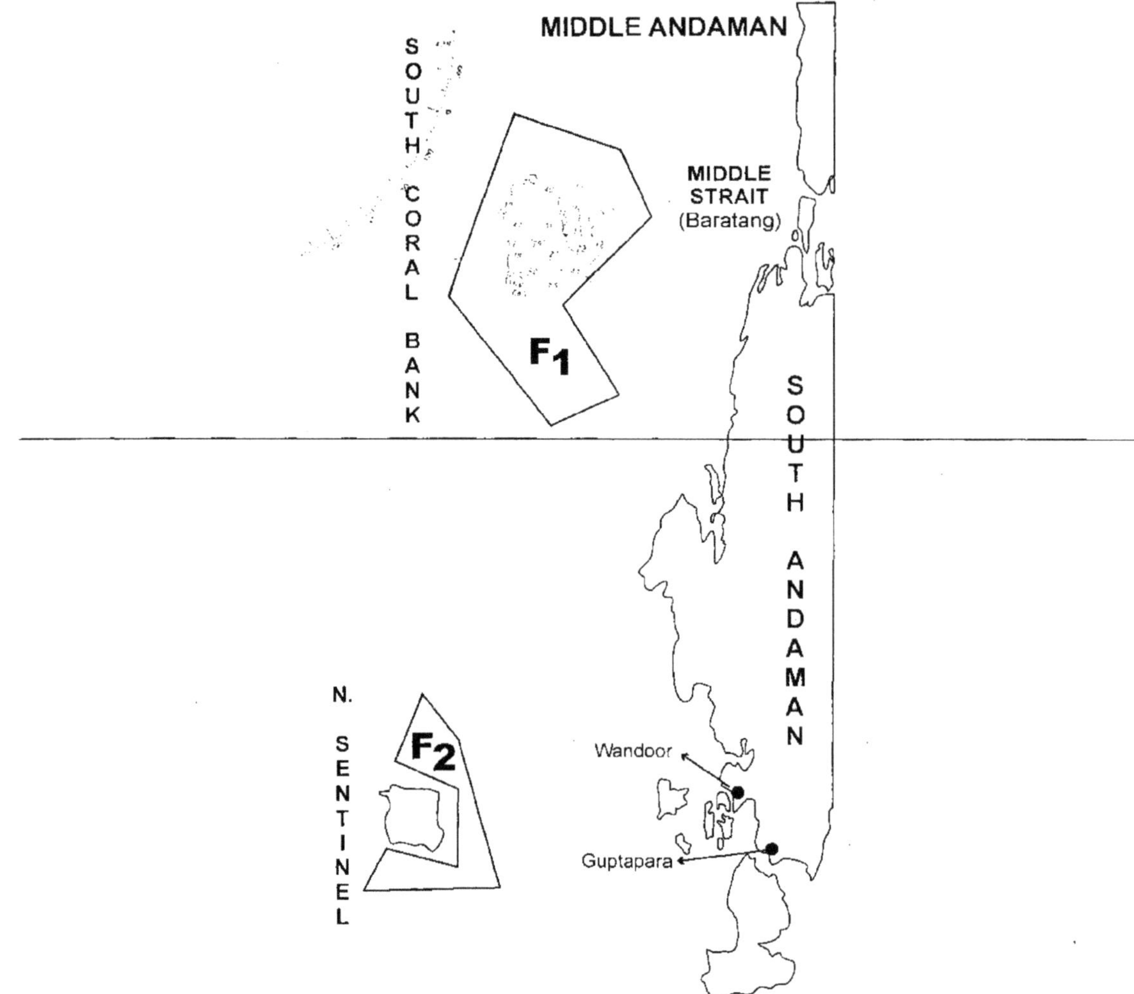

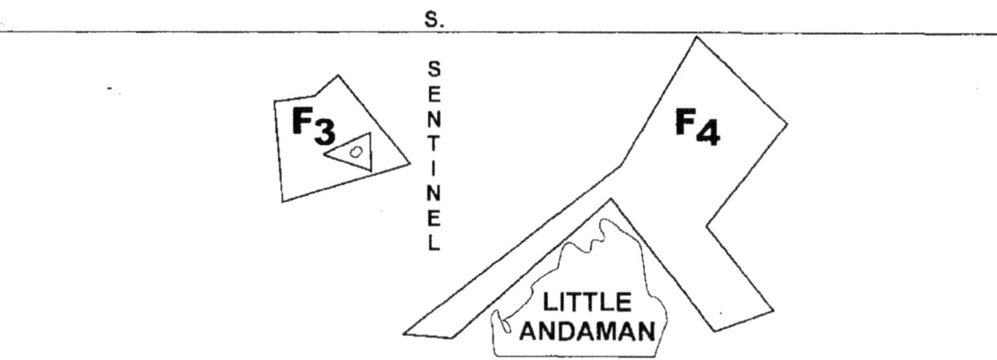

Fishing ground WFZ-SA.

PROCESSING INDUSTRY AT PORT BLAIR

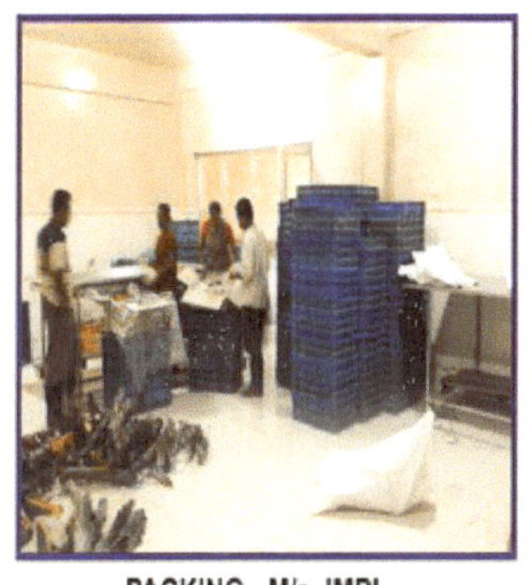
PACKING - M/s. IMPL

MACKERALS

PERCHES & SNAPPERS

PERCHES & SNAPPERS

FROZEN CONSIGNMENT - M/s. IMPL

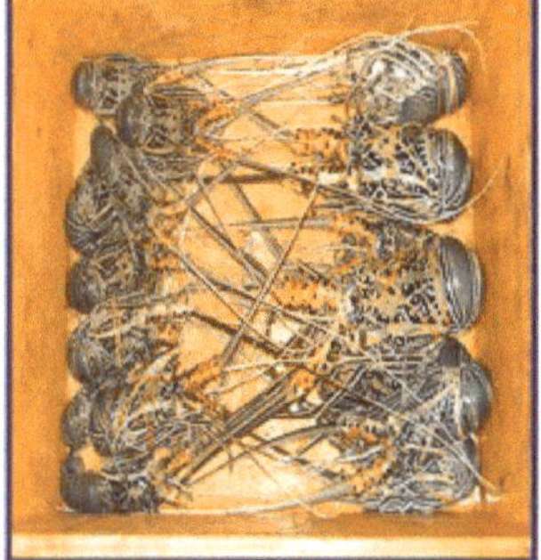
LOBSTER FOR ICE PACKING

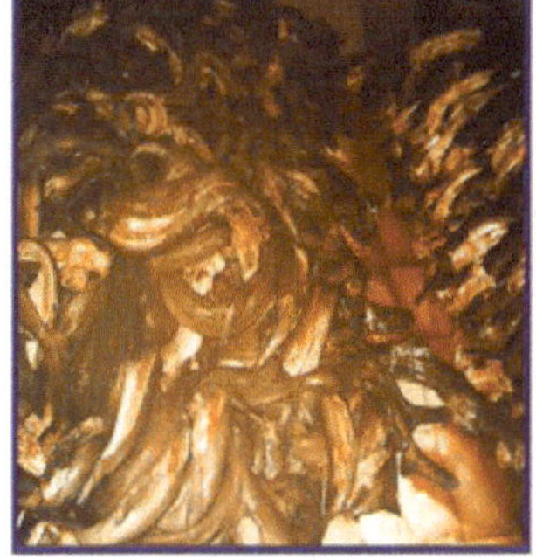
DRESSED DEEP SEA SHARKS

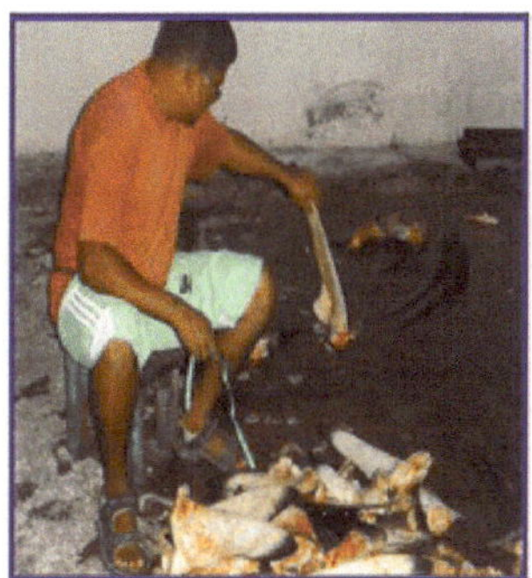
GUTTING OF SHARKS

Surface Sharks under process

Perches

GOI/UNDP/GEF Project
Management of Coral Reef Ecosystem of Andaman & Nicobar Islands

A
Comprehensive Analysis
OF THE

CORAL ECOSYSTEM Vis-a-Vis RESOURCES EXPLOITATION

AROUND
ANDAMAN & NICOBAR IS.

By

Arif M.Mustafa
Asst. Fisheries Devp. Officer,
MB-12, GAFOOR MANZIL (ANNEXE)
Aberdeen Bazar, Port Blair – 744 101.

&

Dr.S.Dam Roy
Director of Fisheries
A&N Islands
Department of Fisheries,
Port Blair – 744 101

January - 2000

A Comprehensive Analysis OF THE

CORAL ECOSYSTEM Vis-a-Vis RESOURCES EXPLOITATION

AROUND ANDAMAN & NICOBAR IS.

INDEX

S.No.	Contents	Page No.
1.	**CENTRAL IDEA**	1
2.	**INTRODUCTION**	2
	2.1 Origin	2
	2.2 Salient features	2
3.	**CORAL REEF DOMAIN**	3
	3.1 Background	3
	3.2 Diversity	4
	3.3 Extent	5
	3.4 Type	5
	3.5 State	5
	3.6 Productivity	7
	3.7 Status	7
4.	**CORAL REEF POTENTIAL & EXTENT OF EXPLOITATION**	9
	4.1 Components	9
	4.2 Fishes	10
	4.3 Shells	13
	4.4 Lobster & Crab	14
5.	**DEPENDENCE ON LIVING MARINE RESOURCES**	15
	5.1 Food	15
	5.2 Profession	15
	5.3 Catch	16
	5.4 Women	16
6.	**POACHING**	17
	6.1 Explosive fishing	17
7.	**UNIQUENESS OF OUR CORAL REEF ECO-SYSTEM**	18
	7.1 Classical Integration	18
8.	**CONSERVATION**	19
	8.1 Beginning	19
	8.2 Theory	19
9.	**SUGGESTIONS**	21
10.	**REFERENCES**	26
	Tables	23-25
	Acknowledgement	29

1. CENTRAL IDEA

Synthesis of available informations tells that a decade ago, the offshore coral reef of Andaman & Nicobar Islands were found to be comparatively well preserved, rich and virgin in Andaman sea region, whereas contemporary findings had than indicated depreciation of inshore coral formations. Later investigations indicate mounting multidimensional pressure in totality on this fragile eco-system. Our traditional coastal fisheries flourishes upon this bio-niche, the primary producer i.e. the fisherman community & aborigines and the local inhabitants as consumer depend on its living resources. The traditional subsistence level reef fisheries is fastly becoming an export oriented commercial venture. Present level of reef fish harvest is estimated to be about 4,000 tones a year, whereas much legislated shellfishery shows scrape landing.

Considering our natural production, which is of a lower order we shall evolve a practical resources sustainable management strategy through synergetic effort without further delay.

2. INTRODUCTION

2.1 *Origin*

Arched Oceanic archipelago of Andaman & Nicobar in the north-east Indian ocean is a mid ocean ***orotath cymatogen*** having pronounced linear stretching & bending. The island chain is believed to have born some 150 million years ago through volcanism due to expanding earth crust as depicted in the fundamental theory of ***contra plate tectonics***. The existing topography have been acquired after subsequent upheavals and submergence, coupled with contemporary biogenic deposits, this atypical genesis gave it a long coast line of about 1962 Km. and a narrow continental shelf of 34,965 Sq.Km.

2.2 *Salient features*

Existing geomorphology interdict normal oceanic circulation, stratification & mixing of water masses, presence of tectonic belt live volcano in east raises the deep water temperature comparatively in that region a few degree above normal, recurring burst of two monsoon per annum adds huge quantity of fresh water & sediment; the synergetic resultant is a unique rhythm of meteorological & oceanographical features, which even after 100 years of study and analysis is yet to be understood, this complex and shrouded eco-set up is susceptible to loose the inherited equilibrium of existence if the regime and its associated fabrics of life is unwisely raided for the satisfaction of human greed. The territory is endowed with diverse, complex interlinked & interdependent bio-potential habitats; the most amazing, appealing & fragile among them is perhaps the 'Coral reef ecosystem'.

3. CORAL REEF DOMAIN

3.1 *Background*

The genesis of such biogenic reefs are estimated to be 35,000 years old (Wilkinson,C.1998)[1] since than the progressing evolutionary events has caused repeated mass destruction of corals through geological past. Present day coral reefs found in over 100 countries covering about 6,00,000 Sq.km. of area began their growth from the living remnant thriving upon steep submerged slopes about 6,000 to 9,000 years ago on the lapse of 80,000 years long glacial retreat episode (Hallock,P.1997)[2] this ecosystem may have as many as three thousand living species in absolute harmony. We have just began to understand the marvels and mysteries of this submarine, complex, heterogeneous niche.

Today we know that 10% of earth's coral reef area has already been lost forever, 30% is likely to be degraded beyond recovery within 10-20 years another 30% may also be damaged after 20-40 years

(Wilkinson,C.1998). All the scientific investigations carried so far are indicative of multidisciplinary human induced disturbances in the sub-marine photic ecosystem arising out of unregulated economical ventures and over population. However, looking at the past geological events it could be safely assumed that this primitive, tiny, fragile, boneless polyp in conjunction with his associates are genetically endowed with unexhaustable ability for resources mobilisation towards self recovery on the advent of optimum ecological conditions.

3.2 *Diversity*

So far about 186 varieties of stony reef building corals ***scleractinians*** have been documented (a.Reddiah,1977; b.Pillai,1978; c.Anon,1987; d.Subba Rao,1999)[3] ***Porites*** **&** ***Favia*** are the chief reef builders, isolated monospecific stands of ***Acropora*** is frequent; beside stony corals many other benthic ***cnidarians*** (non-scleractinians) known as soft corals are seldom seen either in association with stony coral reef or away from it, commonly occurring soft coral colonies are ***Millepora*** (Fire coral), ***Heliopora*** (Blue coral), ***Tubipora*** (Red Coral), ***Isis*** (Marble Coral), ***Antipathes*** (Sea Fan), ***Cirrhipathes*** (Sea whip), ***Muricella*** (Red Gorrgonian), ***Suberogorgia*** (Black coral), beside them ***Sarcophyton***, ***Lolophytum***, ***Xenia*** and ***Dendronephthya*** are also found; the soft coral regime deserve much more taxonomical exploration.

3.3 *Extent*

Remote sensing findings are indicative of about 1000 Sq.km. of coral reef coverage, whereas the traditional projection based on random ground survey tells of 2000 Sq.km. i.e. 6% coverage of shelf area (Mustafa,A.M.1987)[4], a conclusive estimation is yet to be made.

3.4 *Type*

The islands coral reef are loosely classified as of ***fringing*** type all around and ***barrier*** on west of Andaman island, this western formation (Swell,1935)[5] is not a true barrier (Sarano,F. 1989)[6] there are in fact three separate coral banks lined ≈ 21 Kms. away from the coast; further the Ritchie's archipelago is said to have ***channel*** reefs and ***Knoll*** formation, present type nomenclature is based on metaphor and analogy. A proper morphometric type classification based on Maxwell (1970) terminology is yet to be done.

3.5 *State*

Like any other coral reef ecosystem, elsewhere in the world, the reefs of A&N Islands are being subjected to varied anthropogenic stresses; leading to serious degradation & ultimate loss of the ecosystem. In Andamans siltation appears to be the single leading cause (Mustafa,A.M.1990)[7] followed by trampling, anchorage, and extraction for trade which once used to be there; it is estimated that about 18% of the total coral coverage has been lost due to these reasons (Mustafa,A.M.1987)[4]. In addition to human induced stresses the natural causes such as cyclone, invasion by Crown of thorn star fish, fungal infection and bleaching has also been recorded. The French research vessel ***Calypso*** co-related the destruction of Middle coral bank to cyclone during 1977 (Sarano,F.1989)[6] later during December 1987 pronounced cyclonic impact was observed upon the near shore ramose coral reefs on the western coast of Chidiyatapu near Rangachang village (Mustafa,A.M.1990)[8]. An abrupt population explosion of crown-of- thorn star fish (***Acanthaster planci***) was noticed at Wandoor during December,1988 (Mustafa, A.M.1990)[8]. Five varieties of massive corals namely ***Porites lutea***, ***Porites lichen***, ***Montipora tuberculosa***, ***Goniopora*** sp. and ***Goniastrea*** sp. are regularly found to have necrotic patches due to profuse colonization by a hyphomycetous fungus ***Scolecobasidium*** sp. which is a known fish pathogen; the infected coral tissue coenosarc and coenosteum are killed in due course, the infected coral skeleton show brown to black zonation, (Chandralata Raghukumar, Personnel Communication,1990). The mysterious phenomenon of widespread selective bleaching has also been observed during May,1998.

3.6 *Productivity*

Primary production for a coral reef near Port Blair is reported by Nair & Pillai,1969 to be of non-autotropic/heterotrophic (Photosynthesis/ Respiration ratio = < 1) nature having a total annual production of only 1200 gC/m^2 (Nair,R.P.V.1970)[9] ; however in terms of tertiary production based on a flat rate model of 15t/ Km^{-2}/year (Munro,1984)[10] the total biotic production in theory could be 24,600 tones/year from 1640 sq.km. of normal reef area.

3.7 *Status*

The renowned marine naturalist Cap. ***Jacques Cousteau*** with his associates on board Foundation Cousteau's research vessel 'CALYPSO' made 70 hours of underwater observations over 12 dive sites in Andamans during 1989 and recorded that the "Indian waters are preserved, rich and virgin" compared to the east coast region of Andaman sea which belongs to other nations. He further suggested that "some drastic rules must be taken before the industrial fisheries, recreational fisheries and tourism become active". Cousteau's specific five point suggestions than made are as follows :-

" (1) To keep some places absolutely virgin so that marine life reproduce, grow without human interference, and by that way, can recolonise exploited areas. That means industrial or recreative fishing, spear fishing, collecting of any sort are strictly forbidden. That also means rangers are present to avoid poaching. Some places we explored are particularly suitable for that purpose : NARCONDAM Island, SOUTH SENTINEL Islands, INTERVIEW Island and all the places where mangrove forests are well developed as on the RICHIES Archipelago east coast.

(2) Minimum size for the fish and crustaceans to be caught. The minimum size must be bigger or atleast equal to the size of sexual maturity. So fish could reproduce once before they could be caught, and the fish population is regularly renewed by that reproduction.

(3) Avoid fishing during reproduction time, particularly spawning time.

(4) Muddy or sandy area could be "seed" with simple wood and branching structures as artificial reefs. These reefs would offer a hard substrate for algae and sessile animals and shelter for small fish to grow peacefully. Areas with artificial reefs could advantageously be exploited according to a rotation system.

(5) It is very important that collecting by the means of dynamite, and poison is strictly forbidden. Dynamiting that is well developed all around Asia, is the worse fishing method.

If all these measures are taken, Andaman archipelago people could enjoy for a long time good fisheries and rich, various and wonderful marine fauna."

However, today's conclusive perception is indicative of mounting human induced stress upon this "fragile symphony of inner space" so called by Cousteau himself.

4. CORAL REEF POTENTIAL & EXTENT OF EXPLOITATION

4.1 *Components*

Spatial and temporal heterogeneity of bio-forms upon a reef could be as large as 3,000 in nos. larger intricate associates or macrofauna which include (1) Fishes (2) Mollusca (3) Crustacea and (4) Echinodermata are of direct human interest. Percentage composition of major macrofaunal group in the coral reef niche of Andaman & Nicobar islands is reported to be Mollusca – 30%, Curstacea – 20%, Worms – 16%, Fishes – 10%, Echinodermata –

8% and others 16% (Rao,G.C.1999)[12]. The most valued in order of priority are Fishes, Mollusca, Crustacea and Echinodermata.

4.2 ***Fishes***

Fishes of the coral reef are socio-economic crown for any island territory they exhibit striking diversity both in terms of species number and the range of morphologies. So far more than 500 varieties have been indexed from A&N islands (Rao,D.B. Personnel Communication,1999), these waters are teeming with not only edible but also ornamental icthyo beauties, luring commercial and industrial fisheries. Existing level of exploitation for edible varieties is moderate, the fish landing went up with increase in population from 44 tones during 1950 to 28,000 tones during 1998, the 39 years landing profile manifest 8% to 17% of demersal stock contribution annually, the demersal stock exploitation is mainly through hook and line, this also include non-reef dwellers caught through bottom trawl and shore seine. The exploited coral reef fishes belongs to families (1) ***Lutjanidae*** (Snappers) (2) ***Serranidae*** (Groupers) (3) ***Pomadasyidae*** (Grunts) (4) ***Scaridae*** (Parrot fishes) (5) ***Holocentridae*** (Squirrel fishes) (6) ***Mullidae*** (Goat fishes) (7) ***Chaetodontidae*** (Butterrfly & Angelfishes) (8) ***Acanthuridoe*** (Surgeonfishes) and ***Balistidae*** (Trigger fishes). Presently there are 2508 nos. of licensed fishermen with a fleet strength of 1762 (Table-I).

Since 1996 the large predatory fishes inhabiting coral reef areas have gained significant importances with the on-set of frozen live fish export from Port Blair to South-east Asian countries. Andaman Fisheries Ltd. (AFL) Garacharma a subsidiary company of A&N islands integrated development corporation (ANIIDCO) came into commercial production during 1996 followed by M/s Islander Marine Products Pvt. Ltd., Hamfregunj, both 100% export oriented units (Table-II). The AFL continues to maintain export of frozen coral reef fishes whereas it appears that M/s IMP have diversified. The raw material from AFL is mainly procured from the landing centers located on the west coast of South Andaman viz. Guptapara, Wandoor and Loha Barrick, the main fishing grounds are the area North of Little Andaman (Leeboard ledge and Duncan passage) upto South coral bank west of Port Anson in Baratang/Middle Andaman.

During 1998-99 about eight tones of wild caught live perches has been exported by M/s Samudra Marine Products, Junglighat, a company which originally launched its venture with the objective of cage culture, entered into wild harvesting from west coast of Mayabunder, Long island and Havelock island. One battery of there holding cages for acclimatization of stock was visited on 30th December'98 at Sound island in Middle Andaman. The cages were told to contain more than three tones of large demersal predatory fishes comprising of (1) ***Lutjanus*** (2) ***Epinecephalus*** (3) ***Cephalopholis*** (4) ***Variola*** (5) ***Aethaloperca*** and (6) ***Lethrinus*** species

Major targeted varieties for export are the members of Family ***Lutjanidae*** and ***Serrranidae***; typically for ***Lutjanidae*** it is almost a monospecies fishing for ***Pristipomoides typus*** {Ver. Sharptooth snapper (English), White snapper (Commercial), Mrigal (Local)} caught in shallow waters at an average depth of 50m whereas many varieties of ***Serranidae*** from among 43 recorded so far (Rajan,P.T. Personnel Communication,1999) are exploited; fishing season begin with the lapse of North-east monsoon by mid November and lasts upto mid May till the out burst of South-west monsoon.

Quantum jump in shark fishing to reap the fins alone is being observed,; indiscriminate harvest and criminal waste of the whole fish after removing the fins by migrant fishermen groups from mainland is getting established throughout the territory. Main fishing bases are at Mayabunder, Little Andaman & Nancowry.

With regard to ornamental fish resources a cross sectional trade survey was carried .by CBI Netherlands during 1988, the survey revealed availability of 153 marine & 13 brackishwater fishes, all having ready market abroad. (Tomey, W.A.1989)[13].

4.3 *Shells*

Molluscan the largest shareholder macrofaunal component of A&N islands coral reef have brought traditional fame to this territory for its shell wealth, commercial exploitation seems to have began since 1929 when a Japanese company was licensed to exploit ***Turbo*** & ***Trochus*** at the rate of 500 tones in a season of 3 months. Later the shell fishing was regulated through A&N Island fishing Regulation 1938 and Fishing rule 1939 upto 1954, thereafter A&N Island shell fishing rules 1955 was enacted wherein the whole area was divided into 9 geographical zones. The right of fishing was through auction, subsequently it got repealed and A&N shell fishing rule-1978 was enacted limiting the harvestable limit to 15 tones per zone per season of 6 ½ months. The last amendment was promulgated during 1994 through which the right of shell fishing is now granted only to registered shell craft industries by the Director of Fisheries through license, the quantity of shells to be exploited is kept reserved as a discretory decision of the licensing officer.

There is no authenticated statistical data available for the period prior to 1994 to establish the quantum of shells harvested because till 1993 the right to auction and accord permission for exploitation was with the respective Deputy Commissioners of Andaman & Nicobar districts. However with the introduction of licensing system since 1994 Director of Fisheries has been designated as authorised officer and the landing reports for the last five years is now available (Table-IV). Hitherto it is believed by many that the shell resources of A&N islands have been overexploited and pronounced depletion is observed throughout.

4.4 ***Lobster & Crab***

With the establishment of export industries lobster resources also attracted the attention and they too got a place in the export list (Table-III). They are hands picked or caught through spear. The exploited varieties are (1) ***Panulirus*** and ***Thenus*** inhabiting coral and rocky reefs. Contemporary addition is the mud crab ***Scylla*** sp. caught from shadow zone coral area (described elsewhere), lagoons and backwaters. Major fishing grounds are in North Andaman (Paschim Sagar, Mohanpur & Kalighat) Middle Andaman (Tugapur, Uttara & Kadamtalla), South Andaman – (Chouldari & Loha Barrick) and Little Andaman. The peculiar aspect is that only a fraction of professional fishermen are engaged in mud crab fishing, bulk of the catchers are agriculture farmers and forest workers. Kalighat, Mohanpur area in Diglipur North Andaman appears to be the largest crab landing center where it is estimated that about 70 persons are engaged. (Table-III).

5. DEPENDENCE ON LIVING MARINE RESOURCES

5.1 *Food*

The entire tribal and aborigine population heavily depends on all the edible marine resources which has become an essential component of their daily nourishment, for others which include pre and post independence settlers and migrants it is believed that 90% of the present day population is non-vegetarian fish eater.

5.2 *Profession*

Statistically during 1995 we had 11,062 Nos. of fisher community members dependent on fishing, for the new millenium the hypothetical population projection based on scattered random sampling assessment incorporating annual increase due to influx and birth is suggestive of about 20,000 fisher population. The non tribal migrant fishermen of Nancowry Island (Katchal and Kamorta) are assessed to be the richest having average monthly income of Rs.3,800/- per head per month followed by the fishermen of South Andaman (Gupta para and Junglighat) earning about Rs.3,500/- the fishermen of Little Andaman (Hut Bay) appears to be third in terms of income by making Rs.3,100/-. Nancowry and Little Andaman are known for fish drying, the dried fish when marketed at Port Blair/mainland fetches good price, for South Andaman prosperity at Gupta Para and Junglighat is attributed to export house and mechanisation respectively.

The declining seashell industry hardly supports 70 persons in totality, similarly the up-coming crab & lobster fishing supports some 100 people. The shell collectors and crab & lobster fisher are not the true dependent on fishing however a substantial share of their livelihood is by means of this trade.

Through fisheries 'management' about 1,132 persons, under 'survey' about 147 and by 'research' about 42 people by virtue of earning member's employment in the concerned department depends on fisheries.

Summarily today about 21,491 islanders solemnly and solely depends on fisheries.

5.3 *Catch*

Shoaling Mackerel and Sardine when put together are the maximum caught seasonal varieties exploited by traditional gill netters, followed by round the year harvestable Perches caught through hook & line, harpoon and arrows.

5.4 *Women*

In urban areas road side marketing and door to door hawking for fresh fish is done by fisherwomen folk, it is assumed that in the Port Blair city about 40% of fisherwomen population do undertake marketing. In rural areas the fisher women actively participate in fish drying.

6. POACHING

6.1 *Explosive Fishing*

Since independence thousands of poaching fishing vessels have been apprehended. More than 80% of the confesticated vessels are found to contain only coral reef fishes. It is assumed that these poachers might have been undertaking coral reef destructive fishing methods such as 'Blast fishing' and 'Cyanide poisoning' because majority of the arrested vessels are found to have only a few dozens of large scoop nets and many sets of modified diving masks connected to onboard fitted large compressed air reservoirs for group diving. It is not possible to get their fish holds filled with large coral reef area predatory fishes utilising scoop nets and diving masks when no other active or passive form of gear is found to be onboard. Although there is no documented evidence in support of this assumption, I would like to narrate two personnel incidents which are indicative of such unfair activities in our waters. During mid 1997 while at Car Nicobar two Nicobari tribals of Chowra island approached for underwater crackers to kill fish because some years ago once they witnessed such act while accompanying outside 'Doust' (friend) visitors on a shallow reef South of Chowra island and East of Batti Malu island; similarly while at Mayabunder during March 1998 a group of scientists from the Russian Academy of Sciences, ANS Institute of Ecology & Evolution, Moscow after diving expressed their opinion of possible dynamite fishing N-E off Sound Island in the vicinity of Elfin patch.

7. UNIQUENESS OF OUR CORAL REEF ECO-SYSTEM

7.1 *Classical Integration*

Classical synchronisation of coral reefs with mangrove forest is found to be widespread in these islands; the unconscious forces of creation have amazingly blended the two most productive marine eco-system at many places, perhaps invoking the inherited trait of adaptability in many organism. In Alexandra Island at Marine National Park Wandoor it is found that underneath the advancing mangrove forest at sub-soil level there is deposition of corals and molluscan shell and as such soil pH in that area is in the range of 7.2 to 7.4 (Dam Roy,S.1995)[14]. Such shadow zones show meek mingling of inhabitants, one among many is the mud crab ***Scylla serrata*** which although not a true components of coral reef niche is seldom found there.

8. CONSERVATION

8.1 *Beginning*

It was during 1987 when in a symposium entitled Management of Coastal ecosystem and oceanic resources of Andamans, organised by Andaman Science Association at Port Blair an attempt was made by the first author to highlight the level of human induced damage to corals of A&N Islands, its significance and potential. The presentation fortunately got the attention of many govt. and non-govt. organisations, since than, many in-line scientific contributions have been made; certain legislative measures adopted for the welfare of coral reef regime, notable one is the amended regulation for trade in live ornamental fish sharks and invertebrate, exploitation, trade and export to mainland India; and establishment of a public marine aquarium.

8.2 *Theory*

Hypothetically in theory the good coral reef area of A&N island is producing about 24,600 tones of biota per year. The coral reefs are also known as 'nutrient deserts' because with a few exception majority of the available nutrients are locked as biomass comprising of many year classes, which could be harvested. Any harvest is a deduction from the nutrient bank which if not harvested is otherwise available for recycling. Smith & Stimson (1979)[15] are of "the opinion that the coral reefs have low standing crop of exploitable species" this argument is supported due to limited fish recruitment upon a reef by juveniles after pelagic phase further they are cryptic and heterogeneous in composition. Grigg (1979)[15] suggests exploitation of only 10% of standing crop. Gomez.E.D. & Yap.H.T.(1985)[15] are of opinion that reef fishes may sustain only subsistence consumption. For Pacific islands the range of coral reef primary production is found to be 2900 gm.C/m^2/Year (Kohn & Helfrich – 1957) to 4200 gm.C/m^2/Year, based on it the quantum of exploitable fin fish in that area is estimated to be 3-5 tones/Km^2/Year (Marshall.1979)[15]. On the contrary the level of our Primary production is quite low hence in view of the above argument the safe exploitable stock level for various reefs of A&N islands will undoubtedly be of lower magnitude.

9. SUGGESTIONS

It is indeed a mammoth task to achieve the objective of a prudent sustainable exploitation policy to protect our coral reef regime, so as to utilise the available resources for very many more years. The policy shall be based on the findings of scientific investigations undertaken in the area concerned. Today as we all are aware no serious effort has ever been made in this direction, the dearth of scientific data deprive us of a realistic approach in this direction.

In A&N islands we have many departments and NGO's working for the welfare of coral & coral reef resources however all the efforts so far put in are independent and fragmented, perhaps loosing much of its significance & utilisation. Considering various island related limitations and constraints it may perhaps be appropriate on our part to mobilise the available resources towards a common goal through synergetic labour to arouse management response conditioned through wide range of scientific, environmental, social & economical factors leading to a management plan and law to control human activity for responsible utilisation, care and protection of coral reef resources. Prime response shall be to have detailed informations with regard to :-

(1) Geographical extent

(2) Type

(3) Resources

(4) Status

(5) Exploitation

Informations on these aspects will lead us towards a prudent management plan through :-

(1) Zonation : (Preservation, Scientific, Wilderness, National Park, Recreational and General use)

(2) Fishing Pattern : (Underexploited, overexploited, Harvestable limit, Gear and Season).

(3) Manipulation : (Artificial reef, Culture & Sea ranching).

Table – I Fishing might of A&N Islands

Sl.No.	Asset	Nos.
(1)	Traditional Crafts (Inshore)	
	(a) Motorised	267
	(b) Non-motorised	1486
(2)	Larger Vessels (Offshore)	
	(a) Trawler	08
	(b) Long liner	01
(3)	Licensed Fishermen	2508
	(i) Diglipur	364
	(ii) Mayabunder	229
	(iii) Billiground	68
	(iv) Rangat	255
	(v) Kadamtala	159
	(vi) Port Blair	926
	(vii) Havelock Is.	88
	(viii) Neil Is.	77
	(ix) Little Andaman	164
	(x) Car Nicobar	65
	(xi) Katchal	16
	(xii) Nancowry	32
	(xiii) Great Nicobar	65

Table – II Quantity of coral reef fishes procured by export houses for fresh & frozen consignment

Calendar Year	Qty. (Tones)
1996	81
1997	230
1998	76

Table – III Quantity of other items exported to mainland India (Tones)

Item \ Financial Year	94-95	95-96	96-97	97-98	98-99
Lobster	4.5	0.2	0.9	3.01	2.1
Crab	1.4	11.9	19.4	11.9	20.8
Shark Fin	0.07	1.4	0.5	1.4	2.6

Table – IV Variety wise shell landing & revenue in Andamans (Tones)

Variety	94-95	95-96	96-97	97-98	98-99
Trochus	2.0	0.5	4.4		0.5
Turbo	0.2	0.06	0.02		0.02
Dwarf Turbo	x	0.12	0.2		x
King Shell	0.4	0.06	0.04		0.02
Queen Shell	x	0.03	0.002		x
Nautilus	0.4	0.03	0.1		0.12
Clam Shell	0.1	0.7	0.4		0.3
Chank	x	0.2	0.04		0.06
Nancowry shell	x	0.05	x		x
Lambis	x	0.3	x		0.1
Panja	x	0.4	1.14		x
Cowry	x	0.22	x		0.4
M.O.P	0.2	0.2	1.03		0.05
Misc.	x	x	x		x
Total	3.1	2.7	7.3		1.5
Zone Nos.	3 & 7	4 & 5	7 & 3		5
Revenue (Rs.)	14,705/-	6,445/-	82,505/-		20,275/-

10. REFERENCES

(1)	Wilkinson, C. 1998	The global coral reef monitoring network : reversing the decline of world's reef. ***Coral reefs, Proc. Vth World Bank Conf. Oct.9-11 : 1997***; The World Bank Washington, D.C.: 16-19.
(2)	Hallock, P.1997	Reefs & reef limestones in earth history. ***Life and death of coral reefs,*** by Charles Birkeland, ITP, New York : 13-42.
(3)	(a) Reddiah, K.1977	The coral reefs of A&N islands. ***Rec. Zool. Sur.India.*** 72 : 315-324
	(b) Pillai, C.S.G - 1978	Stony corals of A&N islands. ***CMFRI Cochin*** : 1-14
	(c) Anon. 1987	Corals of A&N islands – a status report ***CARI, Port Blair*** : 1-29
	(d) Subba Rao, N.V.1999	Report on state of environment. ***UNDP Project, ZSI, Port Blair***.
(4)	Mustafa, A.M. 1987	Endangered coral reefs of bay islands and their ornamental fishes. ***Proc. Symp. Mang. of Coastal ecosystem and oceanic resources. ASA*** : 60-65
(5)	Swell, R.B.S.1935	***Memories Asiatic Society of Bengal.*** 9 (8) : 461-540

(6) Sarano, F. – 1989 Observations En Plonge De La Fauna Marine Des Iles Andaman. ***Foundation Cousteau, Paris*** : 1-45

(7) Mustafa, A.M.1990 Increasing environmental stress on the coral reef ecosystem around South Andaman ***JASA, 6(1)*** : 63-65.

(8) Mustafa, A.M.1990 Coral reef degradation, conservation and management in Andamans-an overview. ***J.Sci. & Tech. Islands on March – 1990*** : 104-106.

(9) Nair, R.P.V. 1970 Primary production in the Indian seas ***Bull. Cent. Man. Fish Res. Inst. Mandapam Camp***. 22 : 1-56.

(10) Munro, J.L.1984 Coral reef fisheries and world fish production. ***ICLARM, News letter*** 7 (4) : 3-4

(11) Mustafa, A.M.1991 Sex and spawning of corals at Port Blair. ***J.Sci. & Tech. Islands on March – 1991*** : 69-70.

(12) Rao, G.C.1999 Marine ecosystem around A&N islands. ***UNDP, Project report, ZSI, Port Blair*** : 1-33

(13) Tomey, W.A.1989 Technical trade co-operation CBI/India. ***Project 5 Ornamental fish. Report phase – 6 Netherlands***

(14) Dam Roy,S. – 1995 Mangrove ecology of Alexandra Island and Manjeri area of S.Andaman. ***JASA, 11 (1&2)*** : 58-61.

(15) Gomez, E.D. & Yap, H.T. 1985 Coral reefs in the Pacific their potentials and their limitations. Regional seas, Environmental & resources in the Pacific. ***UNEP report no.69*** : 89-106

ACKNOWLEDGEMENT

With profound joy and deep pleasure the first author expresses his indelible thanks to DR. S Dam Roy, Director of Fisheries, A&N Islands for utmost encouragement in the preparation of this document.

Much gratitude and thanks to Dr N.V.Subba Rao (Emeritus Scientist) UNDP Project Manager for this opportunity and much needed generous support.

Reprint. *Proc. National Symp. Agri. Envit. & Forest : Towards Reconciliation ; Andaman Science Association (CARI, ICAR); 1999 ; p.20 & 113-119.*

METAMORPHOSIS OF ARTIFICIAL REEF PINNACLES AT PORT BLAIR, SOUTH ANDAMAN.

ARIF.M. MUSTAFA
Department of Fisheries, A&N Islands.

The paper deals with the creation of first ever man made shallow water artificial reef of its kind in Andamans and perhaps in India in the vicinity of a degraded coral reef at Sisostris Bay, Port Blair and presents a record of macro level quantitative biotic colonisation and settlement occurring upon it over a period of 28 months. An attempt was made through this experiment to rehabilitate the entire cove through the cumulative impact of created structure and coral implantation upon it. The succession of biotic aggregation was observed to be extremely fast, population explosion upon the three pinnacle was multifold, survival of implanted live coral colony was found to be 43%. This successfully tested technique of direct human intervention to arouse the mysterious life rendering power of ocean for the benefit of mankind could be an excellent tool for rehabilitating the degraded shallow water marine environment around the island.

METAMORPHOSIS OF ARTIFICIAL REEF PINNACLES AT PORT BLAIR, SOUTH ANDAMAN

ARIF.M.MUSTAFA
Department of Fisheries
Andaman and Nicobar Islands, PortBlair 744 101.

INTRODUCTION :

Inspired through the reports on fish stock aggregation (Anon - 1982; Bergastrom, M-1983; Maclean,J. -1986; Vega, M.J.M. - 1988; Anon -1991) in a short span of time upon various modules of Fish Aggregating Devices and fast colonisation of benthic FADs/artificial reef (Charles H.Turner - 1969; Peter Benchlay - 1988) a pilot project entitled 'Rejuvenation of Corals and fish aggregation through artificial reef formation at Sisostris bay, Port Blair, South Andaman was conceived and proposed by the author to the Department of Ocean Development, New Delhi, during 1993 with the objective to test the effectiveness of such man made structures towards eco-enrichment in Andamans primarily through natural metamorphosis on and in the vicinity of FAD/ARN secondarily through the implantation of small live coral colonies upon ARN to access the possibility of accelerated cumulative rehabilitation of biota in the degraded cove. The project was accorded immediate approval and execution began during 1994. Fishes are known to aggregate around solid object like floating logs, old nets, jetties, rafts and ship wrecks etc. The fishermen perhaps the world over since long made use of this peculiar fish behaviour which is known as 'Thigmotropism' through the deployment of various man made pelagic column and demersal structures, these artificially made structures are known as Fish Aggregating Devices which could be made up of anything such as palm leave, tree branch, bamboo, old tyres, stone, cement, FRP or old automobiles. The demersal FAD is termed as artificial reef and perhaps the first ever such reef was created by sinking four ships off U.S.Coast during 1935 followed by the construction of first experimental reef during 1958 at Santa Monica Bay California. The doomed war machines of second World war amid Bismarck Sea are now found to have been transformed into a living monument.

During 1968 an artificial reef made up of 800 concrete blocks produced 10 times more catch per unit area compare to two natural areas in 9 m depth in Virgin island(Oren, O.H., 1968). During 1968 four ships alongwith 15,000 tyres were used off South Carolina which enhanced fish availability (Wilkinson, W.A. 1980).

GEOMORPHOLOGY OF THE SITE :

Sea area of about 25 hectares in Sisostris bay, Port Blair (Fig.1) was earlier allotted to the Department of Ocean Development for some other project. In the same area during 1992-94 cage culture project (Mustafa, A.M.1995) was undertaken. The area for about a Century was one of the major fish landing centre of Port Blair till the beginning of this project and was heavily fished through hook and line, spear, cast net and shore seine net (Gulaty, R.K.-1957). The area was fairly good in having many marine organism upto 1978 when CMFRI conducted an indicative survey of the mariculture potential in A&N Islands (Silas.E.G-1983), the expert committee on marine aquaria and related aspects constituted under IDA found the site unpolluted and suitable for marine aquaria/oceanarium/dolphinarium (Anon - 1989). However, the coral cover over the small fringing reef was found to be degraded due to over collection and siltation (Mustafa, A.M. - 1987). Based on the routine visual, hydrological and fishing observations made for three years of cage culture project this area was proposed after base line survey.

Sisostris bay located 92° 45' EL & 11° 40' NL on the eastern edge of Aberdeen is a natural cove. It has a beach about 0.5

Km long the sand deposit is good towards break water whereas the area opposite to college is devoid of sand. Degraded coral patches are more on the eastern side above low water mark. The crescent projection from east towards Aberdeen Jetty covers about 1500 m² is the coral reef, which had less than 1/4th live coverage, maximum depth toward Ross Island is 16 m. The reef edge have a 70° slope. Sand grain size in the range of -1ø to 3.75ø was found to be 0.75ø (0.59 mm), tidal inundation taking flood and ebb tide into account is little more than 1.5 m.

Day time visual macro biotic population in an area of about 6500 m² was found poor.

OBJECTIVE :

The above mentioned project have following specific goals :-

(1) To evaluate the effectiveness and efficiency of FADs and artificial reef under island eco-system and to test the validity of an hypothetical concept of rehabilitating an unpolluted but degraded coral reef .

(2) To observe the magnitude and intensity of fish aggregation and natural metamorphosis of the erected structures in terms of macro biotic aggregation.

(3) To implant small (< 6 cm) coral colonies upon the artificial reef and to determine the possibility of survival on relocation.

(4) To access the positive impact upon the adjacent prevailing natural coral reef.

MATERIALS & METHOD :

For base line survey procedure describe under UNEP handbook 25 were followed. Observations were made during low tide. Ten 100 m transect count beginning parallel to Aberdeen Jetty break water keeping 0 at the low water mark extending outward towards Ross Island making count for 2.5m on either side i.e. covering 500 m² over each transect was recorded. In addition, 4m circle observation through a 4 m rope over the same transact at 10 m, 50 m and 100 m length was also made simultaneously.

It was decided to create (1) Three nos.of artificial reef nucleus (demersal fish aggregating devices) having different geometry made up of locally available quarry boulders each boulder weighing about 50 Kg. (2) Five nos. Of FRP. FADs, out of them one pelagic, three semi-pelagic and one demersal, former four to be anchored on sea-ward side of three boulder ARN and later one embedded upon central ARN (3)A series of 1x1x1 m concrete module in conjunction with old automobile tyres laid in an arc, compressed landward in the cove to have regularly-irregular continuity among three ARNs.

Three ARNs were created during October-November'94 by dumping quarry boulders from barge, an anchored afloat guider was made available while dumping the boulders to obtain desired shape. Few scattered boulders were later brought to shape through diving. The one demersal FAD was embedded upon centrally located ARN-II

ARN-I, located 92° 45' 22" EL & 11° 40'18" NL, one side resting upon a 6m high degraded natural coral reef, bottom sandy, made up of 189 MT of quarry boulder, height 8 m, basal circumference 20 m, it has the shape of a longitudinally half cut inverted funnel and have a 70° slope, depth 10 ± 2 m, surface area » 112 m².

ARN-II , located 92° 45' 15" EL & 11° 45' 15" NL stands upon sandy bottom, made up of 112.5 MT of quarry boulder, height 1.5 m, breadth 4 m, length 6 m, rectangular in shape. On sea-ward side one rectangular (H 1.4m x W 1m x L 2m) almirah type (both sides open) FRP FAD is embedded, depth 6 ± 2 m, surface area » 54 m².

ARN -III, located 92° 45' 10" EL & 11° 40' 14" NL over sandy bottom, made up of 62.5 MT of quarry boulder, height 2 m, basal circumference 16 m, paramydial in shape and have a 50° slope, depth 5 ± 2 m, surface area 408 m².

Due to certain administrative reason further structures are yet to be deployed.

The A&N Administration was approached to exempt the artificial reef area from fishing accordingly a total ban on fishing in the area was immediately imposed (Annon- 1995).

ARN-I was implanted with small (<6 cm) stony coral colonies, those found attached to detachable substratum were collected from adjoining area and transported over ARN-I. All possible care was adopted to maintain the original angel of orientation of the colony to sea bed while transporting and implanting upon ARN-I. These mini colonies with original substratum were firmly wedged in between the stony crevices of ARN-I.

Since the initially created structures were not large enough to apply the standard visual fish census procedure of Dartnall and Jones,1986 a floating 1 m^2 quadrat made up of PP rope having iron angel over edges and fitted with counter floats was fabricated and suspended at two sites through rope, anchor or stick support some distance above the pinnacle 3-4 days before the actual census. The fish count was made for the individual quadrat area, then projected for the total surface area of pinnacle after taking averages for groups. The enumeration over each quadrat was for a minimum of 15 minutes in one stretch. In this procedure there is a possibility of the same fish passing over the quadrat repeatedly which was inevitable and thus must have given rise to some enumeration error.

OBSERVATIONS :

Soon after the creation of three ARNs daily observations through SCUBA was made for about 30 days, but at times it got interrupted due to difficult sea condition. However, regular monthly visual census continued upto September'95, thereafter observation was made during April'96 and April'97. For the first one week there was no perceivable change in any of the pinnacle, by second week visual changes were noticed first tiny brown algae (Phaeophyta) were found upon ARN-II & ARN-III, whereas ARN-I developed a green pigmentation by the end of second week. Simultaneously fishes were seen hiding within the rock crevices, by the end of third week innumerable shell spat settlement was found upon ARN-I.After a month fishes were found hiding within the crevices and hovering upon all the ARNs. During second month two anemones (Radianthus sp) appeared upon ARN-I, initially only one have clown, clown fish egg mass was found by the end of second month. As the third month began shell and barnacle spat settlement was visible over ARN-I & III alongwith coloured sponge outgrowth, whereas ARN-II developed the ear plant Padina sp. all over except FRP portion which by now have many fishes. Pronounced aggregation was after five month when a quantum jump in fish population was observed upon all the pinnacles. ARN-III have many rock lobsters by this time. Dense squid shoal was found staying upon ARN-II by July'95 with 14 squid egg mass clusters attached to FRP component. After six month many sea snake (Lacticuda sp.?) were frequently seen in the deeper areas of ARN-I. By seventh month accumulation of sessile organisms of phyla. Echinodermata, Mollusca, Arthropoda, Coelenterata and Porifera were found flourishing upon ARN-I and ARN-III whereas ARN-II was covered with ear plant. For another three month the biotic density and composition went on increasing. Visual enumeration for fishes was possible only upto fourth month for later months an estimated quantity was recorded.

The increasing aggregation of residential and transitory fish stock made census extremely difficult hence for the convenience of accountability the fishes were broadly grouped in accordance with their relation to ARN Pinnacles (Ukkrrit Satapoomin - 1994) as follows :-

Type-A Fish living within the crevices, such as scorpion fish (Pterois spp), Grouper (Epinephelus spp), Dottybacks (Pseudochromis spp.) etc.

Type-B Fish found swimming close to ARN and gets into crevices when threatened, vacating the shelter as soon as the threat is over, such as Damsel and clown fishes (Chrysiptera spp. Chromis spp.

Dascyllus spp.) Pomacentrus spp. Abudefduf spp. Amphiprion sp.) Cardinal fishes (Apogon spp.) etc.

Type-C Fishes hovering around and touching or getting into projections while passing over such as Butterfly fish (Chaetodon spp.), Moorish idol, Bat fish (Platex sp.)Snapper (Lutjanus spp.), Parrot (Scarus spp.) Wrasses (Thalassoma sp, Gomphosus sp.), Trigger etc.

Type-D Fishes living around the base and on the ground around pinnacle such as Monocle bream (Scolopsis sp.)whiting (Sillago sp.) etc.

Type-E Pelagic hovering school of fishes and their juveniles, such as carangidae, Fusiliers (Caesio sp.) live bait (Spratelloides sp.) etc.

The initial fish aggregation was fast and intense, hence a random suspended quadrat count technique was adopted, the procedure has been explained elsewhere. Which is indicative of immense quantitative and qualitative enhancement in fish stock.

The surface metamorphosis specially upon ARN-I was found remarkable over a short span of 1½ year, the broad spectral colonisation of marine biota was amazing. The ARN-II showed little of faunal settlement it went floral through dense Padina sp. growth, whereas ARN-III have 40% of its surface area covered with Oyster (Crassostrea sp.)

Out of 14 implanted coral colonies four died during the first month, another four were found gradually bleaching the next month, all the eight dead colonies were removed. Rest six were found surviving upto April'97.

RESULT & DISCUSSION :

All the three pinnacles behaved differently with regard to biotic colonisation possibly due to their location and depth within the cove. The ARN-I is at the outer edge of a natural coral reef which offer better eco-conditions, ARN-II is a little above the lowest low water line in the centre of the cove having minimum height, the suspended silt load in the water column above it appears to be more compare to ARN-I & III and the dispersal of silt from the area also appears to be retarded, ARN-III is close to brake water in an area much agitated compare to ARN-I & II.

The qualitative and quantitative fish aggregation in such a short span of time is remarkably amazing. The accumulation of other living organisms and plants is comparable to any other artificial reef built elsewhere in the world. The survival and growth of 6 out of 14 implanted colonies is indicative of the possibilities of 'rejuvenation' of degraded coral reef having suitable eco-conditions.

This technique if adopted judiciously around the degraded coral reef areas of A&N Islands will definitely lead to multifold enhancement of standing fish stock regeneration of corals and large scale rehabilitation of degraded area with marine biota. Such human creations are gradually transformed into living monuments through mysterious influence of the sea and could well be subjected to exploitation after optimum gestation.

ACKNOWLEDGEMENT :

The author deeply express sincere thanks to Dr.S.A.H.Abidi, Director-cum-Vice Chancellor, CIFE, Bombay for suggesting this project, the Secretary to Government of India, Department of Ocean Development, 12 C.G.O.Complex, Lodhi Road, New Delhi, and his team of learned officials for sanctioning this project and their permission to utilise the data for Ph.D work. However, after premature reversion from DOD the A&N Islands Administration did not accorded permission to continue the research, hence a part of the data is presented here.

Thanks are also due to the able Scientists of the National Institute of Ocean Technology, IIT, Madras for their field level technical support in the selection of pinnacle site, its geometry and construction.

REFERENCES

(1) Gulaty, R.K. 1957 Fishing in Andaman Waters, The A&N information -1957.

(2) Turner, Charles, H. 1969 Man made reef ecology. Fish bulletin.46, State of California, : 1-221.

(3) Anon 1982 Fish Aggregating Devices. Bay of Bengal News.No.6.

(4) Bergstrom, M 1983 Review of experiences with and present knowledge about fish aggregating devices. BOBP/WP/23.

(5) Silas, E.G. 1983 Mariculture potential of Andaman and Nicobar Islands. CMFRI Bulletin 34.

(6) Maclean, Jay 1986 Who's working on artificial reef? NAGA 9 (2) : 22

(7) Mustafa, A.M. 1987 Endangered coral reefs of bay islands and their ornamental fishes. Proc. Symp. Mang. Of coastal ecosystem. ASA : 60-65.

(8) Peter Benchley 1988 Ghosts of war in the South Pacific. National Geography 173(4): 424-456.

(9) Vega, M.J.M. 1988 Who's working on fish aggregating devices? NAGA 11(4) : 16

(10) Anon 1989 Marine Aquaria. Report of expert committee. DOD, New Delhi.

(11) Mustafa, A.M. 1990 Increasing environmental stress on coral reef eco-system around S.Andaman JASA 6(1) : 63-65.

(12) Anon 1991 National Workshop on artifical reefs on the west coast of Thailand. Bay of Bengal News. No.41.

(13) Helen Newman et.al. 1994 Transplanting a coral reef : A Singapore community project. Coastal management in tropical Asia. Sept.'94 : 11-14

(14) Munro. J.L. et.al 1995 Artificial reefs in Philippines ICLARM. Conf. Proc.: 1-56

(15) Mustafa, A.M. 1995 A report on experimental coastal sea cage culture at Port Blair. Tech. Report No.02/95 - ANCOD, DOD.

(16) Anon 1995 A&N Gazettee (Extraordinary) No.96/95/F.No.5-1(318)/93 Dev.I. A&N Administration.

ABERDEEN
JETTY
ARN-III
SESOSTRIS BAY
ARN-II
ARN-I
PILLER
(CONSPIC)
S A N D
BUILDING
(CONSPIC)
SOUTH POINT
ROSS ISLAND

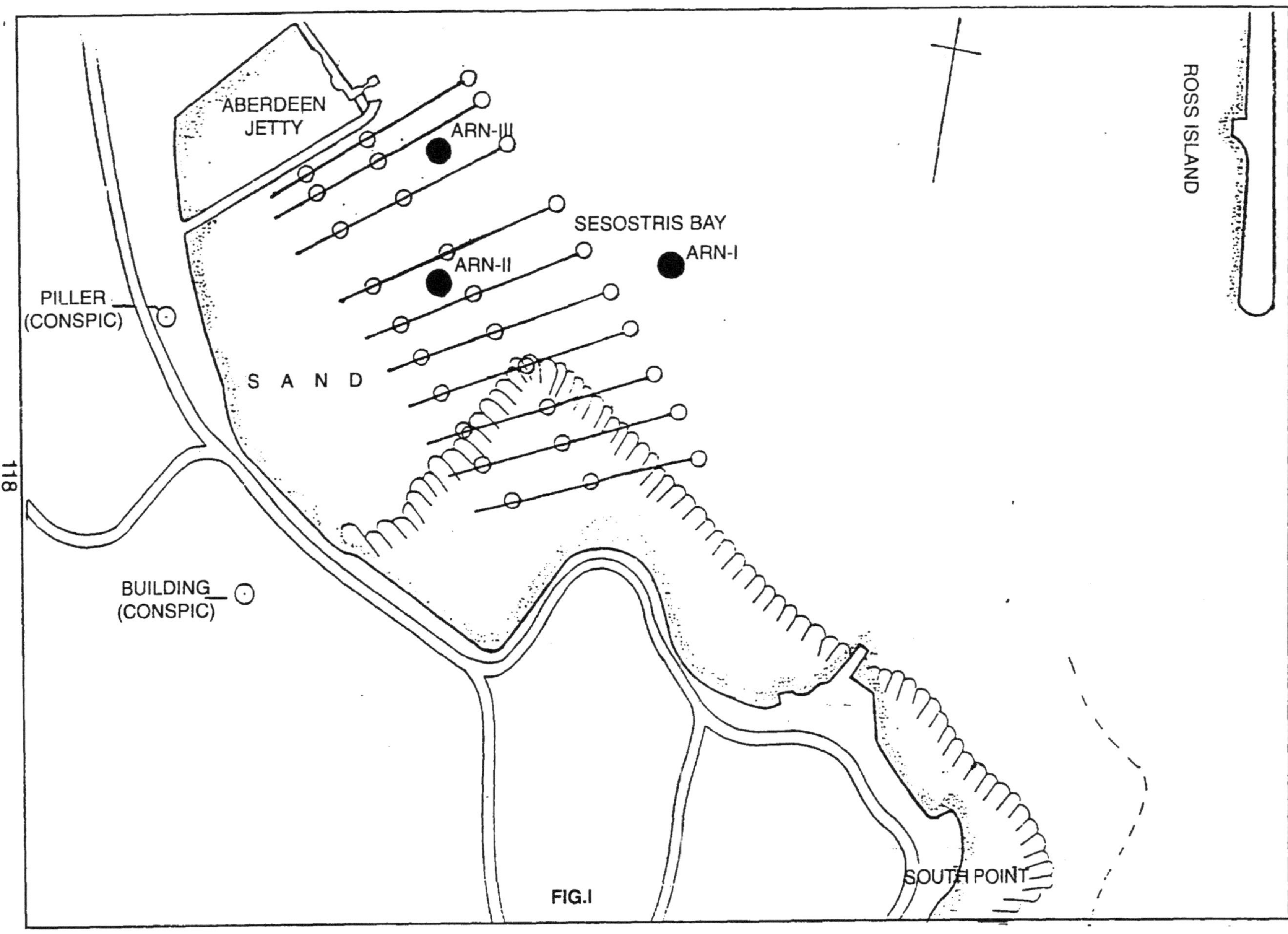

FIG.I

Stone ≈ 189 MT
H = 8 m
Cin (b) = 20m
Area = 112 m^2

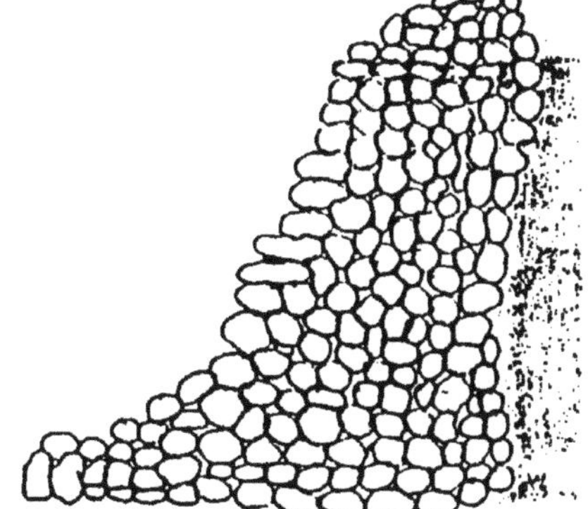

6m high degraded coral reef.

ARN. I
Side Elevation

FRP
1.4 x 1 x 2

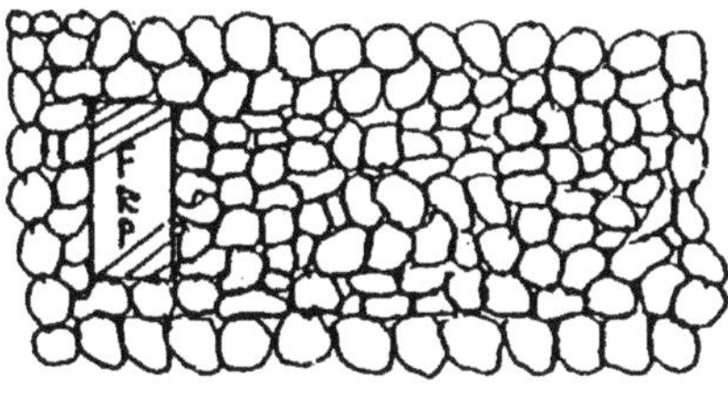

Stone ≈ 112 MT
H = 1.5 m
B = 4 m
L = 6 m
Area = 54 m^2

ARN. II
Top Elevation

Stone ≈ 62 MT
H = 2 m
Cin (b) = 16 m
Area = 408 m^2

ARN. III
Side Elevation

Artificial Reef
Unplanned accidental formation
a)
b)
c)
d)
e)
f)
g)
h)
i)
j)
k)
l)
m)
n)
Modules for planned artificial reefs

Base of Granite for Pinnacle No.2 & 3 at Sisostris Bay

3rd day recruitment over FRP structure in Pinnacle No.2 at Sisostris Bay

10th day recruitment over Pinnacle No.1 & 3 at Sisostris Bay

GOVERNMENT OF INDIA
A & N CENTRE FOR OCEAN DEVELOPMENT
(DEPARTMENT OF OCEAN DEVELOPMENT)

NEAR ABERDEN JETTY
P.O. ABERDEEN BAZAR
PORT BLAIR- 744 104
A & N ISLANDS.

A
REPORT ON

EXPERIMENTAL COASTAL SEA CAGE CULTURE AT PORT BLAIR

JANUARY – 1995

For bibliographic purposes this document may please be cited as follows :-

Arif M. Mustafa, 1995 - A report on `Experimental Coastal sea cage culture at Port Blair.'

(Tech. Report No.02/95-ANCOD)
A & N Centre for Ocean Development,
Department of Ocean Development,
Port Blair - 744 104.

-oOo-

A C K N O W L E D G E M E N T S

The 'PCC' team earnestly desires to express their gratitude to :-

Prof. P. Rama Rao, Secretary to Govt. of India, Department of Ocean Development for entrusting this micro-level project to ANCOD.

Shri J.V.R. Prasada Rao, Joint Secretary, **Dr. S.A.H.Abidi,** Director, **ER. B.N. Krishnamurthy**, Director and **Dr. B.R. Subramanian,** P.S.O. of D.O.D., New Delhi for their constant encouragement and untiring effort in accomplishing this project.

Dr. Arun Purlekar, Dr. Z.A. Ansari and Dr. Anil Chatterji, Scientists, BOD, NIO, Goa, for their scientific support and guidance at the beginning of this project.

Dr. G.C. Rao, Officer-in-charge, Z.S.I., A & N Regional Station, Port Blair, for critically going through the manuscript and improving upon it.

Dr. D.V. Rao and Shri P.T. Rajan of Z.S.I., Port Blair for identifying the icthyofauna utilised in the present experiment.

Dr. A.R.P. Sinha, Head, Department of Botany, JNRM College, for identifying the marine algae.

-oOo-

CONTENTS

Sl. No.		Topic	Page No.
1.	-	Introduction.	1
2.	-	Objective.	1
3.	-	Budget.	2
4.	-	Assets.	2
5.	-	Experiment site.	2
6.	-	Floating cage development.	3
7.	-	Survey & selection of fish.	12
8.	-	Mono-culture.	16
9.	-	Systematics & biology.	16
10.	-	Observations.	27
11.	-	Polyculture.	34
12.	-	Eco-impact.	34
13.	-	Hydro-eco-status.	37
14.	-	Recommendations.	44
15.	-	Bibliography.	45

-oOo-

1. Introduction

In order to explore the possibilities of culturing the locally available marine fishes in floating cages under Andaman conditions, a pilot project entitled "Polycage culture" was initiated by this centre during June 1992 at Port Blair and completed in July 1994.

2. Objective:

The specific objective of this project was to:-

1. Establish technical feasibility of culturing marine fishes through floating cage culture in the coastal waters around A & N Islands and its demonstration.

2. Design a floating cage suitable for subsistence level cage culture utilising locally available material through experimental cage development.

3. Identify a few commercially important varieties of marine fishes from Andamans water suitable for cage culture, out of the wide spectrum of around 800 varieties of fishes recorded from A & N islands, through field trials.

4. Screen/select new varieties of fin fishes which are not being cultured elsewhere, to avoid duplication.

5. Evaluate the growth, survival and mortality pattern of the prosperous varieties under semi-captive state, in the prevailing hydro-ecological conditions.

6. Evaluate preliminary eco-impact of cage culture.

Experimental floating cage set up at Sesostris bay

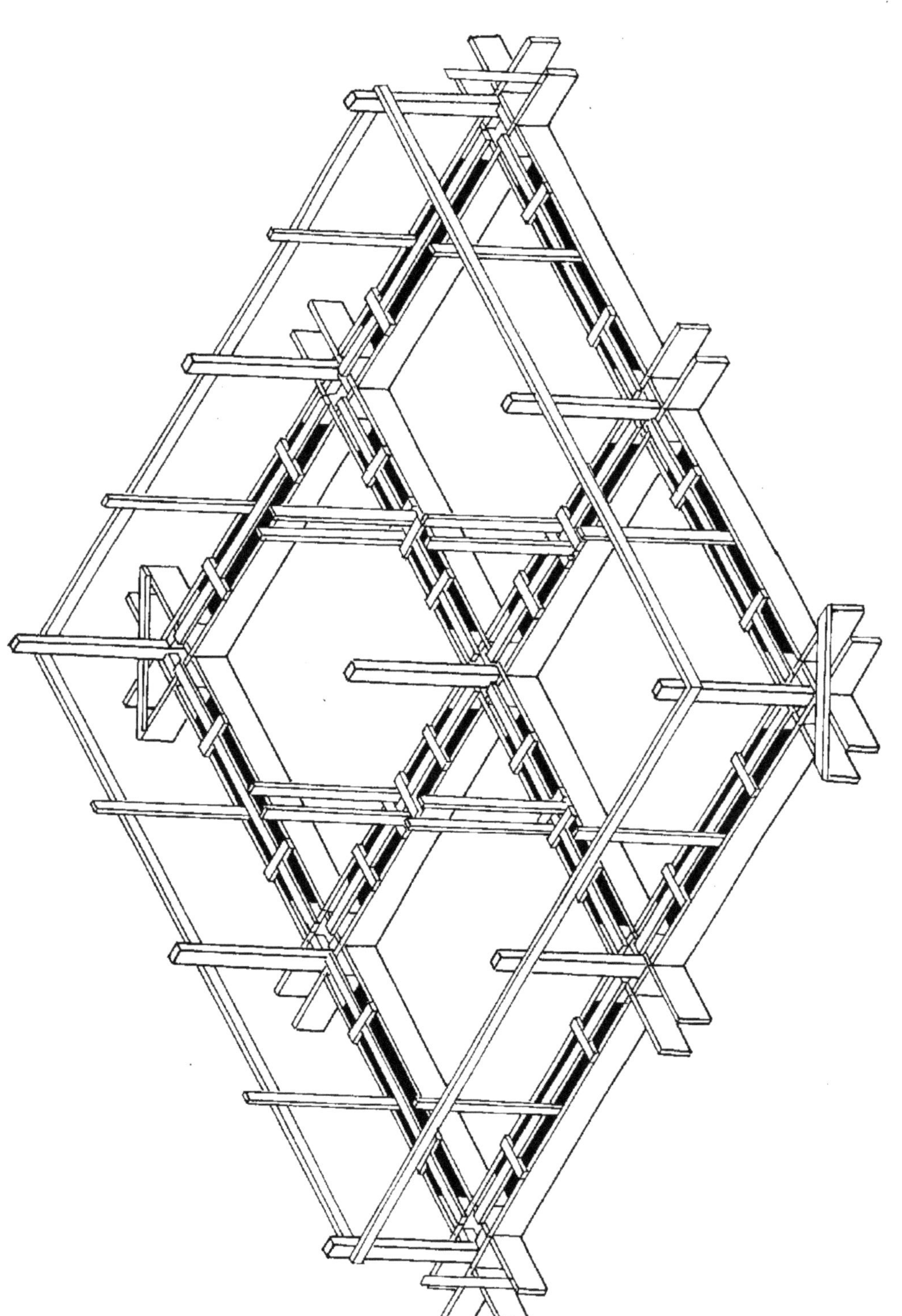

(Fig.1. ISOMETRIC VIEW)

Experimental Fish Culture Cage

(Not to scale)

A & N CENTRE FOR OCEAN DEVELOPMENT

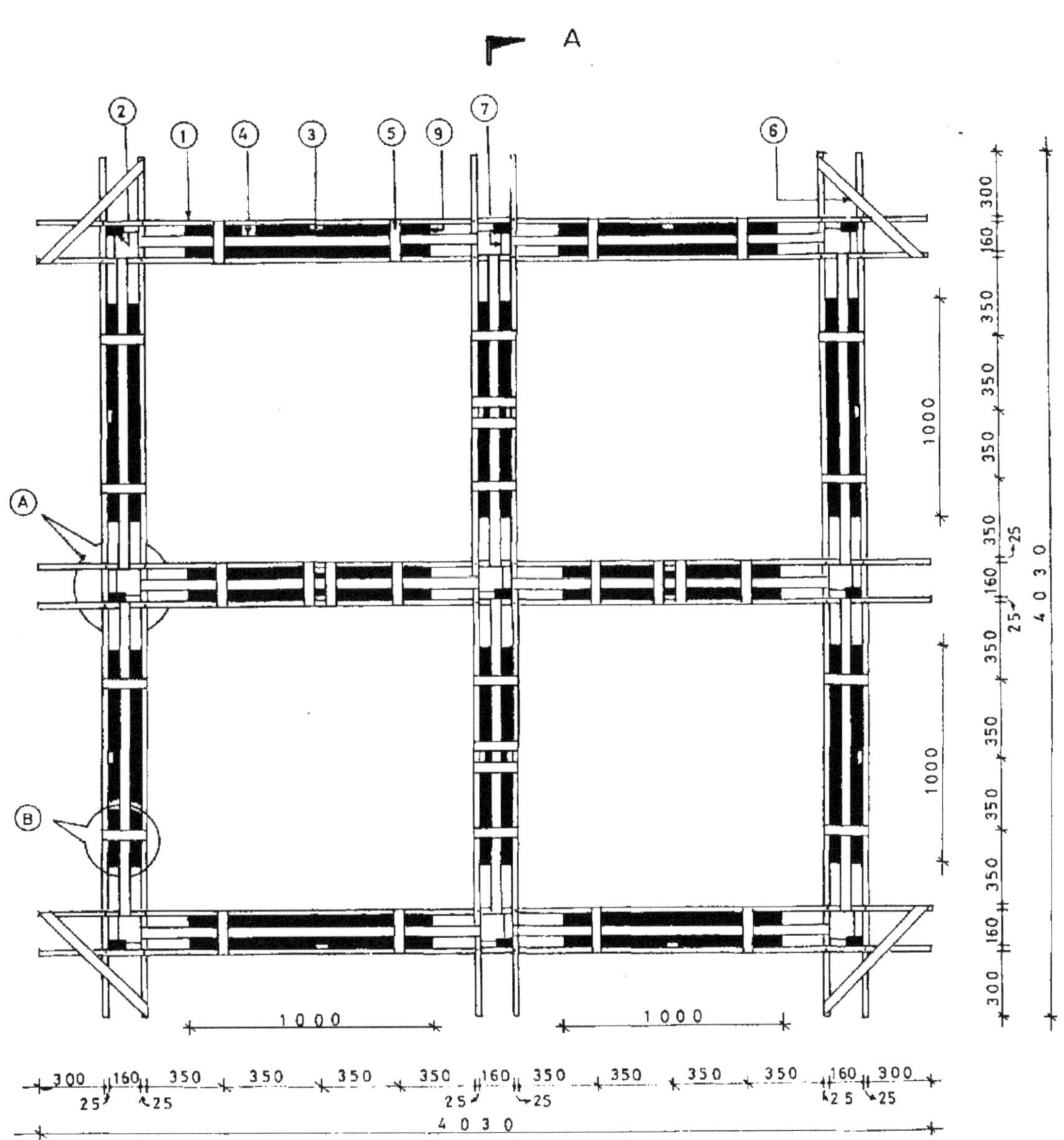

(Fig. 2. Top View)

Experimental Fish Culture Cage

(Not to scale)

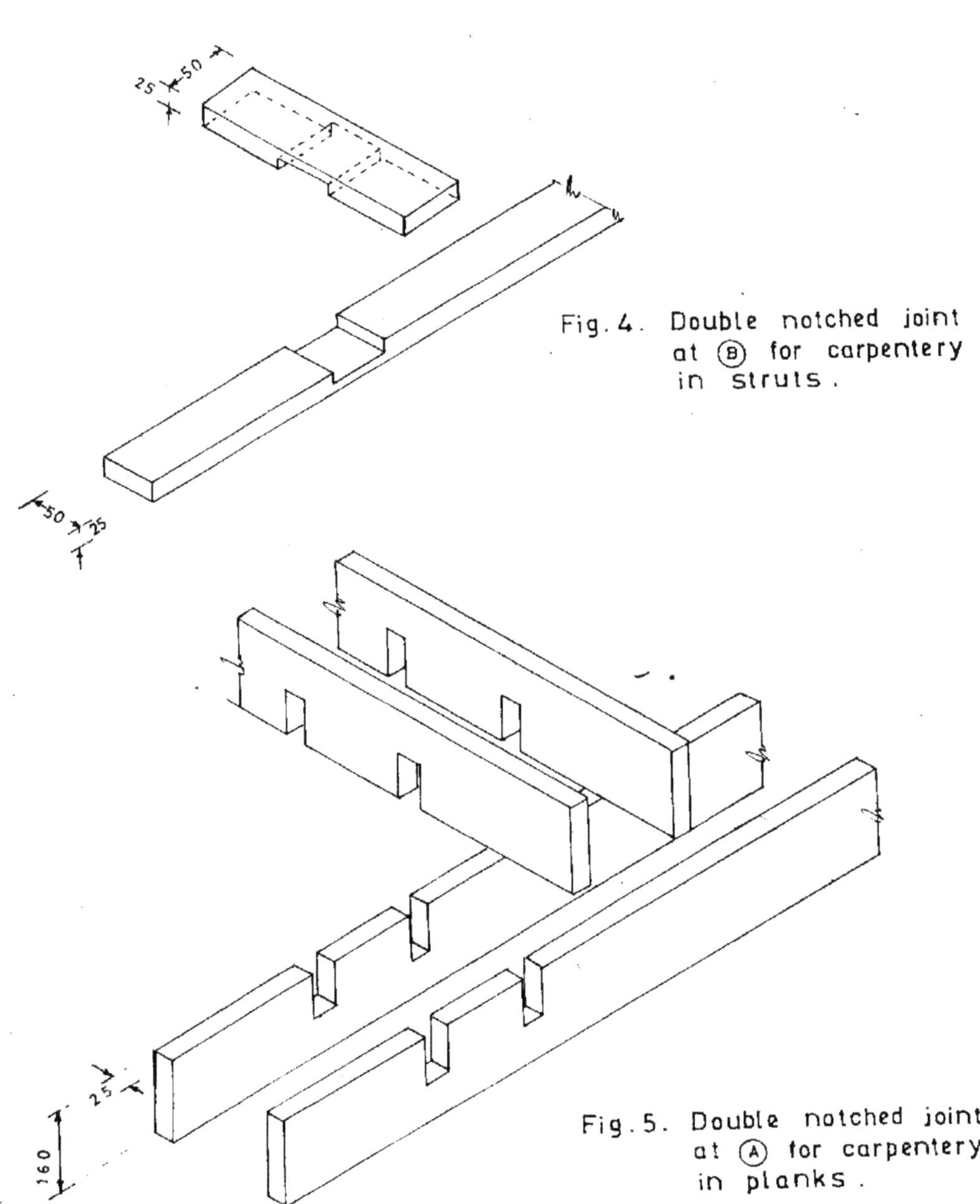

Fig. 4. Double notched joint at (B) for carpentery in struts.

Fig. 5. Double notched joint at (A) for carpentery in planks.

Note :- All dimensions are in mm.
Refer Fig. 2.
(Not to scale)

Table - 1. Details regarding 90 days culture period

Sl. No.	Fish Species	Density per m^3	Average size (cm)	Feed	Qty. of feed (%)	Average size after 3 months (cm)	Mortality Unknown (%)	Mortality Cannibalism (%)	Estimated monthly linear increment (cm)	Young ones availability	Remarks
1.	Caesio cuning	50	11	Trash fish	5	12.00	70	Nil	01	Fair	Rejected
2.	Terapon jarbua	50	8	Trash fish	10	11.00	Nil	40	03	Fair	Rejected
3.	Siliago sihama	50	12	Trash fish	5	13.00	80	Nil	01	Fair	Rejected
4.	Abudefduf sexatilis	50	3	Trash fish	10	4.00	Nil	Nil	01	Excellent	Rejected
5.	Scolopsis bilineatus	50	10	Trash fish	5	13.00	2	Nil	03	Fair	Selected
6.	S. ciliatus	50	5	Trash fish	5	8.00	Nil	Nil	03	Fair	Selected
7.	Siganus stellatus	50	6)	Algae +	8	9.00	40	Nil	03	Seasonal	Rejected
8.	S. spinus	50	5)	Trash fish	8	8.00	45	Nil	03	Seasonal	Rejected
9.	S.vermiculatus	50	8)	(8:2)	8	14.00	5	Nil	06	Seasonal	Selected
10.	Epinephelus fario	40	12	Trash fish	10	14.00	8	16	02	Rare	Rejected
11.	E.fasciatus	40	10	Trash fish	10	12.00	12	30	02	Fair	Rejected
12.	E. summana	40	15	Trash fish	10	17.00	20	20	02	Rare	Rejected
13.	E.areolatus	50	10	Trash fish	10	13.20	Nil	Nil	03	Good	Selected
14.	Lutjanus kasmira	80	8	Trash fish	10	12.10	Nil	Nil	04	Excellent	Selected
15.	L.quinquelineatus	80	8	Trash fish	10	12.00	Nil	Nil	04	Excellent	Selected
16.	Cephalopholis leopardus	40	12	Trash fish	10	13.50	30	20	01.5	Fair	Rejected
17.	C.microprion	40	15	Trash fish	10	21.00	Nil	5	06	Fair	Selected
18.	C. argus	40	14	Trash fish	10	16.00	15	22	02	Rare	Rejected
19.	C. miniata	40	12	Trash fish	10	16.00	Nil	2	04	Fair	Selected
20.	C. boenak	80	8	Trash fish	10	10.00	Nil	Nil	02	Good	Selected

due to intense algal settlement each bag was provided three nos. of adult *S. vermiculatus* from wild, which resulted in a total cleaning of algae in about 7 days and preventing further profuse settlement.

Table - 2 *Cephalopholis microprion* (Density $40/m^3$)

Sl. No.	Month & year	Total length (cm)	Increment during the month (cm)	Approximate weight (grams)	Mortality (%)
Ist set.					
1.	Jan. 1993	21.0	5.0	160	Nil
2.	Feb. 1993	26.0	4.5	-	5
3.	Mar. 1993	30.5	4.8	-	4
4.	Apr. 1993	35.3	5.1	-	Nil
5.	May. 1993	40.4	4.6	600	8
6.	Jun. 1993	45.0	-	710	10

Initial (approximate) bio-weight = 6.2 kgs.
Final (approximate) bio-weight = 28.00 kgs.

Sl. No.	Month & year	Total length (cm)	Increment during the month (cm)	Approximate weight (grams)	Mortality (%)
II set.					
1.	Jul. 1993	16.0	4.0	75	Nil
2.	Aug. 1993	20.0	5.0	150	"
3.	Sep. 1993	25.0	5.0	-	8
4.	Oct. 1993	30.0	4.5	-	9
5.	Nov. 1993	34.5	4.0	-	6
6.	Dec. 1993	38.5	-	515	12

Initial (approximate) bio-weight = 2.8 kgs.
Final (approximate) bio-weight = 20.4 kgs.

Table - 3 *Cephalopholis miniata* (Density $40/m^3$)

Sl. No.	Month & year	Total length (cm)	Increment during the month (cm)	Approximate weight (grams)	Mortality (%)
1.	Mar. 1993	12	03	38	11
2.	Apr. 1993	15	02	-	Nil
3.	May. 1993	17	03	-	"
4.	Jun. 1993	20	04	-	"
5.	Jul. 1993	24	02	-	"
6.	Aug. 1993	26	01	200	8
7.	Sep. 1993	27	02	-	4
8.	Oct. 1993	29	03	-	3
9.	Nov. 1993	32	02	-	6
10.	Dec. 1993	34	01	600	16
11.	Jan. 1994	35	02	-	12
12.	Feb. 1994	37	-	650	8

Initial (approximate) bio-weight = 1.4 kgs.

Final (approximate) bio-weight = 25.8 kgs.

Table - 4 *Ephinephelus areolatus* (Density 40/m^3)

Sl. No.	Month & year	Total length (cm)	Increment during the month (cm)	Approximate weight (grams)	Mortality (%)
1.	Jan. 1993	13.30	1.2	30	2
2.	Feb. 1993	14.50	1.5	-	Nil
3.	Mar. 1993	16.00	1.8	-	"
4.	Apr. 1993	17.80	2.1	75	"
5.	May. 1993	19.90	0.9	-	"
6.	Jun. 1993	20.80	0.9	-	"
7.	Jul. 1993	21.70	1.0	100	"
8.	Aug. 1993	22.70	1.1	-	"
9.	Sep. 1993	23.80	1.2	150	"
10.	Oct. 1993	25.00	1.5	-	"
11.	Nov. 1993	26.50	1.5	250	"
12.	Dec. 1993	28.00	3.0	325	"
13.	Jan. 1994	31.00	3.5	-	"
14.	Feb. 1994	34.50	1.8	-	"
15.	Mar. 1994	36.30	-	508	"

Initial (approximate) bio-weight of the stock = 1.08 kgs.

Final (approximate) bio-weight of the stock = 20.00 kgs.

Table - 5 *Cephalopholis boenak* (Density $40/m^3$)

Sl. No.	Month & year	Total length (cm)	Increment during the month (cm)	Approximate weight (grams)	Mortality (%)
1.	Jan. 1993	10.0	0.8	20	Nil
2.	Feb. 1993	10.8	1.5	-	"
3.	Mar. 1993	12.3	0.9	-	"
4.	Apr. 1993	13.2	0.6	-	"
5.	May. 1993	13.8	0.7	-	"
6.	Jun. 1993	14.5	0.4	-	"
7.	Jul. 1993	14.9	0.9	51	"
8.	Aug. 1993	15.8	1.2	-	"
9.	Sep. 1993	17.0	2.1	-	"
10.	Oct. 1993	19.1	1.5	-	"
11.	Nov. 1993	20.6	1.6	-	"
12.	Dec. 1993	22.2	2.1	-	"
13.	Jan. 1994	24.3	1.7	-	"
14.	Feb. 1994	26.0	-	410	"

Initial (approximate) bio-weight of the stock = 7.2 kgs.

Final (approximate) bio-weight of the stock = 16.2 kgs.

Table - 6 *Siganus vermiculatus* (Density 40/m^3)

Sl. No.	Month & year	Total length (cm)	Increment during the month (cm)	Approximate weight (grams)	Mortality (%)
1.	Apr. 1993	16.08	0.38	75	3
2.	May. 1993	16.46	1.39	-	5
3.	Jun. 1993	17.85	1.50	100	Nil
4.	Jul. 1993	19.35	1.00	-	"
5.	Aug. 1993	20.35	0.8	130	"
6.	Sep. 1993	21.15	0.8	-	"
7.	Oct. 1993	21.95	0.7	-	"
8.	Nov. 1993	22.67	1.0	-	"
9.	Dec. 1993	23.67	1.1	-	"
10.	Jan. 1994	24.67	1.2	155	"
11.	Feb. 1994	26.01	1.2	-	"
12.	Mar. 1994	27.21	1.1	-	"
13.	Apr. 1994	28.31	1.1	-	"
14.	May 1994	29.40	1.1	-	"
15.	Jun. 1994	30.41	-	350	"

Initial (approximate) bio-weight of the stock = 3.0 kgs.

Final (approximate) bio-weight of the stock = 13.8 kgs.

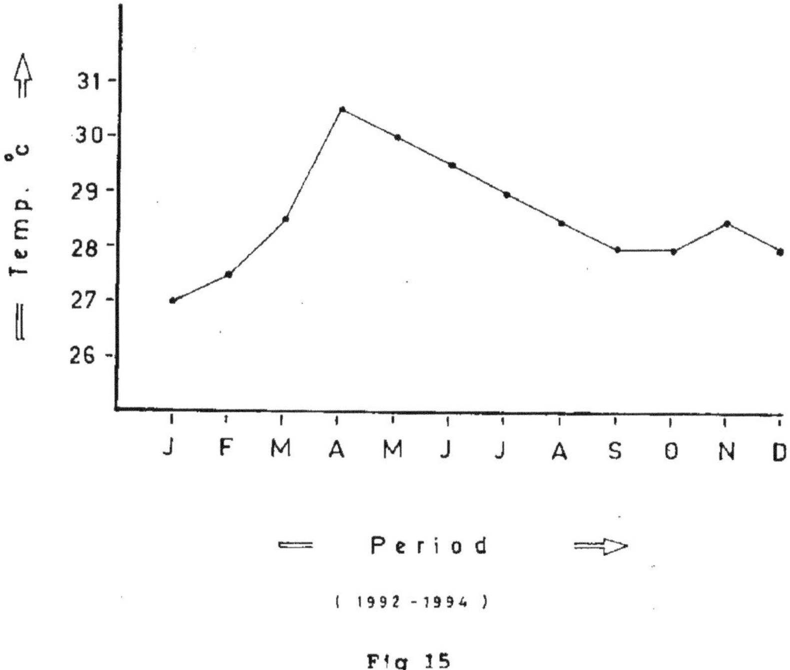

Fig 15

Genral trend of surface temperature

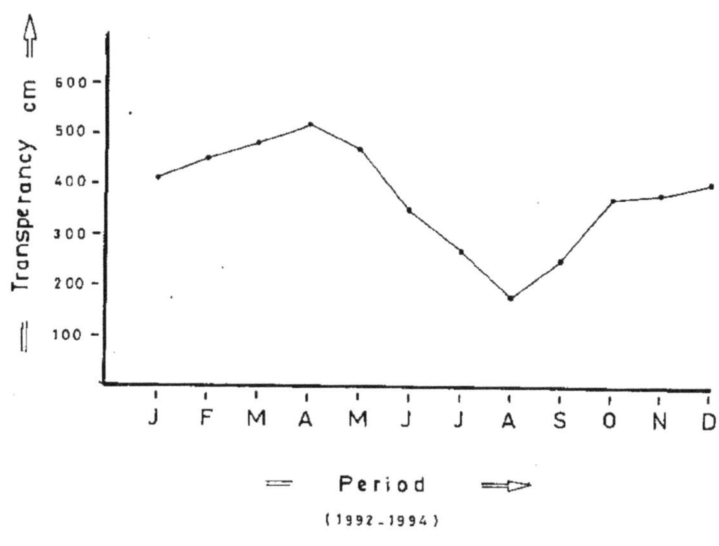

Fig 16

Genral trend of surface vertical transperancy

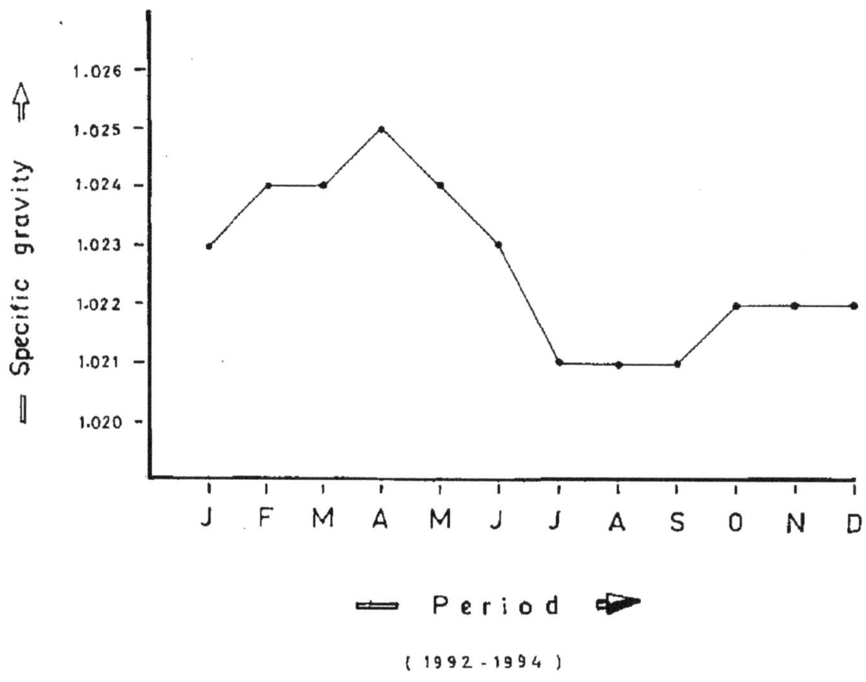

Fig 17

Genral trend of surface specific gravity

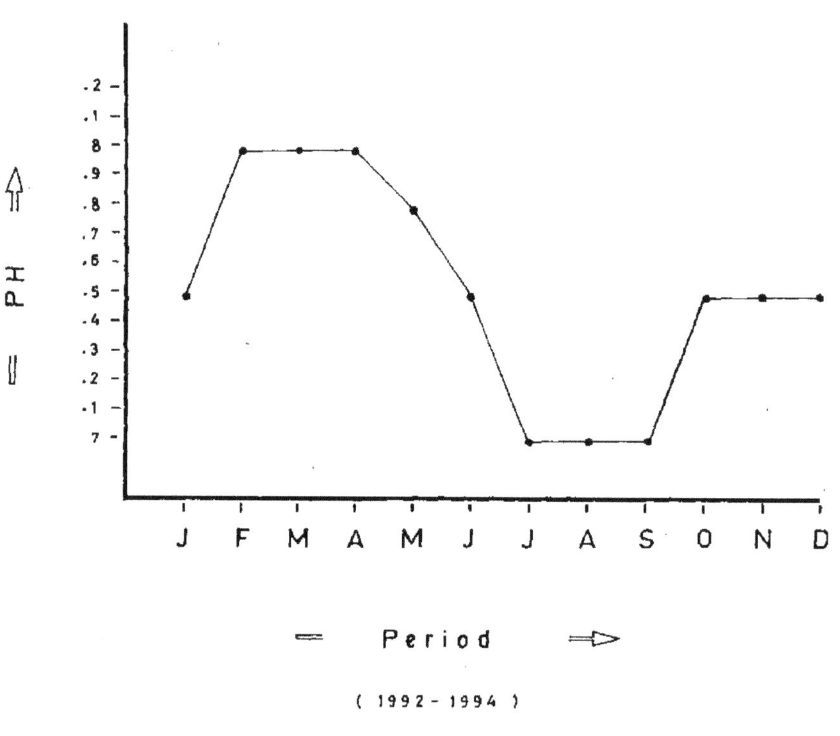

Fig 18

Genral trend of surface PH

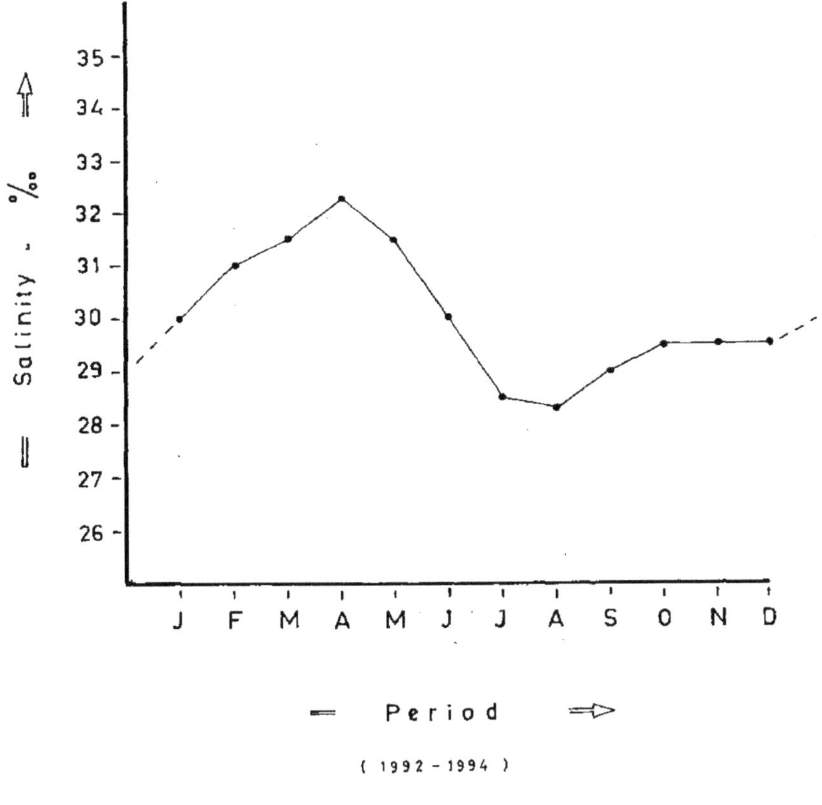

Fig 19

Genral trend of surface salinity.

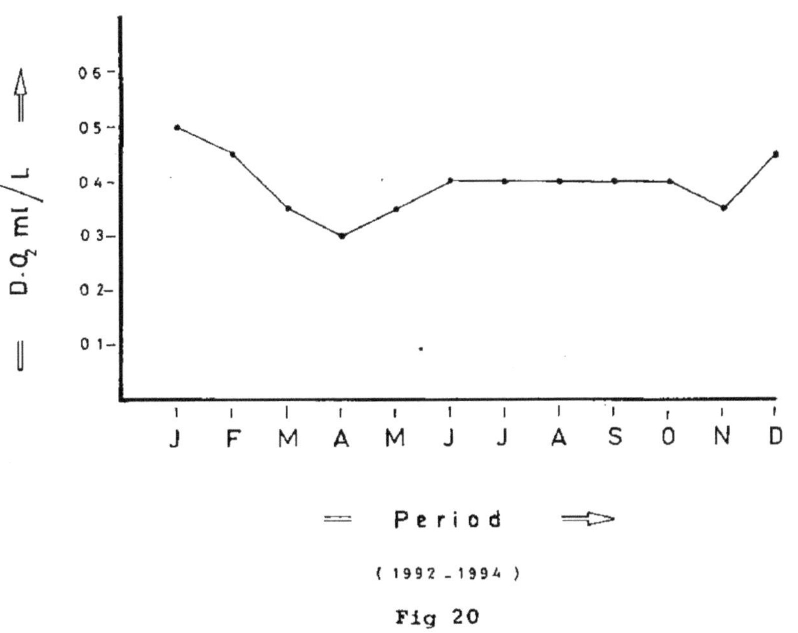

Fig 20

Genral trend of surface dissolved oxygen.

14. Recommendations:

The findings of the first exploratory experimental cage culture presented in this report are indicative of a good commercial venture in this direction. However, the `crude' technology adopted for the present experiment needs to be refined and upgraded through extensive field trials incorporating the unknown or lesser known biological parameters of the promising species. Although, this is a mammoth task, in order to have a better economical prosperity through optimum utilization of vast inherited coastal areas including innumerable bays, lagoons and backwaters endowed by nature, we shall continue further in this direction to evolve a complete technology of our own.

-oOo-

The Project Boat

GOVT. OF INDIA
D.O.D

Epinephelus areolatus

Cephalopholis microprion

Cephalopholis boenak

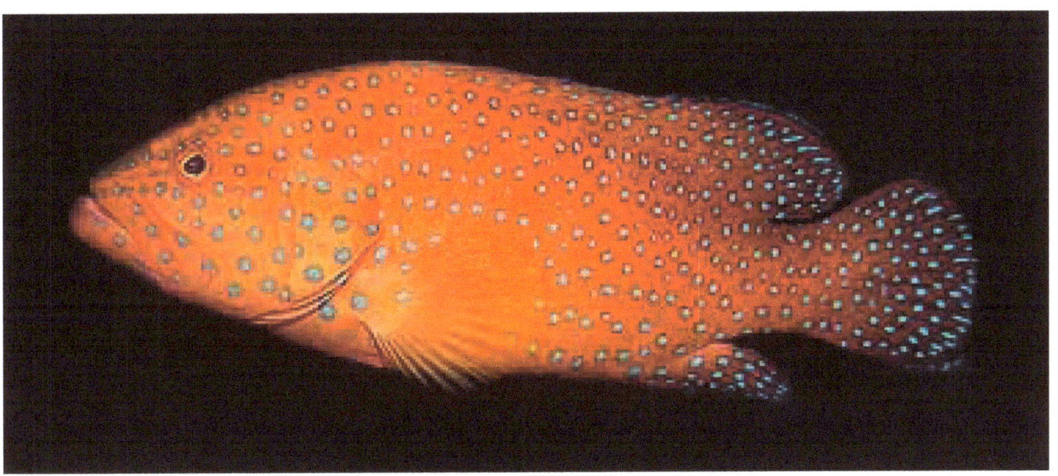

Cephalopholis miniata

Sigamus vermiculatus

GOVERNMENT OF INDIA
A & N CENTRE FOR OCEAN DEVELOPMENT
(DEPARTMENT OF OCEAN DEVELOPMENT)

NEAR ABERDEEN JETTY
P.O. ABERDEEN BAZAR
PORT BLAIR: 744 104
A & N ISLANDS.

A
REPORT ON

BARREN ISLAND VOLCANO

(General Oceanographical Observations)

JANUARY – 1992

<u>For bibiolographic purposes this document may please be cited as follows</u>

Arif M. Mustafa,1992 - A report on Barren Island Volcano - (General oceanographic observations)
(Tech. report No.01/92-ANCOD)
A & N Centre for Ocean Development,
(Department of Ocean Development)
Port Blair - 744 104.

ACKNOWLEDGEMENTS

Thanks are due to the Secretary to the Govt. of India, Department of Ocean Development for granting permission to participate. Also thank the Hon'ble Lt. Governor, A & N Islands with whose help this expendition could take place.

BARREN ISLAND THE ONLY ACTIVE VOLCANO OF INDIA

INTRODUCTION

India has two volcanic islands in the Bay of Bengal namely Barren and Narcondum. The former has been dormant till the beginning of last year and the later is an extinct volcano. Geologically, these two volcanoes belong to the general Sunda group lying on the Neogene Inner Volcanic arc evolved as a result of eastward subduction of the Indian Ocean lithosphere below the SE Asian plate.

The Barren Island Volcano is known for resurgent eruptive volcanism, recent activity of strombolian nature started during March 1991 at a time when Mt.Pinatubo of Phillipine and Mt.Unzen of Japan was also active, beside being an unique phenomenon of the century, this happened to be the first opportunity for the scientific community of independent India to study an active volcano of their own. In this report we are proud to present the first ever coastal ocean related eco-metmorphosis assessment gathered during the active volcanism.

In addition to the two volcanoes mentioned above we also have two prominent sub-merged volcanic sea mounts sitting on the floor of Andaman sea basin, the Alcock SMT is located just 60 km. east of Barren Island and the Sewell SMT 200 km. east of Car Nicobar.

Location:

Barren Island is located at 12° 17' 30" N latitude and 93° 50' 30" E longitude in the northern part of Andaman Sea in the Bay of Bengal. It is about 135 km north east of Port Blair.

Physiography

The uninhabited Barren Island is almost circular in shape, about 3 km. in diameter, 10 sq km. in area and slightly tapering towards the west. It is quite steep with sloping surface raising about 350 m. above mean sea level and has the appearance of a truncated cone from a distance. Due to its steep topography, any conspicuous rocky, sandy and muddy beaches are lacking on the island.

Genesis:

There are couple of theories about the evolution of this island. The most widely accepted one is that, initially an eruption took place in the sea about 1 to 2 million years back during late - to post - Pleistocene period which resulted in the emergence of this island. This phenomenon is considered to have occured in the late Tertiary period and a giant volcanic cone was developed encompassing the whole island. Subsequently, this giant volcanic cone was blown out and the ejected debris and ash materials were deposited on the relict cauldron as pyroclastics, the lava was of `aa', `ropy' and `scoriaceous' types.

Geology:

Geotectonically both the islands of Barren and Narcondum lie on the Miocene to recent Volcanic arc, which proceeds from Mount Popa and Wuntho of Burma in the north and swerving south easterly into the active volcanoes of Indonesia and Sunda region.

It is belived that, during the geological time the existing valcanic cone was developed near the central part of the blown-off cauldron and lava pile of high-alumina olivine basalt was erupted out from the main crater as well as from the three subsidiary vents developed in the north eastern, north western and southern face of the then existing valcanic cone about 50 m. down the main crater and filled up the valley of the amphitheatre. The lava pile finally flowed down into the sea towards the western side of the island.

The last eruption of 1803 was on the relict part of the older valcano and deposited pyroclastics. The SiO_2 content is between 45% to 53.5% (basalt). The older deposited lava sediment is olivine basalt followed by high - alumina olivine basalt whereas the present lava is olivine bearing basaltic andesite. At the prime of present activity the circumference of the crater was about 60 m. at top and 25 m. at base and the depression between 10 to 15 m. deep. The surface temperature of the vent at that time was approximately 100 °C. (Dr. D.Haldar, Personnel Communication). The present volcanic activity which lasted for about nine months from March to November has considerably altered the topography, the height of Valcanic Cone has been reduced from 305 m. to 225 m. above mean sea level and the size of crater has been enlarged many times. The earlier landing site was totally engulfed by advancing lava and at that point a steep about 20 m. heigh lava bed has encroached upon sea, whereas a new crescent cove about 60 m. long has emerged some 200 m. south of previous landing site.

Volcanic history:

According to the records available, during 1795 Capt. Blair observed enormous volumes of smoke and frequent showers of red-hot stones. "Some where of a size to weigh three or four tons, and had been thrown some hundred yards past the foot of the cone. There were two or three eruptions while we were close to it; several of the red-hot stones rolled down the sides of the cone and bounded a considerably way byeond us ... Those parts of the island that are distant from the volcano are thinly covered with withered shrubs and blasted trees".

A few years later Horsburgh records an explosion every ten minutes and a fire of considrable extent burning on the eastern side of the crater. In the next thirty years subterranean forces had considerably diminished in activity and at the end of that period only volumes of white smokes with no flames were to be seen. Drs. Mouat and Liebig, who visited the island within few months of each other in 1857, wrote respectively of volumes of dark smoke, and clouds of hot watery vapour. In 1866, a whitish vapour was emitted from several deep fissures and about 1890, steam was seen to be issuing at the top from a sulphur-bed, which was ..quid and pasty, and a new jet was coming from a lump on the sloping side of the cone; while the sole evidence of activity to be observed was the deposition of sulphur and a escape of steam that often condensed on the surface rocks. There are no authentic records available thereafter. In the third week of March 1991, the volcanic activity was observed first by the Navy, as a follow-up the Inspector General of Police, A&N Islands made an independent aerial survey followed by another survey by Deputy Inspector General of Coast Guard on 6th April, 1991 which confirmed the resurgance of Volcanism at Barren Island. The Lt. Governor made an aerial survey in the first week of May, 1991.

FIRST SCIENTIFIC EXPEDITION:

As the news regarding some geological activity in the dormant volcano of Barren island spread, the Hon'ble Lt. Governor, A&N Islands on a request by one visiting Indian reserach scholar from Oxford University directed to constitute a team comprising of geologist and marine biologist to visit the volcanic island and submit first hand observation report to the A&N Administration. Accordingly, the staff of A & N Centre for Ocean Development, Port Blair, a nescent island centre of the Department of Ocean Development, New Delhi along with the team of GSI geologists and the research scholar from Oxford sailed for Barren Island by mv.Moti under thd command of Cdr. Tilak R.Dewan, Director of Shipping Services on the 11th May 1991, at 9 am. from Phoneix Bay jetty. The sea was calm throughout the journey, visibility was excellent and the Barren Island along with a long streak of smoke cloud moving towards west became visible only a few minutes after crossing the English Island. The team anchored in the vicinity of Barren Island by 5 pm. an OBM fitted boat was lowered and the team moved ahead to land, as we were only about 100 m. away from the shore, strong showering of valcanic glass/dust/ash particles started, which scared some of the team members and party returned without landing. The dormant valcano was active spitting lava from an auxilliary vent located in the north eastern side about 50 m. down the main crater. The discharge was pulsating with rumbling noice. The ejected boulders were returning on the slope and rolling down. The smoke was thick dark having white patches at regular interval. The smoke column was about 900 m. high from the top of the valcano. The surrounding forest was burning and charred smell was spread around.

The team was summoned by the Hon'ble Lt. Governor on 13th May 1991 for an open debate wherein the Hon'ble Lt. Governor ordered for the second scientific expedition on 15th March 1991. Here it may not be out of context to mention that same morning the Hon'ble Lt. Governor was leaving for New Delhi and was scheduled to return by 15th May 1991. Despite his busy schedule he gave opportunity to the scientific community to come for an open debate with him hardly an hour before his departure. Later the Incharge, ANCOD was informed that while at Delhi he called on Prof. V.K. Gaur, Secretary to Govt. of India, Deptt. of Ocean Development and discussed about this phenomenon also.

SECOND SCIENTIFIC EXPEDITION:

The team comprising of 12 persons (Annexure I) left Port Blair by M.V Tarmugli on 15th May 1991 by 6 p.m. the sea was calm with little swell throughout. Visibility was poor at dusk, Barren became faintly visible from about 5 km. The vessel anchored at 6.10 am. about 500 m. away from the lone light becom on western coast at a depth of 300 m. On the left side the island terrain was almost vertical rising straight up from the sea, the landing point on the right side next to the 12 m. heigh navigational light becon was a small sandy patch of about 50 m. length, on the other side sloping terrain was highly uneven rising upto 250 m. Suddenly a whale (*Physeter catodon* ?) appeared in the black water and greated the party with fascinating jet spray and disappeared.

The team landed on shore with the help of a life boat, the

volcano almost lying opposite to the landing beach was dangerously violent continuously throwing out glowing boulders and dense fumes with intermittent rumbling. Surrounding vegetation was partially charred. Entire area was burried under thick volcanic ash and the edge of advancing super heated lava river was about 500 m. from the sea shore. Amidst this catastrophic scenario a weak and trembling cry of a baby she goat (Capra hircus) attracted the attention of the team, the lone starving baby was rescued and sent on board. For a close observation of the cone, we went up the charred hill on the left side climbing upto a height of 200 m. suddenly black rain drenched all, making it difficult to photograph and stay at top.

Marine Science Observation:

Considering the prevailing conditions and limited time the semi enclosed bay where the party landed and some approchable coastal areas were studied. On the north-western side it was a stratified vertical land mass rising upto 100 m. extending encompassing the volcano. The coast is fringed with huge boulders meeting a 50 m. long sandy patch then continuing. The length of the coast surveyed was 1.1 km.

Physical characteristics

10 stations were taken with a depth range of 1 to 5 m. (Fig.2).

1. **Colour and smell**

At first sight in the subdued light, surface water was greyish muddy which slowly turned to blackish brown in the sun. The water smelled of sulphur.

2. **Visibility**

It was measured using standard Sacchi disc and was found to be 10 cm. at noon.

3. **Surface slicks**

The slicks indicating the sub-surface drift was found to appear and disappear at short interval,it was difficult to trace out the entire change which occured during our stay, however,the position as in the morning and evening visualised from the bridge of the anchored vessel is given in Fig. 2.

4. **Coastal current**

Apparently the drift during the stay was from the sandy patch and beyond towards the vertical land mass on the north western side. Two members while surveying were caught in strong surface current and drifted for more than 100 m. before getting hold of submerged rocks. Almost at the tip, a huge whirlpool was seen churning (Fig. 2) apparently fed by coastal conventional current.

5. **Sedimentation**

On the sandy patch the fine sand deposit was more than 60 cm. The top layer was thoroughly mixed and stirred with the black dust cinder giving it an unusual blackish appearance. The submerged area was totally covered with loose lapilli/ dust cinder/ash particles frequently raining therein. Manually it was assessed to be 18 cm. thick. Average dust cinder/volcanic glass partical size was 3 mm. The particles were found sensitive to magnet. Maximum deposit was on station nos. 2,7 and 9.

6. **Surface temperature**

It was recorded for all the 10 stations. Pronounced difference at short interval was observed. Temperature range was 28° to 32 °C.

Stn. No.	1	2	3	4	5	6	7	8	9	10
Surface Temp. °C.	32	32	31	30	30	28	31	32	30	32

Sub-surface temperature

In the absence of a proper instrument, an ordinary thermometer was used and the sea floor temperature was found to be near 50 °C at station nos. 1, 2, 7, 8 and 9.

Stn. No.	1	2	7	8	9
Temperature °C			50		

Surface water salinity

This was measured using salinity refractometer. The salinity ranged from 34% to 36%,readings recorded at all the stations are given below :

Stn. No.	1	2	3	4	5	6	7	8	9	10
Surface Salinity %o	35.5	35.5	34	34	34	34	36	36	36	36

Specific gravity

This was determined through Tetra marine test meter kit and is given below :

Stn. No.	3	4	5	6	8	9	10
Specific gravity	1.022	1.022	1.022	1.022	1.027	1.027	1.027

Chemical characteristics:

1. pH value

It was measured through `pH test kit -CA-1' manufactured by Marine Enterprises Inc. U.S.A (range 6-12) Significant shift in pH value was observed, the phenomenon was pronounced for about 500 m. over an extended tongue towards SE of Barren Island. The observations made for the 10 marked station as shown in Fig.2 was as under:

Stn. No.	1	2	3	4	5	6	7	8	9	10
pH value	6	6.2	6.2	6	6.2	6	6	6	6	6

Normal Sea water pH is relatively constant around 8.0 due to the buffering action of the Carbonic acid system in the photic zone. pH range is usually 7.8 to 8.3. The drop in pH value is attributed to the high sulphur emission from volcano which subsequently disturbed the ionic balance.

The acidic circle prevailed in the immediate vicinity encircling the volcano as shown in Fig. 3. Further the acidic tongue was spread over 500 m. SE of Barren Island as marked over the same figure. Although the value over tougue is quite near to normal still it is an unusual phenomenon.

Biological characteristics

1. Pisceans

No fish with the exception of Rock skipper _Andamia_ sp. was seen. The rock skippers were seen adhering to the rocks in the littoral region, stock was localised, small and scanty; diminished alertness and poor reflex on approach was observed.

2. Mollouscs

Localised, dense stock of rock, mussel, _Crassostrea_ sp. and _Saccostrea_ sp. along the surf beaten rocky coast was seen. Moderate population of _Cellana_ sp. and _Nerita_ sp. was also present on boulders. One specimen each of _Chiton_ sp. and _Patella_ sp. was also seen.

3. Arthropodae

Large number of dead shore crab _Pelocarcinus humei_ (Wood Mason) of family Geocarcinidae were found all along the coast. Many crabs were seen vaccating the shore area and moving landwards climbing up the ash covered mountain range towards south-western side. In order to observe this unusual phenomenon of landward migration, we followed the migrating crabs uphill, they were found upto 250 m. possible in search of a safer and suitable place to live. The migrating stock was large, exhibiting low level of alertness with deminished reflex. Remarkable changes in the body colouration was also observed, the crabs encountered were pale yellow/creamy in colour, no distinct variation in colour as is observed over carapace and chelate legs was seen here. Even on being nocturanal they were wandering in the open at day time.

4. **Coelentrates**

The coral community.was restricted to a small area, near the landing point. This small community was seen growing onto the intertidal rocks. According to the survey carried out on this rocky patch, 9 species related to 5 genera namely Acriporidae, Pociiloporidae, Favites, Favia and Porites were recorded.

Diary of subsequent expeditions:

- 26.6.1991 - Through Coast Guard vessel; the Volcano was dangerously eruptive, the 12 m. heigh light becon was completely engulfed by the advancing lava river, which was further enchroaching upon the sea, the sea-lava interface was producing enormous steam. The synergetic affect of steaming from sea and fire fountain from the crater was an horrifying experience. The surface water temperature about 100-150 m. from the old landing site was 35 °C, water pH still acidic.

- 10.7.1991 - Through Coast Guard vessel; mouth of the spatter cone considerably enlarged. Due to voilent monsoon condition no analytical observations were made.

- 1.11.1991 Through Police boat; a well known German Volcanologist Prof. Dr. Ing. Peter Halbach & Dr. Margret Hälbach were taken to Barren island. The main crater was found to have collapsed, volcano was not emitting any more lava or smoke, the coastal topography was found to be totally changed, 20 m. heigh lava front was found standing much ahead of the earlier landing point, at the sea lava interface the water temperature was 72 °C & pH. 6. Weathering of the condensed magma was in progress, the coastal zone was devoid of any life form, a new crescent shaped landing site about 60 m. long developed some 200 m. south of previous landing point. We disembarked at the new site, the air temperature was recprded to be 38 °C; due to intense heat, constant steaming and emission of sulphur dominated gases we were able to stay only for few minutes.

- 4.1.1992 - Through Coast Guard vessel; Volcano almost silent with the exception of intermittent gas emission, two `driblet' lava cones were found developed on the solidified lava river, solfataric activity was found continuing at some places. Sea surface water temperature at interface was 29 ˚C & pH 7.6, marine algae was found developing over certain rocks, no other marine or terristerial life form with the exception of the crab <u>Pelocarcinus</u> <u>humei</u> was seen, this crab was surprisingly found to be present in good number right from the water edge upto the foot of Volcano, strange creatre !

. * * *

ANNEXURE - I

List of participants of the second scientific expedition to the active Vilcano of Barren Island.

1. Lt. Governor, Lt. Genl.(Retd.) Ranjit Singh Dyal.
2. Secretary to L.G. Mr. Manoj Parida.
3. Director of Shipping Services, Cdr. TR Dewan(IN).
4 Incharge Andaman Division of GSI. Dr. D.R.Haldar.
5. Incharge ANCOD, Mr. Arif M.Mustafa.
6. Research Scholar from Oxford University, Miss Zohra Fatima.
7. Asst. Geologist from GSI, Mr. J.K.Biswas.
8. Zoological Asst. from ZSI, Mr. P.T.Rajan.
9. One Diving Asst. from ANCOD Mr. Raju.
10. Two Photographers.

BARREN ISLAND
N
DENSE MIXED FOREST
183m
Landing Point
FLAT LAND
Volcano
274m
DENSE MIXED FOREST
ANDAMAN SEA
Not To Scale

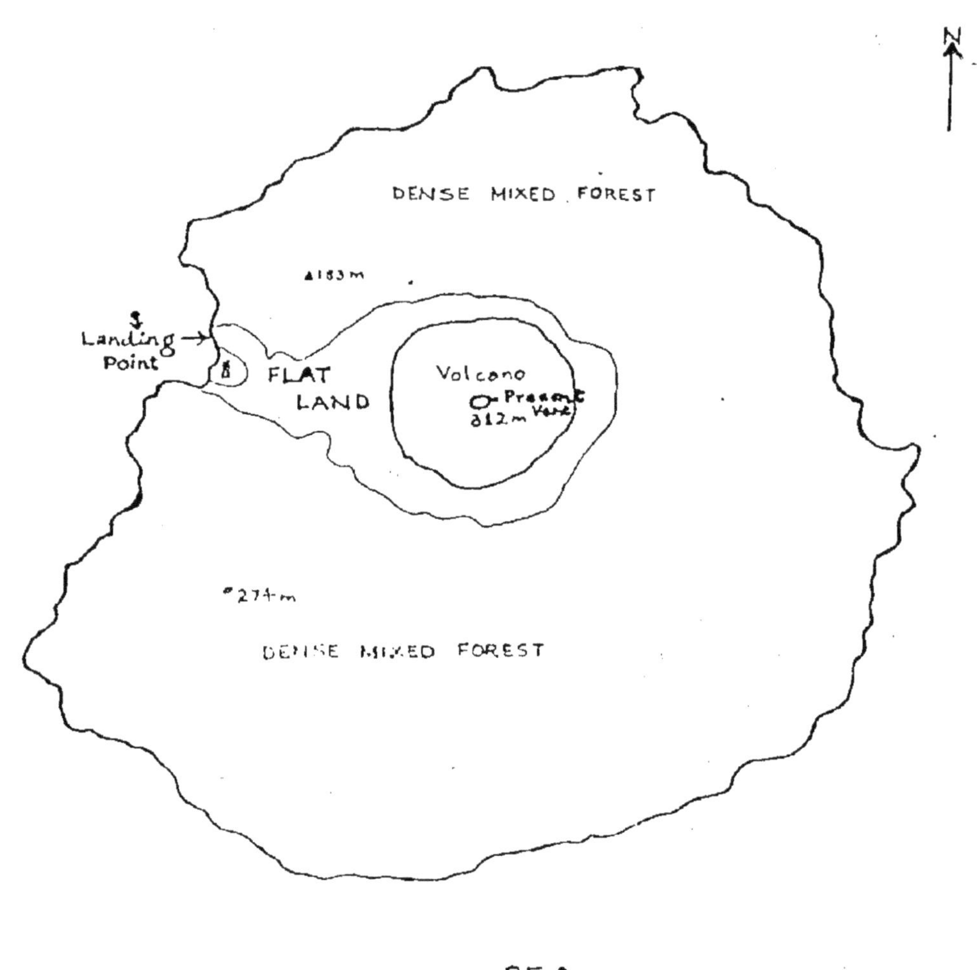

FIG. 1

N
ROCKY
COAST
BOULDER
ZONE
SANDY
PATCH
BAY
S1
7am.
S1
7am.
S2
3pm.
5m.
1.
2.
3.
4.
5.
6.
7.
8.
9.
10.
OLD LAVA FRONT
LIGHTHOUSE
Not to Scale
INDEX
STATIONS
SLICKS
DEPTH CONTOUR
SURFACE DRIFT
WHIRPOOL FORMATION
PROTRUDING ROCKS

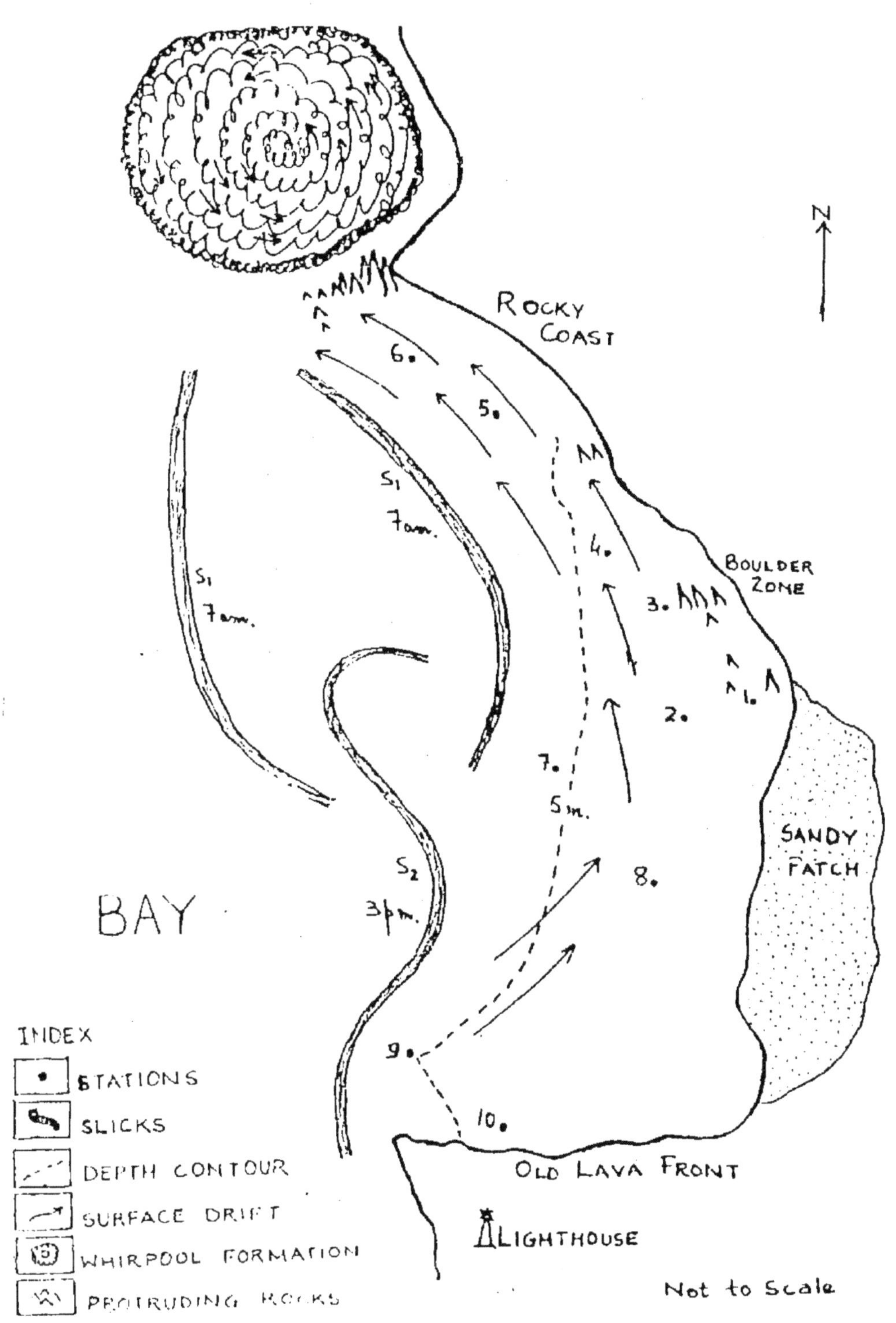

FIG. 2

pH
(6-6.2)
Acid line
12°,17',30".
Previous landing point
Recently developed landing point
Edged rocks
Danger line
Hot Spring
Volcano
Barren Island
S O U T H A N D A M A N
Dense mixed jungle
SOUTH ANDAMAN PROTECTED FOREST
High water line
7.2
7.2
7.2
7.2
7.2
7.2
7.6
7.6
7.6
Acidic tongue

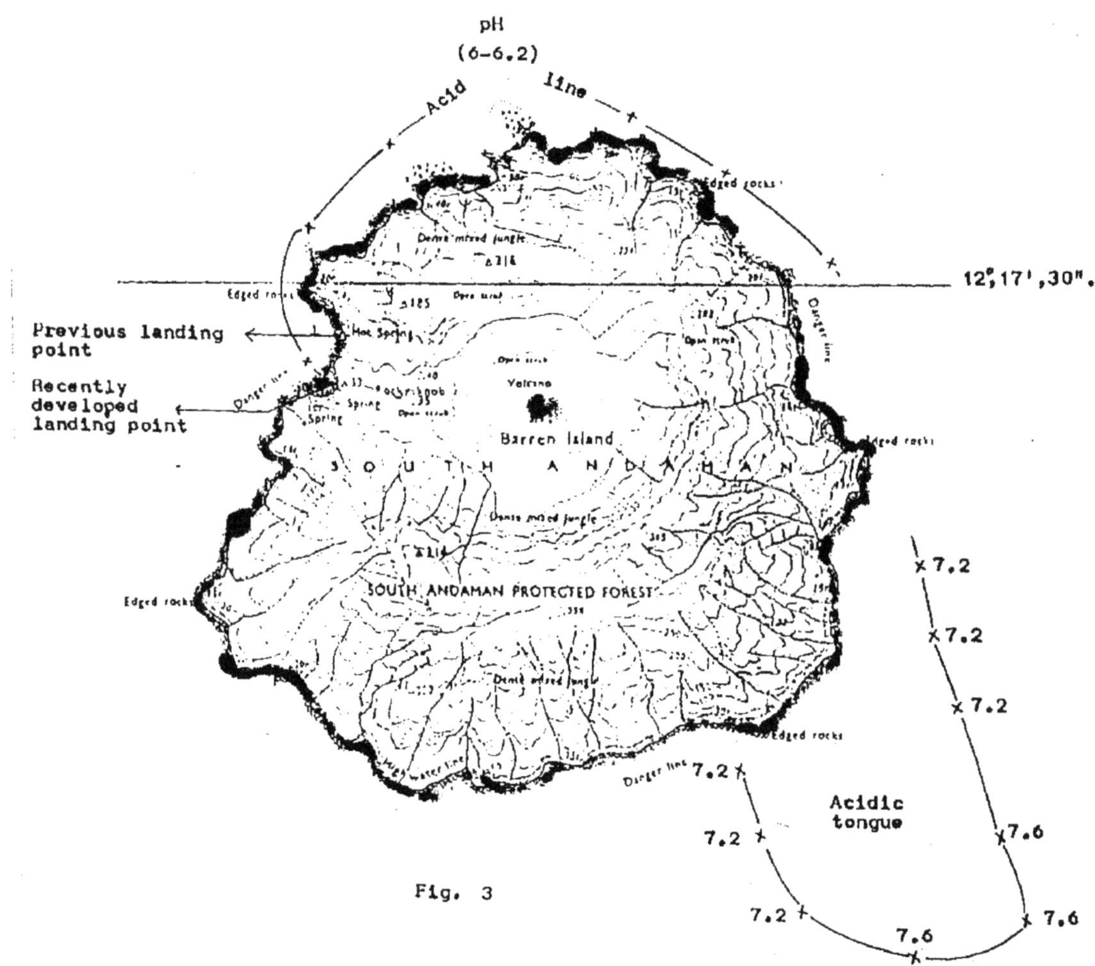

Fig. 3

Reprint : Journal Sci. & Tech. "**Islands on March-1991". Page No. 69—70**

SEX AND SPAWNING OF CORALS AT PORT BLAIR

ARIF M. MUSTAFA* MATHEW C D'SILVA**

INTRODUCTION

Propagation is essential for any living organism to perpetuate its own population. which is achieved through broad spectral reproduction techniques developed and perfected in the course of evolution. A biological multiplication could be sexual, asexual or parthenogenetic depending upon the prevailing physiological and environmental conditions within the limit of genetic freedom of the concerned living being. In the animal kingdom much is known about this essential aspect of life but still much more remains to be discovered and it is more so when the organism is aquatic and lives through shallow coastal waters upto the dark abysal region of the ocean; such as corals representing a primitive group of animals. They are believed to reproduce sexually as well as asexually and present an excellent example to unviel the shrouded mystery of reproductive mechanism. Scleractinians are the stony reef building corals of the World, commonly occuring along continental and insular regions of the tropical seas. They may be either unisexual or bisexual, the reproductive organs namely, the testes and ovaries are formed within the mesoglea of the mesenteries and remain attached encircling the base of pharynx. In many coral species, the gonad development is annual. However, a clear picture of sexual reproduction was not available till the beginning of the last decade. The earlier belief was that the sexual reproduction is a rare event and the main mode of propagation is through asexual 'replication' of the polyp. The sexual reproduction was thought to be internal resulting in the ovulation of brooded larvae called planulae. At the same time, the scientific community was not agreeing in toto to this regular and irregular mode of propagation by asexual and sexual reproduction respectively, the argument put forward was being that, why there were many instances of sudden appearance of large number of planulae and mass annual ripening of gonads in many reef builders? As to help solve these questions a team of devoted coral biologists made some intensive underwater observations in the Great Barrier Reef off Australia and by early 1970 came to the conclusion that the corals exhibits an annual sexual reproductive cycle wherein the fertilisation is external.

RECENT FINDINGS

Encouraged with these results large scale field and laboratory observations were made and it was finally concluded that mass spawning of corals is taking place on the Great Barrier Reef. The findings of wallace et al (1986) reveal that "spawning events are spread over the third to sixth nights after full moon. Lunar/tidal cycles determine the date of spawning, which occurs during the period of least difference between successive high and low tides. This is a period of very low water exchange over the reef and probably is important to the spwaning corals as a time during which the high concentration of eggs and sperm necessary for good fer-

*Department of Ocean Development, A & N Field Station.
**Director, Piscean Enterprises, Bombay.

tilization rates is maintained for the longest time possible. Shortly before releasing their reproductive products, the polyps in the corals can be seen setting. The area around the polyp-mouth becomes distended by the presence of eggs and sperms, which have been gathered within the polyp, most often into a compact ball. Since coral tissues are semitransparent, the brightly coloured gamete ball can be seen within the swelling. The coral can remain in this setting state for about an hour. Then suddenly in some cases rapidly in others, the bundle is pushed through the polyp-mouth and released, Gamete bundles begin to stream upwards from the colony, to join those released from other colonies nearby. In corals with separate sexes, clouds of eggs or sperm are released". These findings unveiled a japanese mythological belief also regarding the occurance of pink slicks appearing in their coastal waters in late spring or early summer said to be the menstrual waters of the dragon princes and was called' **punitsu**'. Further it was confirmed that gametogenesis begins early in the year and the ripen gonads are lodged within the mesenteries, the entire colony could either be male, female or hermaphrodite depending on the species. Male and female gonads develop as carrot-shaped or rounded bundles of sperm and as strings of eggs respectively, ultimately taking the shape of a ball.

ANDAMAN OBSERVATIONS :

The Andaman and Nicobar Islands are known to have a fairly rich coral fauna but such a reproductive phenomenon has never been recorded. During a survey programme for Crown of Thorns star fish infestation on corals, the author came across many ripen gonads in scattered colonies of Faviidae and Acroporidae around Port Blair and in the light of recent reports a constant and close watch was maintained on some colonies located in a radius of about 2 km. at North Bay, off Ross Island and at Sisotris Bay· These reefs were selected due to their proximity to the town. The surface water temperature showed a rising trend after a drop in January, the average surface water temperature recorded at that time being 28.7°c. The major constrain in this work was the non-availability of an underwater camera to keep a recrord of the anticipated events of three small colonies of **Favites abdita** which were brought to shore and maintained in a plastic pool. Ultimately on 25th February, 1989 i. e. on the thrid day of their captivity exactly at 6.40 PM. IST all the coral colonies started simultane ously releasing gametes. Pink balls were shooting up and bursting, it was all fascinating. In order to have photographs of the event, the coral colonies were transferred to a large glass beaker amid the happening and photograph taken (Figs. 1 & 2) Immediately there after, we dived at the nearest Sisostris Bay reef but the water was not very clear making the observation a bit difficult. However, slow spawning was noticed in two species of Acroporidae, while bigger colonies of Faviidae were found empty. We snorkled upto 8.00 p m. but it all stopped much earlier. The surface water of plastic pool turned pink and was full of small granules. These contents were released back into the sea. Thus, these findings constitute the first spawning record of stony corals in Andamans and confirm the findings made by Australian scientists on Great Barrier Reef. The intresting point to be noted here is that there was a Solar eclipse on 26th Jan. 1989 followed by Lunar eclipse on 9th Feb. 1989 and 25th Feb. 1989 was being the 18th day of the new moon.

ACKNOWLEDGEMENT

The author is thankful to Dr. G. S. Rao Officer-Incharge, Zoological Survey of India, Port Blair for going through the manuscript.

REFERENCE

C.C. Wallace, R C. Babcock, P.L. Harrison, J. K. Oliver and B. L. Willis - 1986. Sex on the reef : mass spawning of corals. Oceanus; 29 (2) : 38-42.

Spawning of Corals -
Annual Male Female Gamates release

J. Andaman Sci. Assoc. 6 (2) : 177-180, December, 1990

Linuparus andamanensis, a new spear-lobster from Andamans

ARIF M. MUSTAFA*
Department of Fisheries, Port Blair-744 104

Hitherto the spear lobster of the genus *Linuparus* of the family Palinuridae is known to have three species viz., *L. trigonus* Von Siebold, *L. sordidus* Bruce and *L. somniosus* Berry & George. The genus has a fairly large distribution in the seas around the world, but has not yet been reported from Indian waters. Geographical distribution for the three known species is as follows. *L. trigonus* : East Africa, Philippines, Australia, China, Taiwan and Japan. *L. sordidus* : South China Sea, New South Wales and Australia and *L. somniosus* : East Africa. The genus is mainly characterised by the inflexible and straight antennal flagella and the frontal horns fused to a broad 2 or 4 spined median projection on the anterior margin of the carapace between eyes. Examination of the material of spiny lobster collected by the author in Andaman Sea showed that the species did not agree with any of the known species. Hence, it is assigned to a new species *L. andamanensis* and described in the present paper.

The abbreviations used in the text for principal measurements of the new lobstar using the terminology of Fischer & Bianchi (1984) are : BL - Body length from the anterior marginal spine of carapace to the posterior tip of telson, CL - Carapace length from the anterior marginal spine of carapace to the median anterior margin of submarginal posterior groove, TL - Tail length from the median anterior margin of submarginal posterior groove to the prosterior tip of telson, TF - Tail fan from the transverse grove of VIth abdominal segment to the prosterior margin of telson, AP-Antennular peduncle and AF-Antennular flagella. All measurements are given in centimeters.

Linuparus andamanesis n. sp. (Figs. 1-2)

Material examined : 3 specimens, 2 male, 1 female, BL 26-38. 5.

Type locality : (i) Off the east coast of South Andaman 92° 52′ EL, 11° 37′ NL, depth 415 m, rocky sandy bottom, 27.4.85, 1 male, an incidental catch by the author on bottom set long line for deep sea dog sharks (ii) Off the west coast of South Andaman down to Little Andaman 92° 19′ - 30′ EL, 11° 12′ - 55′ NL, average depth 400 m, muddy bottom, 30.1.1990, 1 male and 1 female, bottom trawl catch by F. V. Champion.

Diagnostic features : Colouration dorsally ventrally and reddish, laterally creamy.

Holotype : Male, loc. (i), BL - 38.5, CL - 14.5, TL - 24, TF - 7, AP - 9, AF - 25.

Allotype : Female, loc. (ii), BL - 30, CL-11, TL-19, TF-5.5, AP-7.5, AF - 17.5.

Paratype : Male, loc. (ii), BL-26, CL-10, TL-16, TF-5, AP-6.5, AF-16.5.

Distinctive characters of the new species are given below :

1. Submarginal prosterior groove of carapace much wider medially than laterally (Fig. 2A).

* Present Address : Senior Technical Assistant, Department of Ocean Development, A & N Islands Field Station, Port Blair-744 101.

Fig. 1. *Linuparus andamanensis* n. sp. A - male, holotype, B - female, allotype, C - male - paratype.

2. No pleopods on first abdominal segment of female.
3. Epistomal ridges coarsely granulated, anterior median tooth not well developed, paired normal spines occur on either side of epistomal groove (Fig. 2C).
4. Smooth chitinous margin of male genital aperture throughout its length (Fig. 2D).
5. Frontal horns small and paired (Fig. 2B).
6. Vth abdominal segment furnished with single median granulated ridge, whereas VIth with two mid-lateral ridges along

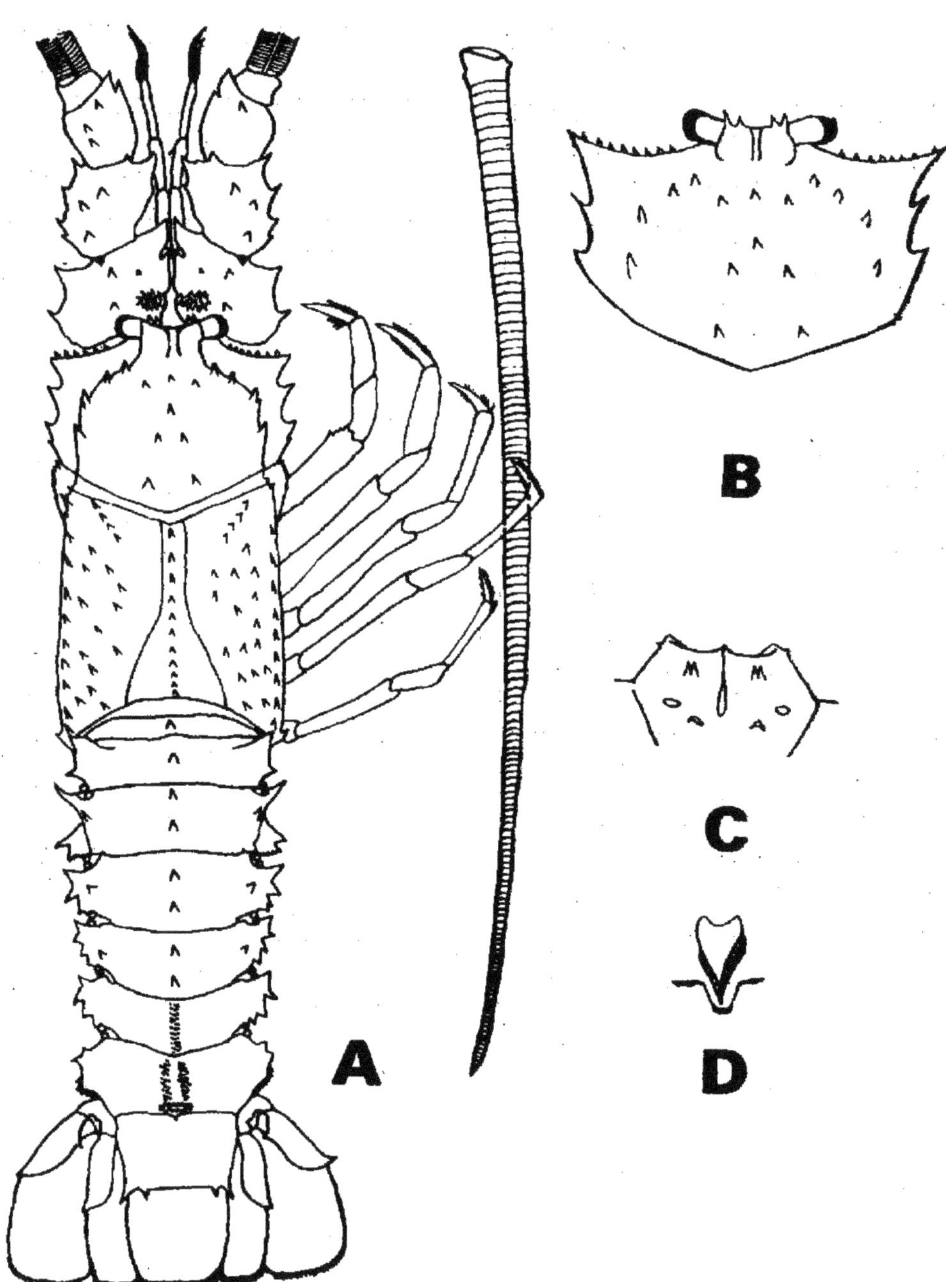

Fig. 2. *Linuparus andamanensis* n. sp. male A - adult, dorsal view, B - Carapace, dorsal view, C - epistomal area' D - genital aperture.

with a small transverse ridge posteriorly (Fig. 2A).

7. Anterior half of telson calcareous with proximal paired marginal spines, posterior half membranous (Fig. 2A).

Among the known species of genus *Linuparus*, the material examined comes closer to *L. somniosus* in having a medially wider and laterally narrow submarginal posterior groove on carapace, but the absence of pleopods on Ist abdominal segment of female is the character exibited by *L. sordidus* and *L. trigonus*. Smooth chitinous margin of male genital aperture disagrees with *L. sordidus* and *L. trigonus*. Presence of normal paired spines on either side of epistomal groove along with coarsely granulated epistomal ridge have an ill-developed antero-median tooth bringing it nearer to *L. sordidus*. However, the smooth chitinous margin of male genital aperture throughout its length clearly differentiates the new species from all the known ones. With the description of this new species, a key to all the four species is deducted here following Berry and George (1972).

Key to the species of *Linuparus* White.

1. Submarginal posterior groove of carapace as wide medially as laterally.

   No pleopods on first abdominal segment of female. ... 2

   Submarginal prosterior groove of carapace much wider medially than laterally.

   (i) Vestigial pleopods present on first abdominal segment of female ... *L. somniosus*

   (ii) No pleopods on first abdominal segment of female ... 2 (ii)

2. Epistomal ridges coarsely granulated, without an acute well developed anterior tooth

   (i) Chitinous margin of male genital aperture with toothed median border and entire lateral border ... *L. sordidus*

   (ii) Chitinous margin of male genital aperture smooth throughout its length ... *L. andamanensis* n. sp.

   Epistomal ridges feebly granulated, with an acute well developed anterior tooth. Chitinous margin of male genital aperture toothed throughout its length ... *L. trigonus*

ACKNOWLEDGEMENTS

The author is grateful to Dr. T. C. Khatri, Lecturer in zoology, J.N. Govt. College, Port Blair, for help in the preparation of this manuscript and to Dr. M. Kathirvel, Senior Scientist, MRC of CMFRI, Madras, in providing necessary literature on the spear-lobster. Thanks are also due to Dr. G. C. Rao, Scientist-in-Charge, Z. S. I., Port Blair, for going through the manuscript.

REFERENCES

Berry, P. F., and R. W. George, 1972. A new species of the genus *Linuparus* (Crustacea : Palinuridae) from South-east Africa. *Zool. Meded.*, 46 (2) : 17-23.

Fischer, W. and G. Binachi, 1984. *FAO species identification sheets for fishing area 51*, Vol. V.

Linuparus sp.
Deep Sea Lobster

J. Andaman Sci. Assoc. 6 (1) : 63-65, June 1990

Increasing environmental stress on the coral reef ecosystem around South Andaman

ARIF M. MUSTAFA

Department of Fisheries, Port Blair-744 101

The renowned coral reef ecosystem sprawling upon the continental shelf of the Andaman and Nicobar Islands in the Bay of Bengal has lost much of its domain in recent years in shallow coastal waters (Mustafa *et al.*, 1987). Pillai (1983) pointed out siltation as one of the major causes for the degradation of shore corals in these Islands. The condition was further aggravated by human interference, disturbing the balanced ecosystem through deforestation, trade collection, tourism, navigation and mining. An attempt was made in the present study to ascertain the factors responsible for the distribution of coral reefs in these islands. For this purpose, the physical parameters as siltation and turbidity and the biological factor as the predatory crown of thorns starfish, were studied.

The two fringing reefs one located near Port Blair at Sisostris Bay on the east coast and another at New Wandoor on the west coast of South Andaman, were selected for study. The Sisostris Bay is located at 92° 45′ EL and 11° 40′NL, having a small fringing reef of 1500m^2 of which 3/4th of the reef flat towards western shore is dead. The live front has the coral *Porites* sp. as a dominant variety over reef crest and beyond, the huge colonies located towards the South Point measure 4-5 m in diameter. The reef has also branching corals as *Acropora* sp. and *Pocillopora* sp. mixed with a few scattered plate corals. Fish fauna inhabiting the reef is poor and comprises of Electric blue damsel (*Pomacentrus melanochir*), Yellow damsel (*Abudefduf sulphur*), Vagabond butterfly (*Chaetodon vagabundus*) and Long nosed file fish (*Oxymoncanthus longirostris*), etc.

The fringing reef off New Wandoor (92°, 36′EL and 11° 35′NL) stretches 2 km along the coast from the northern tip of Alexandra Island lagoon upto Andromeda Point with an average width of 500 m. The reef supports a large number of stony corals, minimonospecific stands of *Acropora nobilis* and *A. stoddarti* towards the southern side opposite to Grub Island and near Andromeda Point. The reef crest has the capping of massive colonies of *Porites* sp. Non-reef builders as *Gorgonia* sp. are also present. Reef slope is not pronounced. Fish fauna is better represented compared to Sisostris Bay reef and consists of Rainbow butterfly (*Chaetodon trifasciatus*), Vagabond butterfly (*C. vagabundis*), Threadfin butterfly (*C. aurega*), Lined butterfly (*C. lineolatus*), Banner fish (*Heniochus accuminatus*), Sharp nosed puffer (*Canthigaster solandri*), Stripped parrot (*Scarus* sp.), Undulate trigger (*Balistapus undulatus*), Sailfin tang (*Zebrasoma veliferum*), Lined sweetlip (*Gaterin liniatus*), Lunar wrasse (*Thallosoma lunare*) and Clown (*Amphiprion* spp.). In addition, the calm (*Tridacna* sp.), sea anemone (*Radiantus* sp.) and sea cucumber (*Holothuria* sp.) are also found in good number.

The transparency of sea water was measured by a standard Secchi disc of 20 cm for a period of one year between 10.30 to 11.30

AM. For measurement of siltation, coral plates of 13 cm were wrapped in polythene bags, brought to the surface, silt was washed and its amount estimated by drying and weighing method. For silt estimation 12 samples were randomly collected on 9th and 15th September, 1988 at the withdrawal of first monsoon. The crown of thorns starfish was sighted on the coral reefs through toe snorkelling.

The transparency of sea water ranged from 190 to 530 cm with a gradual increase from December 1987 through February 1988 and a gradual decrease in the following months, showing its minimum value in August. An increasing trend was again observed from September to December. The average silt deposition in various samples of coral plates (*Montipora* sp. and *Acropora* sp.) was found to be 1.46 mg/cm². The first abnormal population of crown of thorns was sighted off New Wandoor on 2nd December, 1988. These observations were later extended to Grub, Red Skin and Tarmugli Islands. These starfishes occurring in groups numbered 5, 18, 44, 46, 96, 102, 108 and 132. The size measured *in situ* for 32 specimens ranged from 26 to 33 cm.

In the course of evolution of the animal kingdom, corals have developed an intimate commensalism with marine algae without which their flourishment in coastal waters is not possible. For their synthesis, energy is supplied through solar radiation. The transparency of sea water has a direct bearing on the development of algae and in turn of the corals. The poor visibility in these waters from 190 to 530 cm indicates the hindrance in penetration of light. Gradual increase in transparency from December onwards coincided with the pre-monsoon period. With the onset of monsoon and the increase of rainfall, the Secchi disc reading fell upto 190 cm in the month of August possibly due to heavy precipitation brought by the run-off from the catchment areas in both the reefs (Table 1).

Table 1. Transparency and rainfall from December 1987 through December 1988 at Sisostris Bay, Port Blair

Month	Transparency in cm	Rainfall in mm
December	430	285.5
January	490	000.4
February	530	024.7
March	470	029.5
April	470	087.9
May	400	274.3
June	390	263.0
July	260	602.8
August	190	422.9
September	320	748.5
October	360	274.4
November	390	561.6
December	470	043.1

The run-off in deforested areas has more silt compared to the forest protected catchment areas. The quantitative analysis of the deposited silt of New Wandoor clearly indicated a heavy deposition of the silt. The run off from adjoining fields sprinkled with artificial fertilizers enriched the nutrient content of the coastal waters, resulting in a bloom of the phytoplankton and an increase in the population of the crown of thorns starfish (Birkeland, 1982). The greenish water during the monsoon showed that the phytoplankton flourished more during that season than in post or pre-monsoon season. Both these factors i.e.

organic matter and dense phytoplankton are considered favourable for a rapid development of crown to thorns, which was rare between 1978 and mid-1988. It was only on 2nd December 1988 when hundreds of the crown of thorns starfish were observed feeding on corals over a strip of about 2 km off New Wandoor. The starfish is nocturnal in habit (Lucas, 1986), but the over cast sky with some drizzling at noon might have confused the animal to come out from their hiding places. Later observations of the animal recorded in small but dense groups mostly on plate corals, at Grub, Red Skin and Tarmugli Islands also supported the large scale sightings of off New Wandoor. However, this population explosion of the crown of thorns was not fully understood, although it appeared that the particular pattern of rainfall was responsible for their sudden outbreak (Birkeland, 1982).

The main causes for this heavy run-off from adjoining areas are cutting of the trees from coastal areas and nearby hills for human settlement and agriculture, indiscriminate removal of mangroves and sand from New Wandoor, Old Wandoor and the adjoining coast. The death of 3/4th of reef flat at Sisostris Bay seemed to be due to the influx of heavy siltation from Aberdeen Bazar, Atlanta Point, Raj Niwas Hill, Gymkhana Ground, Namak Bhatta and South Point, causing suffocation and resulting in the death of coral polyps. The same factors were also found responsible for the death of marginal corals on reef flat and mid-reef off New Wandoor. These findings are indicative of an increasing environmental stress on the coral reefs around South Andaman and is therefore a matter of great concern for those who are interested in the conservation of this fragile and unique oceanic ecosystem.

REFERENCES

Birkeland, C. 1982. Terrestrial run-off as a cause of outbreak of *Acanthaster planci*. *Mar. Biol.*, 69 : 175-185.

Lucas, J. 1986. The Crown of Thorns starfish. *Oceanus*, 29(2) : 55-64.

Mustafa, A.M., Dwivedi, S.N. Yamini, M. Warwadekar, Abidi S.A.H. and E.K. Raveendran 1987. Endangered coral reefs of Bay Islands and their ornamental fishes. *Proc. Symp. Management of Coastal Ecosystem and Oceanic Resources of the Andamans. J. Andaman Sci. Assoc.*, 60-65.

Pillai, C.S.G. 1983. Coral reefs and their environs. *Bull. Cent. Mar. Fish. Res. Inst.*, 34 : 36-40.

Reprint: *Journal Sci. & Tech.* "Islands on March-1990". *Page No. 104—106*

CORAL REEF DEGRADATION, CONSERVATION AND MANAGEMENT IN ANDAMANS-AN OVERVIEW

ARIF M. MUSTAFA*

ABSTRACT

This paper updates knowledge on the status of coral reef and the causes of reef damage in the oceanic Island of Andamans. Recent scientific investigations are indicative of reef degradation due to magnifying environmental stress on this fragile oceanic ecosystem primarily due to increasing human interaction. Natural causes are cyclone, geological dynamics and biological imbalance such as recent reporting of coral eating Crown of Thorns star fish. Existing legislation is required to be updated to preserve the coral ecosystem for posterity.

INTRODUCTION

It has now become clear that the problems of environmental protection are inseparable from the problems of economic development. They are in fact two sides of the same coin, two aspects of the same goal, one reinforcing and underpinning the other. It is then just a matter of prudent management to approach environmental protection as a dimension of economic development. We desire neither an immaculate and pristine environment for its own sake, nor all out economic development at the other extreme which extracts a heavy price in terms of degraded environment. Much will depend on the prudent management of coral reef resources considering the capacity of the system to generate essential renewable resources within the limit of Island condion. These Islands, compared to most oceanic territories have seemingly good resources and have experienced rapid growth in the rate of exploitation. Such unisectoral overuse is slowly producing various environmental problems.

Important reef degradation problems in the Andaman and Nicobar Islands were identified and reported earlier (Alcock,1893; Pillai, 1983; Mustafa, 1987). Additional documentation on each problem is presented here, in the hope that future directions for monitoring, management and conservation may be indicated in a new dimension.

Human activities causing reef degradation.

1. Tourism:

The aesthetic appeal of these Islands is attracting large number of tourists from within India and from abroad. During 1988-89 about 37,500 tourists including 2438 foreigners visited these Islands. It is estimated that about 99% of tourists comming from abroad and 75% of the domestic tourists get into the sea to have a glimpse of tropical icthyo-beauties and the coral formation. Such unescorted and unguided groups of reef gleaners most of them learning to snorkel, are responsible for large scale destruction of corals in shallow near shore waters due to trampling, walking and breaking of corals for fun. Even the SCUBA and skin divers are seen disturbing and damaging the reef inhabitants. Impact of such indiscriminate activity has caused severe damage in localised pockets. Example of such merciless exploration is seen all around the small Islands open to tourists. Under water survelance and briefing prior to such trips is the

*Supdt. of Fisheries-Cum-Curator, Deptt. of Fisheries
Andaman & Nicobar Administration

only way to minimise the extent of damage due to direct interferance.

2. Collection or reef invertebrates :

The illegal collection and export in decorative corals and shells continue to be a major problem. Such over exploitation of invertebrates makes the reef weak and suseptable to many infestation and disturbs the ecological balance. Many invertebrates help dessipate the accumulating silt. The depletion of Trochus sp. Turbo sp. and many varities of cowries Cypria sp. all over Andamans has been noticed. Strict survelance and sea ranching are the only ways left to protect and replenish the invertebrate resource.

3. Siltation :

Situation appears to be the chief indirectly human induced cause of large scale reef degradation. During 1988 the annual range of transparency at Sisostris bay was found to be 190 to 530 cm. and the amount of sedimentation on laminar corals off New Wandoor was 1.46 mg. $cm.^2$ (Mustafa, 1990.) The French research vessel 'Calypso' has reported that the reefs around Labyrinth group of island and Ritchies Archipelago are partly covered with mud coming from adjoining land (Sarano, 1989). Major causes of siltation are deforestation for settlement and agriculture, land clearing for construction and dredging of soil in the harbour area. The agencies engaged in such works shall be educated in coastal area development works.

Natural Cause of reef degradation.

1. Cyclones and Geological dynamics :

Gradual evolution and adaptation through geological time have allowed a dynamic balance to develop between reef growth and maintenance and the multitutde of destructive processes. Occurance of the latter is discussed here. Immediately after the devastating cyclone of December, 1987 large quantities of freshly broken coral colonies were found heaped and scattered near the shore around southern part of south Andaman. Most significant collection in the intertidal region was on the two kilometre coastal stretch beyond Rangachang upto Chidiatapu. Eight varities of stony corals were recorded from the heap.

'Calypso' reported total destruction of South coral bank and large scale damage to Middle coral bank due to cyclone which occured some twelve years ago i.e. during 1977 or so. Both the coral banks are about 21 km. off the west coast of Andaman Island at a depth of 15 to 20 m. Considering the depth and the type of formation it is assumed that all the three banks on west namely North, Middle and South used to be a barrier reef during glaciation time but got sub-merged due to tectonic movement.

2. Biological interaction :

Threats to the survival and maintenance of coral reefs are also posed by their biological components. Examples are competition of corals among themselves and with other reef inhabitants, boring of coral colonies by various organisms, disease and parasitism. The coral out crops of wandoor Marine National Park are suspected to have 'White band disease' (Wood, 1989) but it requires detailed pathological analysis for confirmation. However a factor which can achieve tremendous importance is predation, such as by voracious coral eating Crown of Thorn star fish, the star fish could feed on live corals at a rate of 300 sq. cm per day (Campbell, 1976). Since Dec., 1988 localised congregation of this star fish around south Andaman has come to notice and adequate preventive measures has been taken by the Administration to lessen the damage. In a single operation during March, 1989 a total of 523 numbers of adult Crown of Thorns were manually destroyed from the fringing reef off New Wandoor. Subsequent observations showed re-occurance of few and then total absence of the star fishes from the area. This phenomenon of sudden appearance in December, 1988 and then a total disappearance during the summer of 1990 on that particular reef is not yet understood.

Discussion

The International Union for the Conservation of Nature in world conservation strategy has identified coral reefs as an "essential ecological life support system necessary for human survival and sustainable development. The coral reefs benefit mankind by :

1. Providing excellant nursery grounds for the growth and development of young ones of many commercial marine fishes.

2. They yield many shallow water fishes, molluscs, crustacears and echinoderms both edible and ornamental providing nurishment and playing a significant role in many maritime nation's economy.

3. Antileukemia, antitumour, antimicrobial and ultraviolet blocker biochemicals of high medicinal importance are extracted from reef inhabitants.

4. Reef glamour support tourism.

5: Reefs protect seaward erosion of coastal areas.

6. Coral reefs provide support and sustenance to other coastal ecosystem upon which people depend.

7. Coral reefs are one of the most productive ecosystem on earth and one reef alone may support as many as 3000 species of living organisms.

For these Islands it is assessed that about 2000 sq. km of the continental shelf around Andaman and Nicobar is covered with corals, in theory it is producing about 24,600 tonnes of biota per annum (Mustafa, 1987), and about 50% of Island fish production is from coral reef areas, but as these Islands are just the tips of sub-merged mountainous range having narrow continental shelf, productive zone is limited. Potential productive areas are adherant mangroove and the coral reef ecosystem. Coral reefs are in fact the 'generators' of life for such oceanic Islands. This coral reef ecosystem is very fragile and easily disturbed when pushed beyond its limits. It is felt that coral reefs have low standing crops of exploitable species possibly due to the cryptic nature of much of the reef fauna and as such could support only subsistance consumption. Grigg (1979) postulates that sustained harvest from coral reef ecosystem may not be much greater than 10% of standing crop. Hence in order to have long term benefit for the Islanders it is advocated to leave the coral reef areas only for subsistance level utilisation so as to allow the replenishment of renewcable resources with in the frame work of a sound and healthy conservation and management policy, thus improving the quality of life for the inhabitants of A & N Islands.

REFERENCE

Campbell, A.C. 1976 The coral seas. *Orbis publishing Ltd. London.* pp 128.

Grigg, R.W. 1979 Coral reef ecosystems of the pacific Islands : issues and problems for future management and planning. *FWS / OBS-79 / 35. p6-17, U S. Fish and Wildlife service.*

Pillai, C.S.G. 1983 Coral reefs and their environs. *Bull. Cent. Mar. Fish. Res. Inst.* 34 : 36-40.

Mustafa, A. M., S.N. Dwivedi, Yamini M. Warwadekar, S. A. H. Abidi and E.K. Raveendran 1987 Endangered coral reefs of bay islands and their Ornamental fishes. *Proc. Symp. Management of coastal ecosystem and oceanic resources of the Andamans. Andaman Sci. Assoc.*, 60-65.

Wood, E. 1989 Coral mortality on reefs in the Wandoor Marine National Park. *Report to INTACH.*

Mustafa, A.M. 1990 Increasing environmental stress on the coral reef ecosystem around South Andaman. (In press).

Sarano, F. 1989 Observations en plongee de la faune marine des iles Andaman. *Foundation Cousteau, Paris.* pp. 42.

Degraded & bleached Corals and the Coral Reefs

REPRINTED FROM

ADVANCES IN AQUATIC BIOLOGY AND FISHERIES

PROF. N. BALAKRISHNAN NAIR FELICITATION VOLUME

PUBLISHED BY

PROF. N. BALAKRISHNAN NAIR FELICITATION COMMITTEE

DEPARTMENT OF AQUATIC BIOLOGY AND FISHERIES

UNIVERSITY OF KERALA, TRIVANDRUM - 695007 INDIA

Advances in Aquatic Biology and Fisheries, 1987, pp. 207–224.

A VIEW TOWARDS A BLUE REVOLUTION IN ANDAMAN AND NICOBAR ISLANDS SEA-PRESENT STATUS AND PROSPECTS

ARIF M. MUSTAFA
Department of Fisheries, Andaman & Nicobar Islands, Port Blair, India

S. N. DWIVEDI
Additional Secretary, Department of Ocean Development, New Delhi, India

&

S. A. H. ABIDI
Director, Department of Ocean Development, New Delhi, and
former Director of Fisheries, Andaman & Nicobar Islands, Port Blair, India

ABSTRACT

The paper outlines the great potential for fisheries development in the seas around the Andaman and Nicobar Islands. History of fisheries development in the territory has been indicated together with informations on the crafts and gears in use. Details of the aquaculture programme undertaken, the species composition and production potential of the resources in the pelagic and demersal regions, have been highlighted. The need for greater governmental support for fisheries development has been stressed.

INTRODUCTION

The archipelagoes in the Bay of Bengal comprising 585 lush green islands sprawling between latitude 6° 45' N & 13° 45' N and Longitude 92° 15' E and 94° E is called the Union Territory of Andaman & Nicobar Islands (Fig. 1).

According to 1981 census the population figures at 1.07 lakhs and it is estimated that nearly 25% (provisional) of the total land area is inhabited. The islands have a coast line of 1962 kms (Anon, 1983) which is about one fourth of the total coastline of India and has an Exclusive Economic Zone (E. E. Z.) of 5,96,554 km^2 (Anon, 1977) which is 30% of the total E.E.Z. of India. The area has achieved special significance with the discovery of natural gas and oil in the offshore region. The living and non living resources potential is very vast and it is almost unexploited with the exception of shell resources. The annual fish landing is around six thousand tonnes. Major developments in oceanic fishery and mariculture with large capital investment, long term planning and appropriate management strategies could boost the economy of these islands and bring a "*Blue Revolution*".

OCEANOGRAPHY

The interest in oceanographic research in the Andaman Sea (area 6.02 x 10^5 km^2, volume 6.6 x 10^5 km^3) dates back to 1869 when Francis Day, a well known army officer and fishery biologist, visited these islands. But the most recent oceanographic investigation was when, the National Institute of Oceanography undertook two cruises (51 and 52) from 29 January to 27 February 1979 and three cruises (66, 67 and 68) from 27 December 1979 to 7 February 1980 on R. V. Gaveshani.

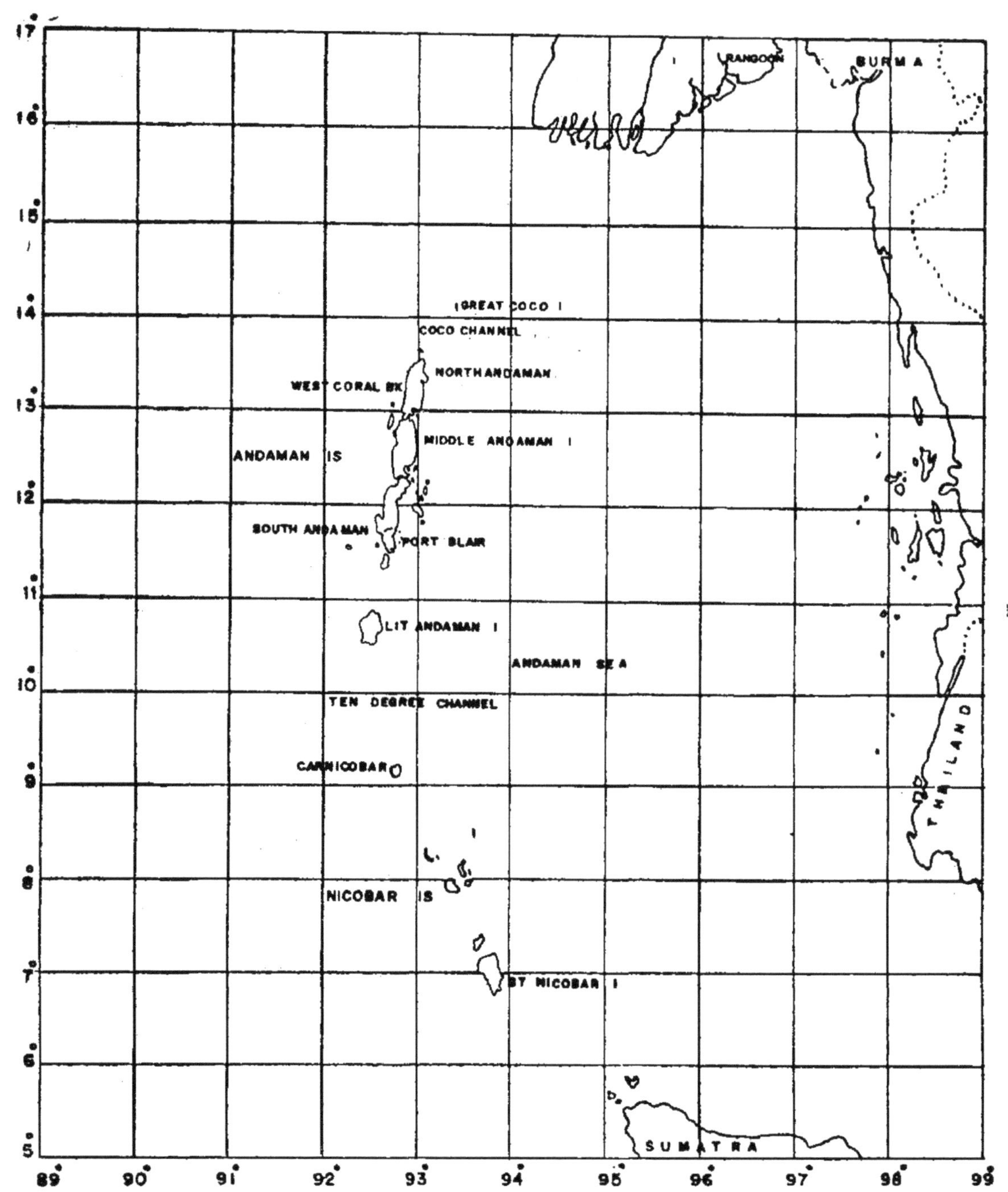

Fig. 1. Andaman and Nicobar Islands

Physical characteristics

The Andaman sea has very uneven bottom topography. It receives very large and variable quantities of fresh water and is partially isolated from the Bay of Bengal by the island arcs. The Andaman sea is connected, through the Strait of Malacca, to the South China Sea. During winter, intense evaporation of surface water occurs in this region. The southern parts of this sea are areas of intense air-sea interaction, well known for the frequent cyclones which originate in this region. Low density surface water enters the Andaman Sea in the north-western region through the Preparis Channel. At the same time in the south-eastern region similar surface waters enter the Andaman Sea from the Suda Sea through the Strait of Malacca. It is reported that the latter spreads westwards through the Great Passage into the Bay of Bengal as it does towards the northeast near the Ten Degree channel. In the area in between, its influence can be seen in the vicinity of the Kamorta-Nancowri islands. Part of the mixed surface water reaches the eastern side of the Ten degree channel and a little South of it, between the Car Nicobar and its southern neighbours (Rama Raju, *et al.*, 1981)

The thickness of the surface mixed layer varies, in general from 25 to 80 m. The sea surface temperature varies from 27 to 28.5°C and salinity from 31.87 to 32.15 ‰ with an increasing trend from north to south on the western section in the case of both and from south to north on the eastern side in the case of salinity.

Presence of a tongue of relatively warm water (28.4°C) starting before Car Nicobar and extending beyond Great Nicobar was observed by Rama Raju *et al.* (1981). At depths of about 1500 m and below the waters on the eastern side of the Andaman islands are warmer than on the western side.

The surface waters around the little Andaman Island flows towards the south-south-west from the north-north-east and deepening of mixed layer on the south-western side of the island results not only as a 'lee' effect but also lends evidence to the fact that the waters on the western side of Andaman island as a whole are slightly different from those on the eastern side.

Chemical characteristics

Sea water is supersaturated with oxygen at the surface and few metres below at most of the places on eastern Andaman sea. But on the western side supersaturation with oxygen was observed even upto 50 m.

The values for oxygen, phosphate-phosphorus and nitrate-nitrogen at few places are relatively higher than those recorded for the corresponding depths in the Arabian Sea. The only exception is silicate-silicon which is lower in the entire Bay of Bengal than in the Arabian Sea. These indicate that the oxidation of organic matter in depths of Bay of Bengal, including the Andaman Sea proceeds at a faster rate resulting in quick depletion of dissolved oxygen. Faster rates of oxidation and oxygen depletion result in the sinking of organic matter in undecomposed or partially decomposed state. (Sen Gupta *et al.*, 1977, 1981). Since the Andaman Sea is a confined physiographic basin, flow to open ocean areas of Bay of Bengal occurs through the channels around and between Andaman and Nicobar

Islands. During January to April the flow through Malacca Strait is in the north-western direction and is the strongest. The surface circulation is controlled by the monsoon system of the northern Indian Ocean. During January to April anticyclonic circulation is established in the Bay of Bengal and the Andaman sea. Circulation pattern indicated the flow of water from Bay of Bengal to Andamn sea through the Preparis Channel. This southward flow leaves the Andaman Sea near 10°N channel. The flow is increased further due to the westerly current originating from Malacca strait. Ramesh Baby & Sastry (1976) have reported that basically the flow consists of cells (cyclonic and anticyclonic of 100-200 km size) which eventually lead to upwelling and sinking.

Productivity

On the basis of the available data Qasim & Ansari (1981) made an assessment of the production at various levels in the Andaman sea, which is as follows:

Primary production

Average primary production of the Andaman Sea (Bhattathiri & Devassy, 1981) was estimated to be 273 mg C/m^2/d. At this rate, the primary production of the entire Andaman Sea (area 6.02 x 10^5 km^2) would be approximately 6 x 10^7 tonnes of Carbon/ Year.

Secondary production

Average zooplankton biomass (Madhupratap *et al.* 1981) of the Andaman Sea, estimated as zooplankton carbon is 288.8 mg C/m^2 Assuming that about 50% of the biomass gives rise to daily rate of secondary production the total production of the next trophic level for the entire area of the Andaman Sea would be of the order of 31.73 x 10^6 tonnes of Carbon/ Year. The 50% figure is based on a series of conversions from biomass values to secondary production which range from 3 to 98% (Mullin, 1969).

Tertiary production

For the open ocean 0.1% of primary production and 1% of secondary production are to be taken for calculating the tertiary production. The mean of the two gave a value of 18.70 x 10^4 tonnes of Carbon/Year. In terms of live weight the figure would be 18.70 x 10^6 tonnes/year (using factor 10). Therefore, 25% of potential yield could be taken as the exploitable stock. This would amount to 4,70,000 tonnes of fish/year (Qasim & Ansari, 1981). Salient oceanographic features of the E. E. Z. of the Andamans & Nicobar regions is given in Table I.

FISHERIES

The major developments in the area of fisheries development in the Andaman Nicobar islands are summarised in Table II. The marine fish landing in the Andaman and Nicobar islands and the relative share of the different capture techniques in the total production are presented in Fig. 2.

Seventh five year plan (1985–90) outlay for fisheries is Rs. 405 lakhs and the major thrust is on the mechanisation of fishing fleet, expansion of cold storage facility, incentive to private enterpreneurs and aquaculture.

TABLE I. **Oceanographic features of the Andaman and Nicobar Islands sea (Source : N.I.O. Goa)**

Sl. No.	*Region*	*Exclusive Economic Zone*	
	Parameter	*Western Section (Bay of Bengal)*	*Eastern Section (Andaman sea)*
1.	Surface temperature	27 ° to 28.5°C	28° to 28.5°C
2.	Thermocline	– 200 to — 250 m	—180 to 280 m
3.	Deep water temperature	3°C at — 2000 m	5.1°C at – 1750 m (over central region)
4.	Isothermal layer	- 44 to — 58 m	—26 to – 50 m
5.	Surface salinity	31.87‰ to 32.15‰	31.20‰ to 32.15‰
6.	Dissolved Oxygen		Less than 0.2 ml at — 1300 m (over central region)
7.	Oxygen saturation value		
	a) At surface	— 3 to + 15%	— 2 to + 8%
	b) At — 25 m	+ 1 to + 14%	— 8 to + 5%
	c) At — 50 m	— 5 to + 10%	— 63 to + 5%
8.	Surface current	2% of the surface wind velocity	
9.	Pollutants		
	a) Petroleum Hydrocarbon	Well within the permissible limit of global range	
	b) Mercury		
10.	Dissolved chemicals	Nearer the coast Nitrate–nitrogen and Phosphate-phosphorus are absent, the absence of the former in general acts as a limiting factor for the productivity of this region.	

The fleet consists of 950 country crafts of traditional origin, 101 O. B. M. fitted boats and 20 in-board mechanized boats of about 12m length, besides the crafts of aboriginals and tribals. This fleet is equipped with 709 gill nets, 26 anchor nets (Boat Seine), 580 cast nets and 19 shore seines, beside innumerable sets of hooks and lines. There are about 2,860 fishermen, out of which 2,225 are full time, 440 part-time and 195 occasional. The inshore fishery has a total capital investment of only about Rs. 1.5 crores. Fisheries Survey of India, Port Blair, is conducting offshore fishing experiments for assessing the demersal stock and for finding out the best method and gear.

The catch is generally sold in retail market by fishermen themselves and seldom sold to middlemen. Marketing is done by vendors from door to door or sold at road side. In case, a nearby market is availbale, catch is manually lifted or transported by bus to the nearest market. In the Union Territory, only one proper fish stall is available through which departmental catch is sold.

TABLE II. **Major events of fisheries development in the Andamans**

Year	*Major event*
1908	Survey by the Trawler "Golden Crown"
1947	"Anda-marine Development Corporation" was set up under the technical guidance of British experts. Unfortunately due to various bottlenecks it worked only for 9 months
1949	A "Fisheries Research Unit" was established at Port Blair.
1950	Licences were issued to Chinese firms for seasonal shell fishing on royalty basis and one Singapore based private firm was permitted to undertake commercial fishing. Both of these contracts were shortly abandoned due to various reasons
1955	"Fisheries Research Unit" was changed into "Fisheries Development Unit"
1971	A unit of "Exploratory Fisheries Project" was set up at Port Blair
1975	"Fisheries Development Unit" was upgraded into "Directorate of Fisheries"
1978	Deep sea fishing was organised by Indo/Tata/Thai joint venture for a period of about two years
1979	a) Mariculture potential survey by the Central Fisheries Research Institute, Cochin
	b) Oceanographic survey by the National Institute of Oceanography, Goa
1980	Proposal to Government of India to set up a "Fisheries Development Corporation"
1983	Inland aquaculture gains considerable momentum with the perfection of the induced breeding technique
1984	Discovery of potential deep sea spiny dog fish shark resources
1985	Intensive mechanisation and establishment of cold chain under 7th plan

The fish is generally marketed afresh and only 15% of the total annual landing is salted or sundried with the usual defects of poor salting, rancidity and high sand content. Average cost per kg. of good quality fresh fish is Rs. 12/-, Rs. 5/- for low quality small fish and Rs. 20/- for dry fish. Average cost for fresh prawn is Rs. 15/- whereas dried ones cost Rs. 20/-. A fraction of dried fish is exported to mainland along with cured shark fins and liver oil.

Against this background the infrastructural and shore based facilities did not develop much and far from satisfactory but are steadily being built up now. One ice and cold storage plant with a rated capacity of 15 tonnes of cold storage and 5 tonnes of ice production per day was erected in 1968 in the vicinity of Aberdeen Jetty, which is the only plant of its kind. Anchorage and berthing facilities are available a lmost in all the ports, except Car Nicobar, which is being constructed. Repair facilities to a limited extent is also available for the fishing fleet at Port Blair with a battery of newly constructed slip ways and a dry dock, besides an old dry dock of the Marine Department. A training centre was set up in 1977 to train the local youth in mechanised fishing; every year 30 candidates are trained for a 9 months certificate course.

Details of traditional, tribal and aboriginal crafts are given in Table III while that of the traditional gears are provided in Table IV.

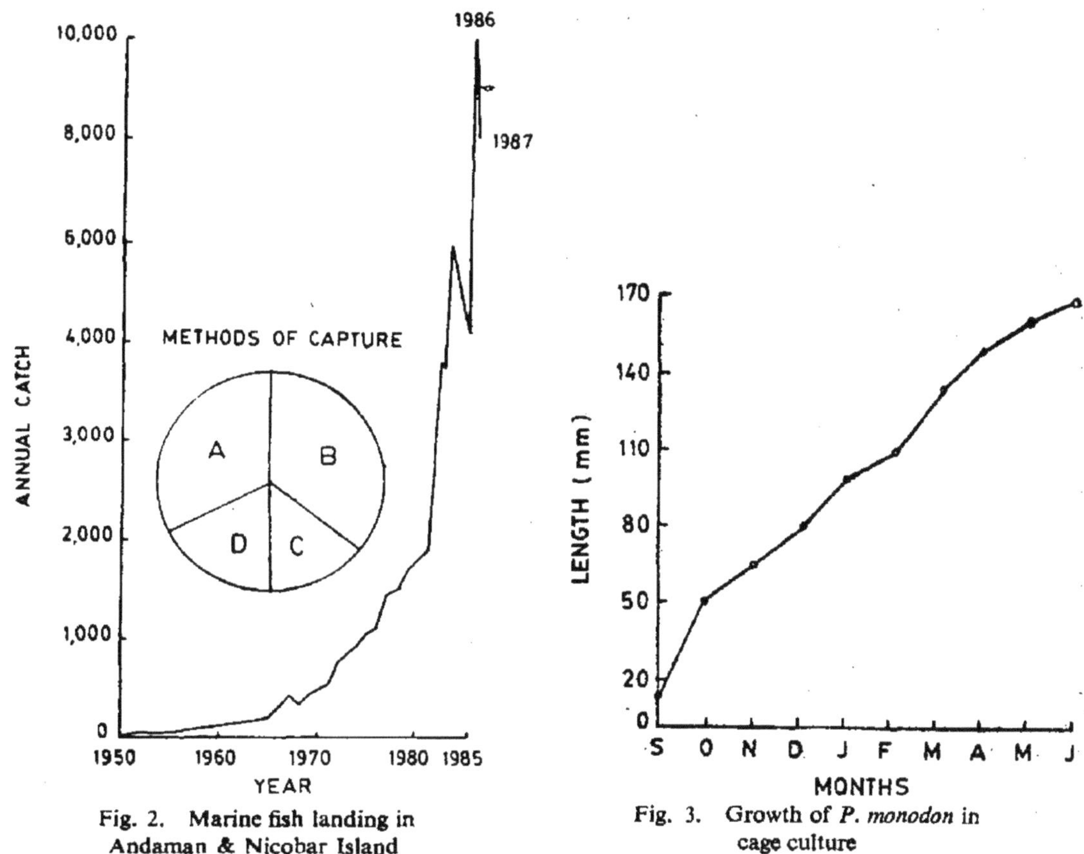

Fig. 2. Marine fish landing in Andaman & Nicobar Island

Fig. 3. Growth of *P. monodon* in cage culture

AQUACULTURE

A brackishwater tank was constructed in early 60's at Tellarabad. The scheme worked for sometime and was then abandoned due to the non-availability of technical manpower. Then, after a long gap, the scheme was revised along with a mariculture pilot project. The brackishwater tank at Tellarabad was reconditioned and cage/pen culture of tiger prawn was undertaken, in September 1979. 5,000 Nos. of *Penaeus monodon* and 100 Nos of *Chanos chanos* seed was airlifted from C. I. F. E.'s brakishwater fish farm, Kakinada to Port Blair, which took about 40 hours. Monthly growth rate of *P. monodon* in cage culture is depicted in Fig. 3 (Abidi *et al.*, 1981 unpublised).

During the period of culture mean average salinity of water was 26‰ and water temperature was between 28 to 32°C. The trend of growth was quite encouraging. Growth of *C. chanos* and *Mytilus viridis* is also found to be good.

In 1985 the fisheries research division of C. A. R. I. has established a one hectare brackiswater farm and the Department of Fisheries has acquired about 2 hectares of coastal land for the construction of a brackishwater farm. However, brackishwater farming is yet

TABLE III. **Details of traditional, tribal and aboriginal fishing crafts**

Sl. No.	*Type of craft* / *Parameter*	*Flat bottom*	*Round bottom*	*Burmese dug-out*	*Tribals*	*Aboriginal*
1.	Material	Wood	Wood	Wood	Wood	Wood *Sterculia villosa*
2.	Planking	Carvel	Carvel	Dug out without side planking	Dug out without rigger arrange ment	Dug out without rigger arrangement
3.	L. O. A. (cm)	655	564	1067	305–1067	400–1500
4.	Draught forward (cm)	50	71	—	—	—
5.	Draught Amidship (cm)	47	89	25	28–30	30–33
6.	Draught aft (cm)	50	71	—	—8	—
7.	Beam amidship (cm)	119	132	119	45–91	35–85
8.	Beam main deck (cm)	102	119	Proper deck absent	Proper deck absent	Proper deck absent
9.	Beam bottom (cm)	79	5	—	—	—
10.	Beam fore (cm)	—	—	71	15–25	30–91
11.	Beam Aft (cm)	—	—	68	15–25	30–91
12.	Fore overhang (cm)	53	66	207	170–210	120–150
13.	At overhang (cm.)	46	61	94	150–190	120–150
14.	Deck	Open	Open	Open	Open 10 to 30 booms of out-rigger serve as seats	Open 4 to 9 booms of out-rigger serve as seats
15.	Man-power (nos)	3–5	4–6	6–16	2–26	3–25
16.	Endurance (hrs)	10:00	15:00	28:00	24:00,36:00	24:00, 96:00
17.	Propulsion	Paddling, and Punting	Paddling, sail and Punting	Paddling, Sail and motorised	Punting, Paddling and sail	Punting, paddling, sail used by Onge and Great Andamanese
18.	Gears used	Gill net, cast net & hook and line	Gill net, shore seine, anchor net & hook and line	Gill net, shore seine & hook and line	Nicobarese-harpoon, spear, cast net & hook and line shompens spears Uni or Multi prong	Bows and arrows, harpoon spear, Primitive
19.	Depth of operation	Up to 20 m	Up to 20 m	Up to 50 m	Up to 300 m	Up to 500 m
20.	Area of operation	North Andaman to Little Andaman	Nort Andaman to little Andaman & Cambell Bay	North/Middle Andaman; Maya bunder & Webi South Andaman Hope Town & Wandoor	Car Nicobar to Great Nicobar	Little Andaman, North Sentinel Island, Western (Medium) coast of South & Middle Andaman and Strait Island

TABLE IV. **Details of traditional gears**

Sl. No.	Gear / Parameter	Pelagic drift gill net	Cast net	Shore seine net	Anchor net (Boat Seine)
1.	Length (m)	300–420	Top meshes 40–86 Nos. bottom meshes 600–2500 Nos	70	39
2.	Depth (m)	3–15	3–5	Near bridles 1.5 at middle 6	3
3.	Mesh size (cm)	3–14	0.5–2.5	0.2–2	0.5–15
4.	Material	Nylon	Nylon	Cotton	Cotton
	If nylon twine size	210/1/2 210/2/2 210/4/2 210/8/2 210/9/2	210/1/2 210/2/2	—	—
5.	Depth of operation	Up to 50 m	Backwater canals	Up to 8m	Up to 10 m
6.	Main catches	Sardine, tuna, mackeral and seer	Anchovies, sardine, silver bellies, & mullets	Silver bellies, anchovies, sardine, perches & misc.	Up to 10 m
7.	Man-power needed for operation	2–5	1–2	10–20	10–30
8.	Area of operation	All over the Andaman & Nicobar Islands (excluding aborigines area)	All over the Andaman & Nocobar islands (excluding aborigines area)	South Andaman: north Bay, Aberdeen Jetty & Carbyn's Cove	North/ Middle Andaman/ Aerial Bay & Mayabunder South Andaman; Jugligha & Haddo

to be popularised. On the other side, freshwater fish culture has gained considerable momentum with the perfection of induced breeding technique (Mustafa, 1985). Presently about 30 hectares of water spread area is under freshwater fish culture and there are around 250 pisciculturists and by the application of proper technique the production has increased from 0.18 tonnes in 1980 to 4.027 tonnes in 1985.

RESOURCES

The rich marine resources in the region have attracted foreign poachers from the beginning of the present century. The first poaching on record took place in 1926 and a fleet of 21 foreign fishing vessels were for the first time confiscated in 1929 (Rao, 1939). Since then hundreds of poaching vessels have been intercepted. Much of the present day information on marine resources has been originally gathered from the analysis of catch and record of such vessels.

The available survey reports manifest that the area is fairly rich in demersal and pelagic exploitable resources. The bottom trawling has been done more extensively as compared to other methods; but due to unfavourable topography it is not possible to locate extensive grounds, where trawling could be done continuously for more than 1½ hours, without encountering obstructions. The average catch per hour of bottom trawling is around 100 kg in depth zone of 20 to 29 m. The period from January to April is most productive. Seasonal fluctuation, variation and the catch composition based on the survey conducted by F.S.I. is shown in Figs. 4, 5 & 6.

So far about 700 species of fish has been recorded from this area of which about 200 varieties broadly classified into 62 groups are found to be commercially important. Exploited fishery resources and their seasonality is given in Table V.

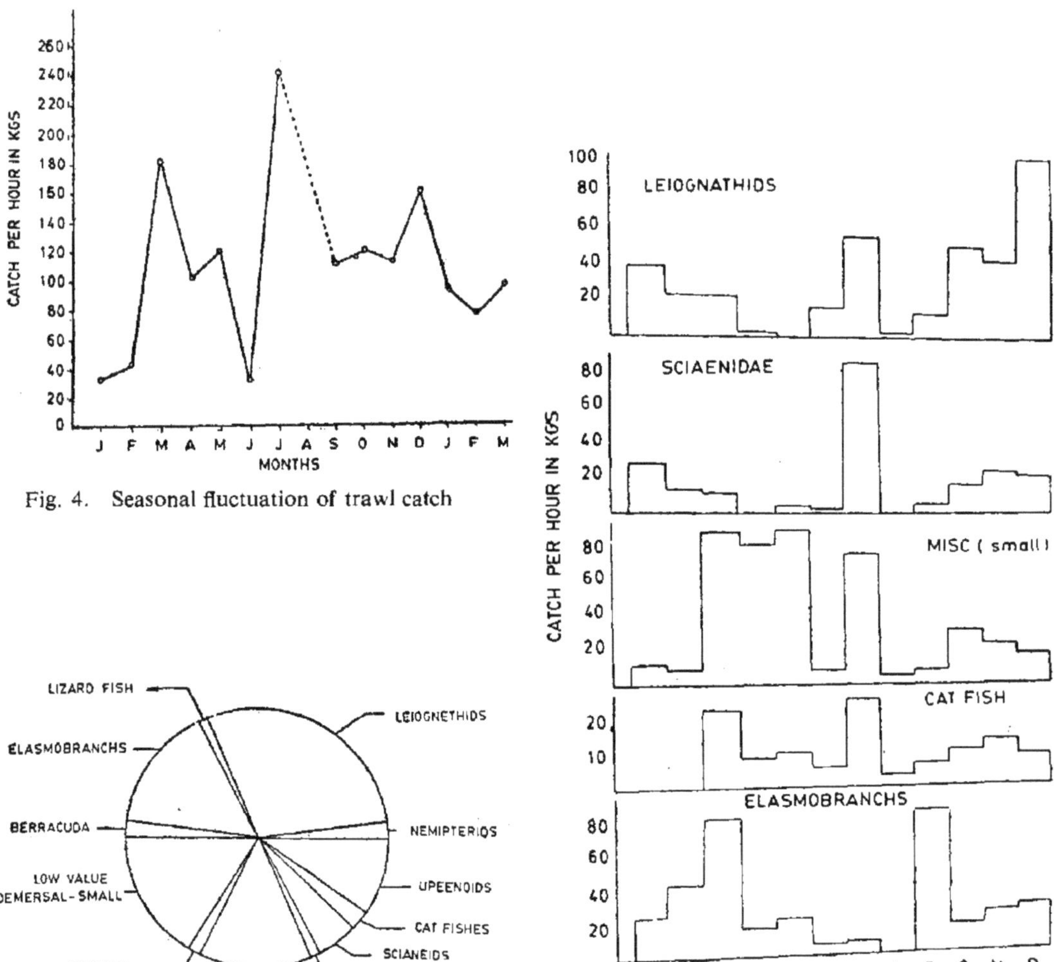

Fig. 4. Seasonal fluctuation of trawl catch

Fig. 6. Average bottom trawl catch composition

Fig. 5. Seasonal variation of main groups
Source : Fisheries Survey of India

TABLE V- **Fishery resources and seasonality (After Mustafa, 1983)**

Resources	*% of catch*	*Peak season*
Pelagic resources		
Sardines—*Sardinella* spp., *Dussumieria* spp., and *Anodontostoma* sp.	13.9	July—December
Carangids—*Megalaspis cordyla, Caranx* spp., *Elagatis* sp., *Decapterus* sp. & *Selar* spp.	9.2	July—December
Mackerels—*Rastrelliger kanagurta* and *R. brachysoma*	9.1	March—June Sept—December
Anchovies—mainly *Anchoviella commersoni* and *Thrissocles mystax*	7.6	June—Nov.
Scombrids—mainly *Scomber commersoni* and *S. guttatus*; some *Auxis thazard*	7.2	March—August
Barracuda—*Sphyraena jello*	2.9	July—Sept.
Tuna—yellow fin (*Thunnus albacares*), bigeye (*T. obesus*), albacore, (*T. alalunga, Euthynnus affinis*) & skipjack (*E. pelamis*)	2.6	Dec—January
Demersal Resources		
Perch-like fishes—especially *Lates* sp., cods (*Serranus* spp. and *Epinephelus* spp.) thread fins (*Polydactylus* spp.) emperors (*Lethrinus* spp.) and grunts (*Pomadasys* spp.)	13.4	August—Nov.
Silver bellies or slipmouths—mainly *Leiognathus equula, L. splendens* and *Gazza* sp.	10.2	June—December
Low-value demersal fishes-croakers (Scianids), goatfish (Upeneoids), threadfin breams (Nemipterids) and lizard fish (Saurids)	5.7	June—December
Mullets (and minor estuarine fish), mainly *Mugil tade, M. cephalus* and *Liza spp.*	5.3	*July—December*
Shark and rays—15 shark spp., 6 rays	2.7	—
Crustaceans—mainly the shrimp (*Penaeus merguiensis, Metapenaeus dobsoni*), and the lobsters *Panulirus polyphagus* and *P. ornatus*	2.4	November and Feb—March

Pelagic fishery resources

It consists of Sardines, Anchovies, Seer fish, Mackerel, Tuna, Marlin, Sail fish. Barracuda and Herrings. Clupeidae fishery sources is mainly represented by *Dussumieria* spp. *Sardinella fimbriata, S melanura, S. sirm, S. longiceps, Harengula punctata, H. ovalis, Anodontostoma chacunda, Megalops cyprinoides, Thrissocles mystax,* Belonids *Lactarius lactarius,* and *Trichiurus haumela.* Major pelagic resources group after clupeids is Scombrids which include the most economical varieties such as *Rastrelliger kanagurta, R. brachysoma, Scomberomorus commersoni, S. guttatus, Auxis thazard, A. rochei, Thunnus albacares, T. obesus T. tonggol, T. alalunga, Euthynnus affinis, E. pelamis, Acanthocybium* spp. and *Gymnosarda* spp.

Tuna resources are little exploited. A stock of about 15,000 tonnes of yellow fin and big eye have been estimated for this area which is based on hooking rate and yield rate (Kikawa *et al.*, 1969). There is no estimation for skipjack stock in this area. although 500,000 to 700,000 tonnes has been given as standing stock for the entire Indian Ocean. From this an estimate of 50,000 tonnes for the E.E.Z. around Andaman and Nicobar Islands could be projected. A catch of 858 kg Tuna, 328 kg seer fish and 223 kg sharks for each operation has been estimated by earlier experts for a 90 M.T. Gross tonnage multipurpose vessel of about 24 mt.

Mid water resources are mainly constituted by Barracudas, Horse mackerels, Marlin, Silver bellies, Sword and sail fishes. The muqiliformes include *Sphyraena jello* as the main variety. Perciformes consists of *Sillago sihama* and more economical varieties as *Megalaspis cordyla, Caranx sexfasciatus, C. melampygus* and *C. armatus, Scomberoides commersonianus, S. tala, Selar kalla, Selar boops, Makaira mitsukurii, M. indica, Istiophorus gladius, Tetrapturus* spp. and *Xiphias gladius*. The fishes like *Decapterus macrosoma, Leiognathus equlus, L. splendens* and *L. ruconius* are widely exploited.

Demersal fishery resources

It consists of perches, snappers and cat fishes and members of the family sciaenidae. Mainly exploitable varieties are percoidei such as *Lates calcarifer, Epinephelus merra, E. elongatus, Lutjanus argentimaculatus, L. lineolatus, Nemipterus* spp., *Pomadasys* spp. and *Johnius* spp.

The elasmobranch resources of the area is totally untouched which consist of about 15 species of sharks and 6 species of rays. Survey reports and day-to-day experience have shown that the potential of elasmobranch fishery is rich and they are available throughout the year. Most abundant varieties of sharks are *Carcharhinus melanopterus, Chiloscyllium griseum, C. indicum, Scoliodon walbeehmi, Hemigaleus* spp. *Pristis cuspidatus, P. microdon* and rays of Genus *Dasyatis* sp. *Aetobatus* sp., and *Mobula* sp. are also encountered very often.

Recently deep sea spiny dog fish shark resources comprising of *Centrophorus* sp. and *Squalus* sp. has been located off Andamans, the liver oil contains squalene ($C_3O\ H_5O$) and this has good commercial potential. Catch per hundred hooks is found to be 94 kgs (Mustafa, 1986).

Estuarine resources

The islands have got innumerable rivulets, intervening creeks, swamps, mud banks and intensive mangrove swamps which are highly productive. It has been estimated that mangrove swamps have an area of about 1 lakh hectares (N.I.O. 1980). Further about 37,916 hectares of marshy low lying areas and mangrove swamps are also available (Anon, 1975). Creeks and rivulets are main fishing grounds for shrimps. The fish fauna consists of *Mugil cephalus, M. dussumieri, M. tade, Chanos chanos, Sillago sihama, Lates calcarifer, Siganus* sp., *Lutianus* sp. and *Serranus* sp.

Molluscan resources

Shell fishery is comparatively better developed although no stock assessment survey work has ever been done. The shell resources are being exploited since 1929 in an organised fashion. There are 9 geographically demarcated shell fishing zones as mentioned below:

1. Cape price to Mayabunder,
   12° - 66.5' N and 13° - 34.5' N. Port-Mayabunder

2. Cape price to Austin Strait,
   12°-54' and 13°-34.5' N. Port - Mayabunder.

3. Mayabunder to Long Island,
   12° - 24' N and 12° - 55' N. Port Blair - Long Island

4. Long Island to shoal Bay,
   12° - 05' N and 12° - 18' N. Port-Long Island.

5. Shoal Bay to Chiriatapu,
   12° - 29' N and 11° - 50.4' N, Port-Port Blair.

6. Chiriatappu to Port Mount,
   11° - 29' N and 11° - 38' N. Port-Port Mount

7. Ritchie's Archipelago,
   11° - 46.5' N and 12° - 19' N. Port-Port Blair.

8. Nicobar Central Group,
   7° - 52' N and 8° - 35' N'. Port-Nancowrie

9. Nicobar Southern group,
   6° - 45' N and 7° - 31' N. Port-Nancowrie

The zones are auctioned every year in accordance with Fisheries Regulations in force. There are about 30 registered shell handicraft small scale units (Anon, 1984) and 7 zones in Andaman group with an annual turnover of about Rs. 29,26,318/-. Average monthly turnover rate of a shell handicraft unit is about Rs. 0.01 to 0.9 lakhs. The composition of trade is 1:3:3 for ornamental.

Miscellaneous Novelties

The shells are either marketed in whole form or diversified articles of various design, colour, shape and size are carved out. Photographic embossing and engraving is also done. The shell industry is mainly based on: *Turbo marmoratus*, *T. petholatus.*, *T. Cochlus*, *T. porphyrites*. The turbo shell is quite heavy and solid in structure with a hard shelly calcareous operculum known as Cat's eye. *Trochus niloticus*, *T. pyramis*, *T. maculatus*, *Pinctada margaritifera*, *P. fucata*, *P. suggillata*, *P. anomioides*, *Cassis cornuta*, *Xancus pyrum fuses*, *Naeutilus pompilus*, *Harpa* spp., *Hamlius* spp., *Murex* spp., *Cypraea* spp., *Conus spp.*, *Tridacna maxima*, *T. crocea* and *T,. squamosa* are the major varieties being fished. A very good stock of edible Oyster *Crassostrea madrasensis*, *C. gryphoides*, *C. discoidea*, *Saccostrea cucullata* (Ramadoss, 1983) and the green mussel *Perna viridis* (Mahadevan, 1983) is also present.

The shelled molluscan are taken as food only by the aborigines of Andaman and tribals of Nicobar Islands. Cuttle fish, *Sepia* sp. and the Octopus stock remain unexploited.

Crustacean resources

There is no organised prawn fishery. Hence no specific craft or gear is employed. Most of the fishing grounds are confined to the mangrove infested lagoon and creeks and November is found to be the major fishing season followed by the period from January to April. Major landing centres are around Port Blair, Rangat and Diglipur and the main fishing ground around Port Blair is the S–W region of Port Blair harbour from Oyster Island onwards, Shoal Bay creek and Homfray's Strait.

So far about 37 species of Penaeid prawns, 28 species of non-penaeid prawn (Family–Palemonidae) and 2 species of Family Sergestidae have been recorded. *Penaeus merguiansis* is the dominant variety accounting for 49% followed by *M. dobsoni* (42%), *M. ensis* (2.5%) and *P. monodon* (0.6%) (Silas *et al.*, 1983).

Other resources

Bottom topography of the area and past experience of foreign fishermen provides ample evidence that the area does have rich lobster beds. *Panulirus homarus*, *P. polyphagus*, *P. orgatus*, *P. penicillatus*, *P. versicolor* and *Thenus orientalis* are reportedly found around A & N Islands (Shanmugam *et al.*, 1983).

Crabs belonging to *Scylla serrata*, *Portunus pelagicus* and *P. sangrinolentus* are occasionally caught as by-catch. World's largest land crab *Birgus latro* weighing about 1.8 to 2.2 kgs is also found in South Sentinel and Great Nicobar Island.

40 species of shallow water sea cucumber has been recorded (James, 1983). Superior varieties like *Holothuria scabra*, *Echinometra mathaei*, *Actinopyga mauritiana* and *Bohadschia vitiensis* were caught, processed and exported to mainland as beche-de-mer. At the moment the collection of sea cucumbers has been stopped by the Administration for conservation reasons.

Ornamental fishes

The coral banks and reefs harbour various forms of aquarium grade tropical fish but so far no attempt has been made to exploit them. Marine mammals, such as *Dolphin* spp. Porpoises, Whales and *Dugong dugong* are protected, however the bio-product of sperm whale *Physter catadon*, the Ambergris deserves special mention which is collected on a commercial scale and exported. The cost of Ambergris at Calcutta and Bombay is around Rs. 50,000 to 60,000 per kg.

Turtles

Hunting of turtles is prohibited under the Wild Life Act. So far about 7 varieties are recorded. *Chelonia mydas* (Green turtle), and *Eretmochelys imbricata* (Hawkbill turtle) are the most common varieties.

Crocodiles

Crocodylus porosus, the most important salt crocodile is found in abundance in the creek and lagoons of North and Middle Andaman and Great Nicobar. It is also protected under the Wild Life Act.

Corals

The collection of corals has also been recently banned by a regulation of the Island Administration. Andaman & Nicobar Islands are affected by erosion, abrasion and dissolution of shore rocks forming reef-flats. Five types of modern reefs exist in these shores. There are windward reefs, channel reefs, bay reefs, coral attols and pitch reefs. *Acropra, Porites, Favites, Montipora, Hydrophora, Cycloseris, Goniastrea, Platygyna* and *Fingia* (Z.S.I.) are the chief reef building genera in these shores. In both the windward and leeward shores of Andaman & Nicobar Islands, three types of oceanographic and depositional environments are distinguished. These are 'shallow mid-reef moderately agitated environment' and 'fore-reef calm environments'. Varying degrees of coral growth occur in these environments.

Sea weeds

55 species comprising of Chlorophyceae (16) Phaeophyceae (17) and Rohdophyceae (22) have been recorded (Solimabi *et al.*, 1981; Gopinathan *et al.*, 1983). Domination of alginophytes such as species of *Sargassum, Turbinaria* and *Pandina* have been observed. Economically important agarophytes such as *Gracilaria* spp., *Gelidium* spp., and *Gelidiella acerosa* are also found.

SUMMARY

Andamans have an area of 8,293 km^2. The total area of the E.E.Z. around the island is about 5,95,554 km^2. This area receives rainfall during both the spell of monsoons and the range between minimum and maximum temperatures vary less than 8°C. The surface waters are productive and have high oxygen content. Primary production rate is around 273 mg C/m^2/day and the estimated production of the Andaman sea would be 6 x 10^7 tonnes of C/Year. The secondary production based on zooplankton biomass is 288.8 mg C/m^2 and the production of entire area of Andaman sea i.e., 6.02 x 10^5 km^2 would be around 31.73 x 10^6 tonnes of C/Year. Tertiary production in terms of live weight has been estimated at 18.70 x 10^5 tonnes/year and the exploitable potential of fish will be around 4,70,000 tonnes/ year. As against this large potential, the present fish landings are around 6,000 tonnes/year. Though these figures of total fish potential may vary depending on different estimates, they definitely indicate that fish production from Andaman sea can be increased manyfold.

The major areas of thrust in the seventh five year plan are: developing marine fishing by Government efforts like setting up of a corporation, encouraging private enterpreneurs and developing infrastructure facilities for storage, marketing and supply of fishing fleet. In addition, manpower training, development of culture fisheries, strengthening of fish seed production centres and setting up of cold chain are being supported. These schemes are further supported by social welfare activities, organising cooperative societies, mechanization of indigeneous boats, provision of facilities like oil and potable water at fisheries centres and settlement of families of fishermen under the new 20-point Programme.

At present, the fishery is confined to coastal areas. Crafts smaller than 12 m length fish in 12 to 40 fathom depth. In A & N Islands there are about 2,860 fishermen who have 1071 country crafts, thus on an average there are two fishermen on each boat. The nets used are simple and primitive. They generally use drift gill nets, cast nets, shore seines. Aboriginals of Andamans use some indigeneously developed fishing implements.

Offshore fisheries survey is conducted by F.S.I., Port Blair base. FSI is operating medium sized trawlers and have surveyed an area of about 3,000 km^2. The long-line fishing is most economical as varieties such as tuna, marlin, shark, carangids, barracuda and seer fish are caught by this method. The available survey reports indicate that the area is rich in demersal and pelagic fishery resources, but the ground is uneven, which restricts the use of bottom trawling. Therefore, the exploitation of pelagic fisheries resources could be done in a big way. It has been pointed out that the FSI base could further be effectively utilised for development of fishery of elasmobranch resources, the pelagic and deep sea sharks.

Traditionally, the Andamans are known for shell fishery. Shell fisheries have been divided into 9 geographical molluscan zones. The fishing rights of these zones are auctioned annually and Andamans have about 30 registered small scale industries which deal with shell fisheries. The shell are either marketed in whole or some dealers prepare diversified articles of various designs, colour, shape and size.

As indicated earlier, bottom topography is uneven, but it has a few rich lobster beds. *Panulirus polyphagus* and *P. orgatus* are common lobsters. The culture possibilities for prawn and mussel is extremely good and this can become an item of export trade in addition to the development of sharks and aquarium fish industry. A glance into the state of art of fisheries of Andaman Islands indicate that the Government of India has paid much attention for the development of fisheries in the Andamans. However, the efforts need to be strengthened. Shore facilities for bunkering, storage and transport should be developed. The Andamans offers good scope for mariculture. Long-line fisheries and the catch of tuna and other pelagic fishes can be increased manyfold. Thus Andaman Islands which form a window towards the Indian Ocean have a rich natural resource potential and have the capability of becoming an important fisheries centre. How quickly this will be achieved will depend upon the Governmental efforts. Adopting a scientific approach and developing appropriate management methods for quick development and rapid expansion of trade can bring about prosperity, happiness and bettar standard of living among the local inhabitants and tribals of the Andamans.

REFERENCES

Abidi, S.A.H., Present status and prospects of future development of aquaculture technology in Andaman, *India Today and Tomorrow.*

Anon, 1976. *Report of the technical team*—pp. 132., A & N Administration.

Anon, 1977. *Indian Fisheries* 1947-1977 pp. 96., C. M. F. R. I., Cochin

Anon, 1978. *E. F. P Bulletin* No. 7, pp. 53, Bombay.

Anon, 1980, *Technical Report* 02/80 pp. 208, N. I. O. Goa.

Anon. 1983. *C.M.F.R.I Bulletin*, 34 pp. 108, CMFRI Cochin.

Anon, 1984. *Basic Statistics*. pp. 274. A & N Administration.

Bhattathiri, P. M. A. & Devassy, V. P. 1981. Primary productivity of the Andaman Sea, *India J. mar. Sci.*, 10 : 243 pp.

Madhupratap, M., Sreekumaran Nair, S.R., Achuthankutty, C.T. & Vijayalakshmi R. Nair, 1981. Major crustacean groups and Zooplankton diversity around Andaman and Nicobar Islands. *Indian J. mar. Sci.*, 10 : 266–269.

Mullin, M. M. Oceanograpny and Marine Biology: An annual review, 7, Ed. by Barnes, H. (George Allen and Unwin., London)

Mustafa, A.M. 1981. *Fish & Fisheries of Andaman & Nicobar Islands*. Dissertation CIFE Bombay, 363 pp.

Mustafa, A. M. 1983. Fishereis of the Andaman & Nicobar Isalnds *ICLARM Newsletter, Phillipines.* 6 (4) : 7-9.

Mustafa, A.M. 1986. New deep sea spiny dog fish shark Reources off Andaman Naga.*ICLARM quarterly, Phillipines.* 9 (1) : 18 p.

Mustafa, A. M. 1986. Observations on the seed production of Indian major carps in Andaman by hypophysation, *And. Sci., Asso.*, *4*.

Mustafa, A. M. 1986. On the occurrence of an unusually large *Catla catla* (Ham.) P.M Variety from Dilthamman tank, Port Blair, *J. And. Sci. Asso.*, (Submitted).

Qasim, S. Z. & Ansari, Z. A. 1981. Food components of the Andaman sea. *Indian J. mar. Sci.*, 10 : 276–279.

Rama Raju, D.V., Gouveia, A. D. & Murthy, C. S. 1981. Some physical characteristics of Andaman sea water during winter. *Indian J. mar. Sci.*, 10 : 211–218.

Ramesh Babu, V. & Sastry, J. S., 1976. Hydrography of the Andaman Sea during late winter. *Indian J. mar. Sci.*, 5 : 176 p.

Rao, H. S. 1939. *Consolidated report on the shell fisheries in Andamans during the year* 1930–35. 130 Zol. Survey of India, Calcutta.

Sen Gupta R., De Sousa, S. N. & Joseph, T. 1977. On nitrogen and phosphorus in the Western Bay of Bengal. *Indian J. mar. Sci.*, 6 : 102–107.

Sen Gupta, R. Moraes, C., George, M. D., Kureishy, T. W. & Noronha, R. J., 1981. Chemistry Hydrograph of the Andaman Sea, *Indian J. mar.Sci.*, 10 : 228–233.

Solimabi Das, B., Mittal, P. K. & Kamat, S. Y. 1981. Bromine and iodine content in sponges and algae of the Andaman Sea, *Indian J. mar. Sci.*, 10:301–302.

REPRINTED FROM :

PROCEEDINGS OF THE SYMPOSIUM

on

MANAGEMENT OF COASTAL ECOSYSTEMS AND OCEANIC RESOURCES OF THE ANDAMANS

17 - 18th July, 1987

Organised by :-

ANDAMAN SCIENCE ASSOCIATION

CENTRAL AGRICULTURAL RESEARCH INSTITUTE
PORT BLAIR - 744 101

ENDANGERED CORAL REEFS OF BAY ISLANDS AND THEIR ORNAMENTAL FISHES

ARIF M. MUSTAFA[1], S.N. DWIVEDI[2], YAMINI M. WARAWDEKAR[3], S.A.H. ABIDI[4], and E.K. RAVEENDRAN[5].

Department of Fisheries, A & N Islands, Port Blair : 744104.

ABSTRACT

Andaman & Nicobar islands have one of the richest coral reef formations, eastern side with fringing reefs and western side barrier reefs. During the present investigation attempt has been made to assess the impact of multidisciplinary developmental activities on this fragile ecosystem. The data collected indicate that coral reefs around the islands are threatened. Out of the total 2000 sq. km. corals in an area of about 360 sq. km. have been intensively damaged in shallow coastal waters. Reefs of these islands are yielding about 24,600 tonnes of edible and ornamental biota per annum. The area is fairly rich in Ornamental fish and invertebrate resources.

About 6,00,000 sq. km. area of our planet earth in the sub-littoral and neritic oceanic region (<-30 Mt.) is covered by coral reefs rich in myriad forms of marine life (Munro, 1984). This area is almost equivalent to the total exclusive economic zone area of Andaman and Nicobar Islands sprawling between latitude 6° 45' N and 13° 45'N and longitude 92° 15' E and 94° 00'E covering land area of about 8293 sq. kms. The continental shelf around the archipelago which is quite narrow spreading for about 3-8 Km. having an area of 34,965 Sq. km. (Anon, 1977), is the abode of India's richest tropical coral reef ecosystem comparable to any tropical rain forest in productivity and diversity. Corals are perhaps the "base" animals in the ecological pyramid of Anadaman sea (Whitaker, 1985). The present paper deals with the destruction of coral reefs in this region and the rich potential of their ornamental fishes.

Coral reefs are extremely beneficial to mankind : (1) they are excellent nursery grounds for the growth and development of young ones of many commercial marine fishes; (2) they yield many shallow water fishes both edible and ornamental, molluscs, crustaceans and echinoderms providing nourishment and playing a significant role in many maritime nations's economy; (3) antileukemia, antitumor, antimicrobial and ultraviolet blocker biochemicals are extracted from reef inhabitants; (4) reef glamour supports tourism; and (5) reefs protect the seaward erosion of coastal areas.

1 Superintendent of Fisheries (Tech.), Department of Fisheries, A&N Islands, Port Blair.
2 Additional Secretary, Department of Ocean Development, New Delhi.
3 Senior Research Scholar, C.I.F.E., Andheri (W), Bombay-59 AS.
4 Director, Department of Ocean Development, New Delhi.
5 Director of Fisheries, A & N Islands, Port Blair.

Materials and Methods

Observations on the prevailing condition of coral out-growths and reefs was made by surface snorkeling upto 5 m. beyond which SCUBA gear was often used upto 10 m. Most of the observations were made during low tide, at times reaching as far 200-300 m. from coast. Observations thus obtained *in-situ* were recorded as excellent (A grade) i.e., totally undisturbed/no physical damages and rarely fished; good (B grade) i.e., slightly disturbed/very little physical damage and seasonally fished; disturbed (C grade); damaged (D grade), and ruined (E grade). For analytical convenience, area was divided into major, minor and micro squares as 3600 sq. nautical mile area (60' x 60'), 100 sq. nautical mile area (10' x 10') and 01 sq. nautical mile area respectively. Base line shore positions were invariably *fixed for* existing berthing jetties at different places with the aid of navigational assets available on board inter-island shipping vessels. In each randomly selected micro square either a "Z" shape combing or 5 spot observations were made and transplotted over map. Informations related to far off and unaccessible areas were collected from experienced masters of ferry vessels and the professional skin divers.

Live fishes were collected from a depth zone upto 4 m. during low tide using Slub gun, encircling fine mesh stake net with pocket at middle and scoop net. Invertebrates were either picked up or brought out from the coral infested area using any one of the above mentioned gear. Live and preserved samples were identified at Port Blair and Bombay.

Results and Discussion

Coral reef types and their extent : The Andaman islands have rich coral reef formations on western side. Along the islands about 21 kms. away from the coast of main island are located North coral bank, West coral bank, Middle coral bank, South coral bank, North and South Sentinel coral banks and Dalrymple bank. The luxuriant barrier reef extends from latitude 10° 26' N to 13° 40'N covering a longitudinal distance of about 320 km. and separated from the coast by a trench about 80 m. deep (Tikader, 1986). On the eastern side, coral reef formation is less and is of fringing type. The Ritchie's archipelago has raised coral benches from Outram Island to Lawrence island, with knolls in their channels. Patch reefs are in abundance in Nancowry. Flat reef about 500 m. wide is reported from East bay in Katchal island (Pillai, 1983).

In these islands three distinct types of reefs are noticed namely (1) wind-ward reef at North Bay (Port Blair), Cinque island and Jolly Buoy island (2) channel reef/lee-ward reef at Camorta—Nancowry area and (3) bay reef at Havelock, Neil and Nancowry Islands.

Porites sp. and *Favia* sp. are the chief reef builders of A & N Islands, the protected bays have better growth of corals as compared to exposed coast and the Nicobars have a richer growth than Andamans. So far about 135 species of corals under 59 genera have been recorded (Pillai, 1983) out of which 110 species under 45 genera are hermatypes and 25 species under 14 genera are ahermatypes. *Fungia* sp. is the only solitary coral. Most common varieties of stony corals are the Acroporidae (*Acropora* sp. and *Montipora* sp. Portidae (*Porites* sp.) Faviidae (*Favia* sp.) and Fungiidae (*fungia* sp.), while the soft corals are widely represented by *Sacrophytum* sp. and *Labophytum* sp. horny gorgonia and sea fans.

Based on the present survey conducted during Nov. '86 to May, '87, many stations were covered all along the island and it is estimated that about 6% of the total continental shelf area of A & N islands is infested by various reef forms such as coral banks, benches knolls, patch reefs and flat reefs covering an area of about 2000 sq. km. out of which about 18% i.e. 360 sq. km. of coral reef areas have been intensively damaged due to various reasons explained elsewhere.

Reef Resources and their productivity. It is estimated that coral reefs (<30 m.) could yield at the rate of 15 tonnes per square kilometre per year (Munro, 1984). Based on the theory the yield from the undamaged coral reef area of these islands could be 24,600 tonnes per year, but this 3° level production could not be harvested fully due to the topographical limitations offered by reef systems where commercial fishing gears could not be used efficiently. The present survey revealed that the entire area is fairly rich with most spectacular icthyc-beauties of coral reefs. There is no single publication dealing with ornamental tropical reef fishes of A & N islands hence a list of the frequently occurring and important species is appended here.

Chaetodontidae—Butterfly fishes : The most dazzling, gracefully agile, small and swift fishes found in all types of coral reefs, some of which are always found in pairs.

Chaetodon plebius	—Blue-spot butterfly fish.
C. ephippium	—Saddle-back butterfly fish.
C. malanotus	—Black-back butterfly fish.
C. klieni	—White-spotted butterfly fish.
C. trifasciatus	—Rainbow butterfly fish.
C. lunula	—Moon butterfly fish

Zanclidae—Toby fish : Only one variety recorded. A very expensive and difficult fish to keep. One of the nature's wonders respected by muslim fishermen.

Heniochus acuminatus—Moorish Idol.

Monocanthidae—File fishes : Extremely attractive, slowly moving invariably inhabiting *Acropora* sp. knolls, giving a unique colouration to the reef.

Oxymoncanthus longirostris—Long nosed file fish.

Centriscidae—Razor fishes swim vertically with head in down position and occur in small shoals in the vicinity of reef.

Aeoliscus strigatus—Shrimp fish.

Platacidae—Bat fishes : It changes body shape as it grows, live among huge stony corals.

Platax pinnatus—Long finned bat fish.

Pomacentridae— Damsel/clown/anemone fishes : The anemone fishes live unharmed amidst the poisonous tentacles of sea anemones—a symbiotic relationship. They defend their territory very much and is an excellent hard fish for beginners.

Amphiprion akallopisos	—Yellow skunk clown.
A. sandaracinos	—Yellow skunk clown.
A. ephippium	—Saddle anemone fish.
A. frenatus	—Tomato clown.
A. ocellaris	—Common clown.

A. sebae —Brown clown.

A. biaculeatus —Maroon clown fish

Dascyllus aruanus —Humbug damsel

D. trimaculatus —Domino damsel.

Chromis caeruleus —Blue-green chromis.

Monodactylidae—Moon fishes: Deep silver diamond shaped fishes.

Monodactylus argenteus—Malayan angel fish.

Labridae—Wrasses; These cleaner fishes are in abundance. They clean anything which comes in their way especially the parasites and fungal infection from other fishes.

Labroides dimidiatus—Bicoloured cleaner doctor fish.

Scorpaenidae—Scorpion or lion or fire fishes: Common around builder type reefs, brilliantly coloured, fins enlarged and filamentous. Spines highly venomous.

Pterois volitans —Peacock lion fish.

Pteropterus antennata —Spot fin lion fish.

Syngnathidae—Seahorse and pipe fishes:

Hippocampus hystrix —Black sea horse

Dunkerocampus sp. —Zebra pipe fish.

Other ornamental resources encountered during the survey were a large number of invertebrates of decorative importance. As the shell fauns of these islands has been described in many old and new publication, they are excluded here. A list of other attrative flamboyantly coloured and displayable animals is furnished below:—

1. *Phyllidia zaylancia* —A rare sea slug characterised by yellow-pink tubercles and black-brown lines
2. *Astropecten* sp. —Common star fish
3. *Linckia leavigata* —Reef star fish.
4. *Ophiocoma scolopendrina* —Brittle star.
5. *Diadema setosum* —Hetpin urchin.
6. *Tripneustes gratilla* —White urchin.
7. *Echinothrix calamaris* —Banded urchin
8. *Metridirum* sp. —Sand sea anemone.
9. *Adamsia* sp. —Hermit crab sea anemone.
10. *Radianthus* sp. —Clown fish sea anemone.
11. *Stenopus* sp. —Banded coral shrimp
12. *Alphaeus* sp. —Pistol shrimp.

Coral reef damages and destruction: Destruction or damage to a coral reef may be due to many reasons. To cite a few significant ones, the Caribbean coral reefs have been considerably damaged by amateur tourist divers and the Sri Lanka's reefs due to intensive removal of ornamental fish. At Kenya, Seychelles and Guam, reef's were spoiled due to commercial collection of corals and shells. Large scale aquarium fish trade, spear fishing, coral and shell collection have also created major reef problems in Indonesia, Phillippines and Pulau Seribu. Even the traditional Japanese fishing gear 'Munro-ami' intensively damaged the reef off Western Palawan-Phillippines and at Paracels and Maccles field bank in South China sea. The use of icthiotoxic and dynamite fishing have damaged the reefs in Tanzania, Guam, Marianas, Palau, Malaysia, Thailand and Bahamas. An abrupt increase in the population of sea urchin *Diadema* sp. caused considerable damage to corals in the Red sea because the predators such as puffer and trigger fishes were killed in large number by scuba divers using spears. Similarly, the gastropod *Drupella* sp. damaged the reefs of Phillippines and Japan due to an imbalance in the ecosystem by depleting one living component. The Great Barrier reef of Australia was once threatened by the out break of the thorny star fish *Acanthaster planci*.

In these islands damage to corals was insignificant till 1960 by virtue of limited collection. But by mid 1970's with an increase in influx of human population from mainland India and the various developmental activities, the damage was magnified and further intensified in mid 1980's. One could easily notice the damage to reefs due to collection by unscrupulous traders and tourists, at Aerial bay in Diglipur to Car Nicobar by the indiscriminate removal of fancy looking corals as *Acropore armata, A. humilis, Montipora cocosensis Seriatopora crassa, Pocillopora damicornis. Merulina ampliata* and *Tubipora musica* (red coral), The badly damaged area around Port Blair is from Corbyn's Cove to Chatham island along the coast, Mayabunder harbour area and the reefs around Car Nicobar and Kamorta islands. Today it is difficult to collect the once common solitary coral *Fungia echinata*, F. *fungites*, F. *danai* and F. *horrida*. The unescorted, unguided amateur scuba and snorkel divers have virtually turned the beautiful coral reefs of Jolly Buoy and Cinque islands into heaps of dead bits beyond recovery. Only soft and boulder type of corals are now flourishing due to intensive trampling and walking. If the trends continue the effects would be devastating in near future.

The damage to coral reefs due to other causes are of a lesser magnitude but still requires attention. They are the deleterious effects of siltation, seen at Hut Bay in Little Andaman which is of recent origin and dead beds of corals at North Bay, Port Blair. The toxic discharge from timber factory at Chatham, WIMCO match factory at Haddo and A.T.I. Bambooflat has caused considerable damage to the surrounding shallow areas. A new timber industry J.T.P. in Middle Andaman has also started damaging the coral out growths. The formaldehyde and arsenic are the most lethal components of this timber industry discharge. Almost all the inhabited islands have jetties constructed at the cost of coral wealth in and around their vicinity. The reef material is also being used as building material in many of these islands. In addition, the increasing threat is from oil spill of over increasing sea traffic and oil drilling rigs. Some mainland based aquarium firms have recently started carrying away ornamental fishes from these islands causing intensive damage to corals and their associated fauna.

Considering these destructive activities, a total ban on the removal of corals and sea cucumber was imposed in 1978 thus enacting "The Andaman and Nicobar islands Shell Fishing Rule 1978". Later the administration has declared some islands including sea shore areas as parks and sanctuaries. Most famous among them is the Marine National Park Wandoor covering an area of 234 sq. km. But, surprisingly no coral bank or demarcated reef has so far been brought under protection. In order to strengthen the island's economy, the fisheries as a whole has to be developed in a rational way keeping allowances for stock replenishment. To a greater extent the fish production is dependent on the well being of coral reef resources and hence for the future overall social and economic development of these islands, the following management measures are advocated.

1. A separate division on priority basis may be started in the Fisheries Department by appointing a coral reef warden and supporting staff to work in close association with research institutions and allied organisations to manage the multidisciplinary activities in and around the reefs.

2. Suitable conservational rules may be formulated as per the provision of Andaman and Nicobar Islands Fisheries Regulation—1 of 1938. The commercial exploitation of ornamental fishes and invertebrates should not be permitted till the enactment of the proposed rules.

3. Collection and removal of live or dead, horny or soft corals needs to be totally banned.

4. Tourists should not be permitted to explore the coral reefs unguided. Department of Fisheries or Forest may arrange to provide marine guides for underwater surveillance.

5. To bring about public awareness, a marine aquarium displaying various sea life forms along-with extension implements explaining the coral reef system and its importance may be established. Areas with good coral growth should be declared as marine parks where the tourists and the public can be taken around to get them acquainted with the reef life.

Acknowledgements

The authors deeply express sincere thanks and gratitude to Prof. M. Devraj, C. I. F. E. Bombay and Dr. B. F. Chhapgar, Scientist B. A. R. C. for their guidance, encouragement and help. The help extended by Mr. Y. B. Mohan ex-Chief Engineer F.S.I., Mr. Mathew D'Silva marine aquarist from Bombay and Mr. Karan Bahadur of Indian Navy during survey is also gratefully acknowledged. The much needed encouragement and co-operation given by Mr. Seva Ram Sharma ex-Chief Secretary and Mr. M. K. Bezboruah then Development Commissioner, to carry out the project needs special mention. The thanks are also due to Mr. Bhagat Singh, Superintending Engineer (Electricity) and Lt. Cdr. Balwant Singh ex-Assistant Harbour Master, Marine Department for extending the facilities in the present endeavour. We are also greatful to the staff of Z.S.I., Calcutta and N.I.O. Bombay for helping in identification of the specimens.

References

Anon. 1977. *Indian Fisheries* 1947-1977. Issued on the occasion of the fifth session of the Indian Ocean Fishery Commission held at Cochin. pp. 96.

Munro, J. L. 1984. Coral reef fisheries and world fish Production *ICLARM Newsletter*. 7 (4) : 3-4.

Pillai, C. S. G. 1983. Coral reefs and their environs, *Bull. Cent. mar. Fish. Res. Inst.* **34** : 36-40.

Tikader, B.K. 1986. *Sea shore animals of A & N Islands*. Z.S.I., Calcutta, pp. 188.

Whitaker, R. 1985. *Endangered Andamans*. Dept. of environment. G.O.I., pp. 51.

J. Andaman Sci. Assoc. 3(1) : 50-52, June, 1987

Observations on the Hypophysation of *Labeo rohita* by Human Chorionic Ganadotrophin in the Andamans

ARIF M. MUSTAFA

Department of Fisheries, A & N Islands, Port Blair

The technique of hypophysation of Indian major carps by the administration of exogenous fish pituitary hormones for the production of fish seed has been adopted in the Andamans since 1983. A number of problems have cropped up in the course of time. These include non-availability of pituitary glands in the Islands; loss of potency and damage during postal transaction from the mainland; cheating by the private gland dealers by adulterating the glands with vegetable seeds; variability of hormonal activity and difficulties in preservation of glands. In order to overcome these constraints it was felt necessary to introduce some other inducing agent of known activity which could be easily available and stored for long times under normal conditions.

The crude human chorionic gonadotrophin (HCG) recently introduced in the market under the trade name of 'SUMAACH' by INFAR India Ltd., was used in conjunction with pituitary extracts in the ratio 3 : 1 to induce rohu in captivity. It's standardised hormonal activity is 30 IU/mg. The studies were conducted in two private tanks of 0.05 and 0.08 ha at Herbetabad. The fish population in these tanks was 450 and 318, the plankton growth 0.02 and 2.2 ml/50 litre, the pH was 8.0 and 7.8 and the mean DO_2 value during the experimental period was 2.3 and 3.2 ppm, respectively. All the fishes were in the age group of 2-3 years, no special care was taken in raising them as brooders. The general condition of fishes from tank No. 1 was not very satisfactory as all of them looked undernourished, lean and less active. The fishes from tank No. 2 were comparatively better. All the males were oozing milt on applying slight pressure over vent whereas the females abdominal bulging and softness was not as satisfactory as in experiments conducted earlier (Mustafa, 1985). The colouration and condition of vent was good in the females.

Three years old alcohol preserved pituitary glands were used, gland extract was prepared as usual. The required quantity was macerated in tissue homogeniser with distilled water, Clear supernatent fluid was taken after centrifugation in an electrical centrifuge for three minutes. Required quantity of HCG was taken after weighing in single pan electric balance, 2% increment in quantity was given for procedural

Table 1. Details of hypophysation

Recipient fish		1st dose				2nd dose						
Sex	Wt kg	Date and time	Amount of HCG used mg	Amount of PG used mg	Volume of the suspension ml	Date and time	Amount of HCG used mg	Amount of PG used mg	Volume of the suspension ml	Response	Eggs produced	% of fertilisation
Experiment in tank No. 1												
F	0.3	14.6.86 (1830)	0.7	0.3	0.1	15.6.86 (0030)	1.8	0.6	0.2	+	5,000	100
M M	0.3 0.3	15.6.86 (0030)	1.4	0.5	0.3	—	—	—	—	+ +		
Experiment in tank No. 2												
F	0.5	14.6.86 (1900)	1.1	0.4	0.1	15.6.86 (0100)	3	1	0.3	+	10,000	100
M M M	0.3 0.3 0.4	15.6.86 (0100)	2.3	0.7	0.3	—	—	—	—	+ + +		

loss. HCG powder was homogenised in a glass tissue homogeniser with required quantity of distilled water. The solution was centrifuged in an electric centrifuge for five minutes, the light yellow supernatent liquid was taken and blended with pituitary extract in a test tube. The blended hormonal solution was re-centrifuged for two minutes in electric centrifuge.

Experiments were conducted during the ideal breeding season alongwith routine seed production programme. Due to the shortage of breeders only two sets were used for this experiment. Atmospheric temperature ranged between 24°C to 27°C. All the injections were intra-muscular at the caudal peduncle above the lateral line and carried out by 1 ml syringe graduated up to 0.02 ml.

In both the experiments on *Labeo rohita* sexes were kept separated upto second injection. Females were administered first and second dose at the rate of 3 and 8 mg per kg body weight respectively. The males were injected once at the rate of 3 mg per kg body weight at the time of second injection to females. The total pituitary gland requirement was calculated and replaced by 75% of crude HCG by weight. Considerable difficulty was faced in properly loading the tissue homogeniser with HCG powder. The details of inducing agents and recepient fishes are given in Table 1. All the fishes responded positively and absolute fertilisation was achieved. The observed positive response confirms earlier works done on carps such as *Cirrhinus mrigala* (Khan, 1938), *Cyprinus carpio* (Ichikawa and Kawakami, 1944) and *Labeo rohita* (Chaudhuri & Singh, 1984; Singh, 1985). The threshold dose of HCG in combination with pituitary extract is reported to be 100 to 200 IU per kg (Chondar, 1980). In the present experiment the effective dose of crude HCG is found to be 250 IU per kg for the female and 70 IU per kg for the males. The use of crude HCG in Andamans could ease the major constraints in procuring and preserving the fish pituitary gland which is used as the conventional inducing agent.

Acknowledgement

The author wishes to express his gratitude and sincere thanks to the Director of Fisheries A & N Islands for providing all facilities and cooperation in the present endeavour.

REFERENCES

Chaudhuri, H.L. and S.B Singh, 1984. *Induced Breeding of Carps.* ICAR, New Delhi : pp 82.

Chondar, S.L. 1980. *Hypophysation of Indian Major Carps.* Satish Book Enterprise; Agra : pp 146.

Ichikawa, M and I. Kawakami, 1944. Acceleration of spawning in the carp by means of pituitary hormones. *Rep. Hyogo Fish Exp. Stn,* 5 : 1-6.

Khan, H. 1938. Ovulation in Fish. *Curr. Sci.,* 7 : 233-4.

Singh, K. 1985. A study on the effect of Human chorionic gonadotrophin on chinese carps and Indian major carps. *High Tech Aquaculture* special bulletin., CIFE, Bombay.

Mustafa, A.M. 1985. Observation on the seed production of Indian major carps in Andamans by hypophysation. *J. Andaman Sci. Assoc.* 1 (2) : 86-92.

The ICLARM Quarterly

January 1986 ISSN 0116-290X

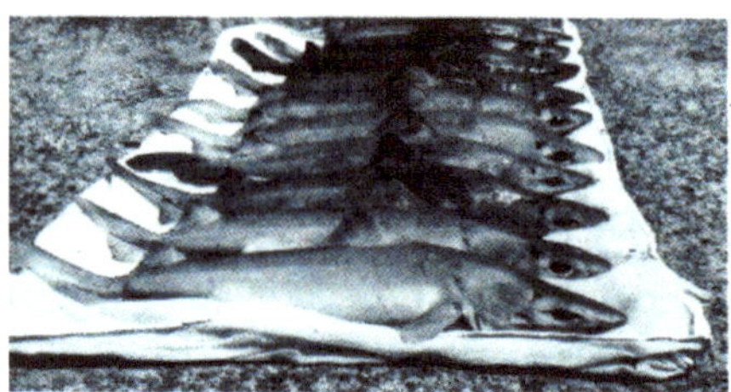

Spiny dogfish sharks (*Centrophorus* sp.) caught during early survey from ground "A" (see map opposite).

New Deep-Sea Spiny Dogfish Shark Resources off Andamans

ARIF M. MUSTAFA
Superintendent of Fisheries (Tech.)
Gafoor Manzil
Aberdeen Bazaar
Port Blair 744 104, India

The oceanic islands of the Andamans with narrow and fairly steep continental shelf have unexploited resources of deep-sea spiny dogfish shark (*Centrophorus* sp.). The Marine Products Export Development Authority of India had recently identified squalene rich deep-sea shark liver oil as an export item from India. The first *Centrophorus* sp. was caught from the continental edge of the Andamans during the third quarter of 1984.

Since then intensive survey beyond 200-m depth has been made by 10-m fishing boats using bottom set longlines. The species is found in the upper continental slope in the depth range of 300-800 m with maximum yield at 360-450 m. For the first time in peninsular India, fishing of *Centrophorus* sp. is picking up which is feasible and economical due to the narrow shelf.

Centrophorus sp. is found on or near the bottom at a temperature range of 5°-12°C. It is ovoviviparous having one to four young in a litter and is perhaps nomadic, moving in small to medium shoals having 20 to 30% of males per shoal.

The liver oil contains about 80-85% of squalene which is extracted and purified in Japan for various pharmaceutical uses. Characteristic features of the oil are as follows: (a) molecular formula of squalene–$C_3O\ H_5O$; (b) squalene content–81.3%; (c) vitamin A–nil; (d) unsaponifiable–75%; (e) iodine value–290; and (f) acid value–1.0. The liver weight is 25-40% of total body weight and each liver yields 50-80% of oil by weight. Extraction is by simple mincing, sun-drying and filtering. The flesh is devoid of peculiar uric shark smell and is acceptable for consumption either fresh or salted.

Japan is a ready market for the oil. The return per unit weight is 25 times more than other liver oil with no or negligible squalene content. Export has started and demand is increasing.

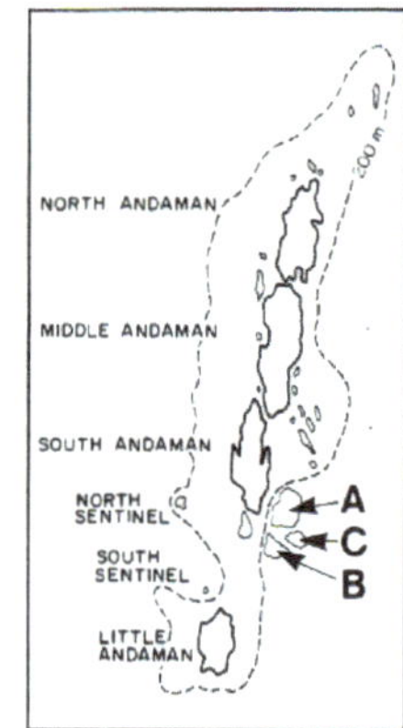

Continental shelf around Andamans, and recently potential deep-sea shark ground.

Potential

Fishing has revealed that the catch for *Centrophorus* sp. is poor only during the dark nights before new moon. Overall, hooking rate is 18% in a broad range of 2-31%, average weight per 100 hooks being 94 kg.

It is estimated that around the Andamans more than 4,000 t of dogfish sharks could be exploited yearly. The estimated catch per year for the recently located grounds marked A, B and C (see figure) alone is 350 t.

A closely resembling variety, *Squalus* sp., generally interferes at a depth of 300-350 m. The squalene content of this variety is comparatively less. •

Fisheries in the Andamans, China, Cyprus, Kuwait and the Philippines

Aquaculture in Nepal and the West Indies

Centrophorus sp.
Deep Sea Shark

J. Andaman Sci. Assoc. 1 (2) : 86—92 December (1985)

Observation on the seed production of Indian major carps in Andaman by hypophysation

ARIF. M. MUSTAFA

Department of Fisheries, Andaman & Nicobar Islands, Port Blair-744 101

ABSTRACT

The paper deals with the induced spawning of Indian major Carps undertaken in the farmers ponds around Port Blair, South Andamans. Alcohol preserved pituitory glands were used for hypophysation. In Rohu out of 12 females and 23 males, 10 females and 22 males responded to the injections. The average egg production ranged between 25,000 and 45,000 eggs per fish with 86 to 94% of hatching rate. In Catla, 1,05,000 eggs were produced by one female with 95% hatching rate and in Marigal the egg production varied from 10,000 to 27,000 eggs with 90% to 92% hatching rate. During the season a total of 4,97,500 eggs were produced by hypophysation and 4,52,000 hatchlings were obtained and distributed to the local pisciculturists.

Indian major carps viz. *Catla catla Labeo rohita* and *Cirrhinus mrigala* were introduced in the Dilthaman tank, Port Blair in 1965 by Prof. R.K. Gulati (Personal Communication). Later on for nearly one decade no serious attempt was made for the development of inland fisheries probably due to limited water resources. In recent past about 250 nos. of minor irrigation ponds have been constructed under various schemes in Andaman District. In addition to the creation of reservoirs such as Dhanni Khari and Jawahar srovar in South Andaman, old water bodies in the islands were also renovated thus providing about 30 hectare of water area for fresh water-fish culture. The demand for Indian major Carps fries slowly built up since 1976 with the re-settlement of refugees from East Bengal. Department of Fisheries arranged to air-lift the fish seed from Calcutta and supplied to fish farmers on no loss no profit basis. Major Carp culture was also taken up by the department for demonstration in a few tanks. Thus, fish culture as a subsistance proposition has gained considerable momentum in the Andamans.

About 205 private pisciculturists alone require more than one lakh of fries per season. Considering the very high cost of fries when air-lifted from Calcutta, the Department of Fisheries initiated induced breeding experiments on major carps by injecting pituitary hormones. First success for *L. rohita* and *C. mrigala* was achieved in 1983 (Anon 1984).

The results obtained in the induced breeding experiments on Catla, Rohu and Mrigal are reported in this paper.

Materials and Methods

All together seven experiments were

conducted. Brooders for these experiments were collected from private fish farmers. Approximately 1 : 2 female-male proportion was maintained by weight rather than numbers. The details of the number of fish used and their general conditions are given under each experiment.

Alcohol preserved pituitary gland suspensions were prepared in distilled water as described by Chondar (1980) and Chaudhuri (1984). Injections were given intramuscular above or below the lateral line on caudal peduncle. For estimating the total egg and the hatching rate the methods described by Chaudhuri (1984) were followed.

The present induced breeding experiment conducted at Andaman is third in the series. They were done at the farmers pond (0.05 ha.) at Caddlegunj and Herbatabad and at the site of the Fisheries Department at Port Blair. At the time of the experiment weather conditions were moderate to heavy raining. Water temperature ranged between 23° and 28°C and the atmospheric temperature 27 and 31°C.

Results

Rohu (*Labeo rohita*). During the spawning season 1985 four experiments were conducted to produce seeds through hypophysation. The details of the receipient fish, particulars of injections and the results of the experiments are given in Table. 1.

In the first experiment, one hour after the second injection splashing of water due to playful jumping of breeders was observed. Ten hours after the second injection 80,500 numbers of fertilised eggs were collected and transfered into hatching hapas. No unfertilised eggs were found in the hapa. On examination all the fishes were found responded positively. 75,000 hatchlings were collected which works out to 93% of hatching rate.

The breeders in the second experiment responded positively to the injections and 1,88,000 fertilised eggs were collected eight hours after the second injection. Hatching rate was 86%.

All the fishes used in the third experiment were not in good condition. All the females had reduced abdominal bulging and with fine ridges all around the internal circumference of vent. Colour of vent was pinkish white; watery jet released when pressure was applied around the vent. In males no free oozing of milt was observed on pressing around the vent. However, on antero-posterior pressing small quantity of milt oozed out as paste. Both the sexes showed prominent signs of absorption. After second injection both the sexes were released in one hapa. eight hours after second injection 25,000 fertilised eggs were collected. One female and two males were found to respond to the injection. 92% hatching was estimated.

Conditions of the breeders used in the fourth experiment was the same as in the third experiment. Out of three females only two spawned and in males all responded positively. Percentage of fertilitation was 98 and out of 49,000 fertilised eggs 47,000 hatchlings were recovered which works out a hatching rate of 96%.

Catla (*Catla-Rohu hybrid* ☿ X *Catla Catla* ♂) In the absence of pure *C. Catla* female, one Catla-Rohu hybrid female and two *C. Catla* males were used in this experiment. Both the sexes were in very good condition. The details of the injections are given in Table 2. Both the sexes responded positively and about 1,05,000 fertilised eggs were collected ten hours after second injection. Hatching ate was 95%.

Mrigal (*Cirrhinus mrigala*). Two experiments were conducted. The details

Table 1. Details of hypophysation on Rohu (*Labeo rohita*)

			Particulars of Injection						Results		
			1st Dose			2nd Dose					
S. No.	Recipient fish Sex	Weight (kg)	Date (Time in hrs)	Qty. of gland used mg	Vol. of the suspension ml	Date (Time in hrs.)	Qty. of gland used mg	Vol. of the suspension ml	Response*	Total eggs. produced. (Av. nos. of eggs per fish)	Hatchlings recovered in nos. (% of hatchling)
1	2	3	4	5	6	7	8	9	10	11	12
Experiment No. I											
1.	F	0.4	4.6.85 (1500)	0.8	0.3	4.6.85 (2300)	2.0	0.5	+		
2.	F	0.3	,,	0.6	0.2	,,	1.5	0.2	+		
3.	F	0.3	,,	0.6	0.2	,,	1.5	0.2	+		
4.	M	0.3	4.6.85 (2300)	0.6	0.3	—	—	—	+	80,500 (26,833)	75,000 (93)
5.	M	0.3	,,	0.6	0.3	—	—	—	+		
6.	M	0.3	,,	0.6	0.3	—	—	—	+		
7.	M	0.3	,,	0.6	0.4	—	—	—	+		
8.	M	0.2	,,	0.4	0.1	—	—	—	+		
9.	M	0.2	,,	0.4	0.1	—	—	—	+		
Experiment No. II											
1.	F	0.9	22.6.85 (1900)	1.8	0.5	23.6.85 (0300)	3.6	0.8	+		
2.	F	0.5	,,	1.0	0.4	,,	2.0	0.4	+		
3.	F	0.5	,,	1.0	0.4	,,	2.0	0.4	+		
4.	F	0.4	,,	0.8	0.3	,,	1.6	0.4	+		
5.	M	1.0	23.6.85 (0300)	2.0	0.3	—	—	—	+		
6.	M	0.8	,,	1.6	0.2	—	—	—	+		
7.	M	0.8	,,	1.6	0.2	—	—	—	+	1,80,000 (45,000)	1.55,000 (86)
8.	M	0.5	,,	1.5	0.1	—	—	—	+		

Table 2. Details of hypophysation on Catla (Catla—Rohuhybrid X Catla Catla)

Sr. No.	Recipient fish Sex	Weight (kg.)	Date (Time in hrs.)	Particulars of Injection: 1st Dose Qty. of gland used mg	1st Dose Vol. of the suspension ml	2nd Dose Date (Time in hrs)	2nd Dose Qty. of gland used mg	2nd Dose Vol. of the suspension ml	Response	Results: Total eggs produced. (Av. nos. of eggs per fish)	Hatchlings recoved in nos. (% of hatching)
(1)	(2)	(3)	(4)	(5)	(6)	(7)	(8)	(9)	(10)	(11)	(12)
Experiment No. I											
1.	F	1.5	4.6.85 (1530)	12.0	0.5	4.6.85 (2330)	18.0	2.0	+		
2.	M	1.2	4.6.85 (2330)	3.6	0.3	—	—	—	+	1,05,000 (105000)	1,00,000 (95)
3.	M	1.0	,,	3.0	0.2	—	—	—	+		

Table 3. Details of hypophysation on Mrigal (*Cirrihinus mrigala*)

Sr No	Recipient		Particulars of Injection							Results	
			1st Dose			2nd Dose					
Sr No	fish Sex	Weight (kg)	Date (Time hrs.)	Qty. of gland used mg	Vol. of the suspension ml	Date (Time in hrs.)	Qty. of gland used mg	Vol. of the suspension ml	Response	Total eggs produced. (Av. nos. of eggs per fish)	Hatchlings recovered in nos. (% of Hatching)
(1)	(2)	(3)	(4)	(5)	(6)	(7)	(8)	(9)	(10)	(11)	(12)
Experiments No. I											
1.	F	0.9	4.6.85 (1600)	1.8	0.2	4.6.85 (2400)	3.0	0.5	+ (—)		
2.	M	0.5	4.6.85 (2400)	1.5	0.2	—	—	—	+	27,000 (27000)	25,000 (92)
3.	M	0.5	,,	1.5	0.2	—	—	—	+		
Experiment No. II											
1.	F	0.8	22.6.85 (1930)	1.6	0.2	23.6.85 (0303)	2.4	0.3	+		
2.	F	0.5	,,	1.0	0.1	,,	1.5	0.2	+		
3.	F	0.5	,,	1.0	0.1	,,	1.5	0.2	+		
4.	M	0.7	23.6.85 (0330)	1.4	0.2	—	—	—	+	30,000 (10,000)	27,000 (90)
5.	M	0.7	,,	1.4	0.2	—	—	—	+		
6.	M	0.7	,,	1.4	0.2	—	—	—	+		
7.	M	0.6	,,	1.2	0.1	—	—	—	+		
8.	M	0.6	,,	1.2	0.1	—	—	—	+		

1	2	3	4	5	6	7	8	9	10	11	12
9.	M	0.5	,,	1,5	0.1	—	—	—	+		
10.	M	0.4	,,	1.5	0.1	—	—	—	+		
11.	M	0.4	,,	1.0	0.1	—	—	—	+		
12.	M	0.4	,,	1.0	0.1	—	—	—	+		
Experiment No. III											
1.	F	0.6	13.7.85 (1700)	2.4	0.3	13.7.85 (2400)	3.0	0.5	+		
2.	F	0.6	,,	2.4	0.3	,,	3.0	0.3	(—)		
3.	M	0.5	,,	1.0	0.2	,,	1.0	0.2	+	25,000 (25,000)	25,000 (92)
4.	M	0.5	,,	1.0	0.2	,,	1.0	0.2	+		
5.	M	0.5	,,	1.0	0.2	,,	1.0	0.2	(—)		
Experiment No. IV											
1.	F	0.7	13.7.85 (1800)	2.8	0.3	14.7.85 (0100)	3.5	0.7	+		
2.	F	0.6	,,	2.4	0.3	,,	3.0	0.5	(—)		
3.	F	0.5	,,	2.0	0.2	,,	2.5	0.4	+		
4.	M	0.6	,,	1.2	0.1	,,	1.2	0.1	+	50,000 (25000)	47,000 (94)
5.	M	0.6	,,	1.2	0.1	,,	1.2	0.1	+		
6.	M	0.6	,,	1.2	0.1	,,	1.2	0.1	+		
7.	M	0.6	,,	1.2	0.1	,,	1.2	0.1	+		
8.	M	0.6	,,	1.2	0.1	,,	1.2	0.1	+		
9.	M	0.6	,,	1.2	0.1	,,	1.2	0.1	+		

* +signs denotes positive and (—) sign denotes negative.

of the reci pient fish, injection and the results are given in Table 3. About 10 hours after the second injection it was found that the females spawned partially, as eggs were oozing out when it was removed from the hapa. Attempts were made to fertilise the eggs by dry method but it was not successful. 27,000 eggs were obtained from the hapa out of which 92% of hatchling were recovered.

The second experiment was conducted in a plastic pool which was provided with aeration and running water supply for about four hours after the second injection. All the fishes responded well and 30,000 fertilised eggs were collected. Hatching rate was 80%.

Discussion

In the present experiment egg production in Rohu works out to 81,78,42 and 41 numbers per gram respectively in the four experiments. Nearly 50% reduction in the egg production the third and fourth experiments as compared with the first and second experiments may be attributed to the fact that these experiments were conducted in the middle of July which is the late spawning period for this species. Chaudhuri (1984) recorded 164 to 185 and Chondar (1980) 200 eggs per gram of body weight. Probable reason for low egg production in the present study might be due to small size fish used for the experiment. Egg production in Catla and Mrigal were also too low as compared to earlier works (Chondar, (1980). However, the hatching rates were comparable with those of earlier works. (Chondar, 1980 ; Chaudhuri and Singh, 1984).

The present study has given encouraging results thus providing a new avenue for fish seed production to meet the requirements of the pisciculturists of Andaman and Nicobar Islands.

Acknowledgement

The author wishes to extend his sincere thanks to Mr. P. Ravindran Nair, Director of Fisheries, A&N Islands for his constant encouragement and providing all facilities for conducting the experiments and to Shri K. Dorairaj, Scientist S-3 CARI, Port Blair for critically going through the manuscript.

REFERENCES

Anon, 1984. Annual General Administration Report 1983-84. (In press). A&N Administration, Port Blair.

Chondar, S.L. 1980 *Hypophysation of Indian Major Carps* Satish Book Enterprise, Agra : pp 146.

Chaudhuri, H.L. and Singh, S.B. 1984. *Induced Breeding of Carps*. Indian Council of Agricultural Research, New Delhi : pp 82

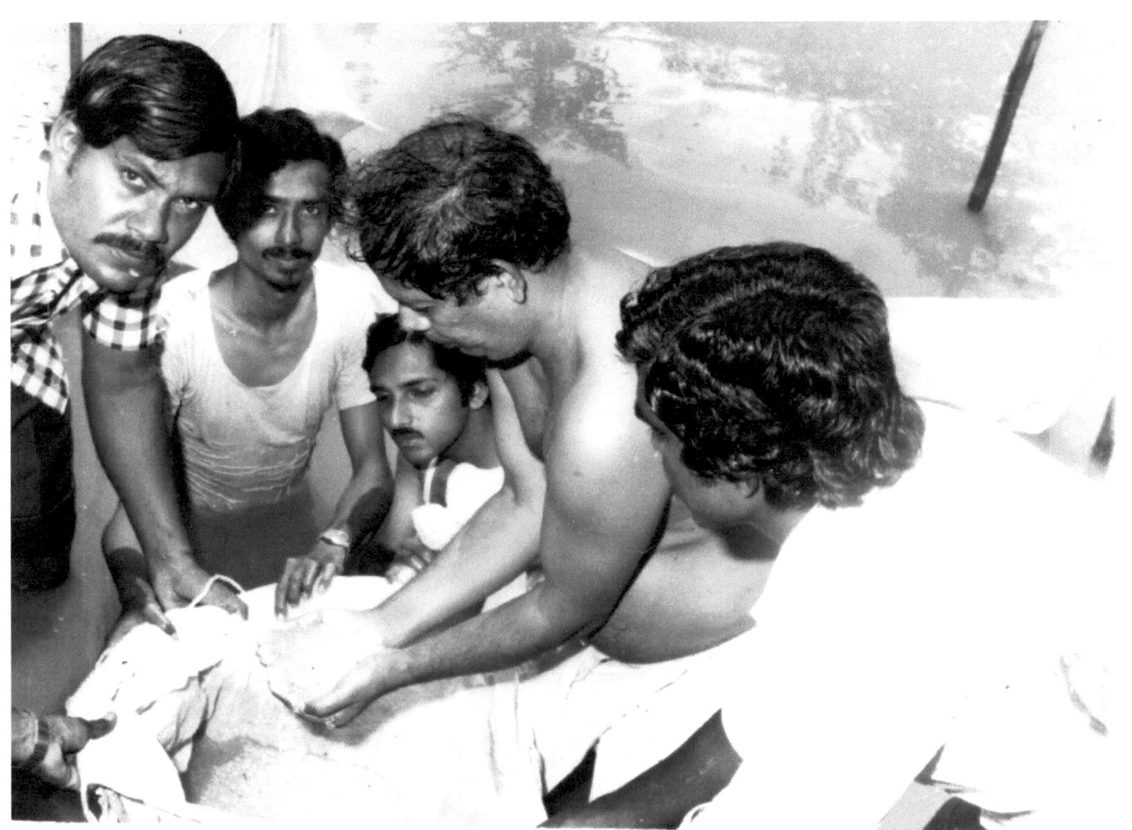

First induced breeding of IMC *Labeo rohita* using fish Pituitary harmone in A&N Islands on 21st June,1983 under the guidance of Dr. K.P.Biswas, then Director of Fisheries

IMC *Catla catla* harvested from Dilthamman Tank during brooder hunt in the year 1984

ICLARM NEWSLETTER

Official Publication of the International Center for Living Aquatic Resources Management

Vol. 6, No. 4 | Metro Manila, Philippines | October 1983

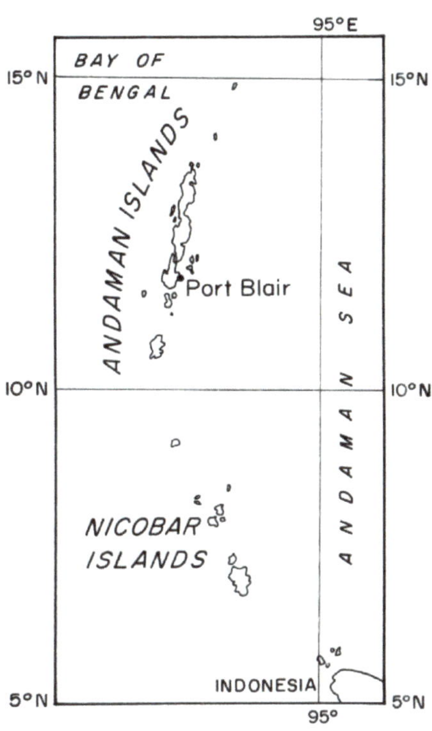

Fisheries of the Andaman and Nicobar Islands

ARIF M. MUSTAFA
Gafoor Manzil, Aberdeen Bazar
Port Blair–744104
Andaman and Nicobar Islands, India

A filamentous stretch of 585 islands, islets and rocks in the northeast Indian Ocean, occupying an area of 8,293 km^2, the Union Territory of Andaman and Nicobar Islands has a coastline of 1,500 km and a continental shelf area of some 35,000 km^2. The shelf is narrow, 3-8 km wide, and fairly steep, with many fringing reefs in the east and barrier reefs on western side harboring rich populations of corals and molluscs. The Union has sovereign rights to a sea area of nearly 600,000 km^2, from which a substantial harvest of fish can be taken annually. The waters are warm but nutrient poor (see box).

Oceanographic features

surface temperature	27-28.5°C
surface salinity	31.2-32.15‰
surface currents	2% wind velocity
nutrients	nitrate-N and phosphate-P negligible near the coast, limiting productivity

Major Developments

The first fisheries regulations were laid down in 1938. However, a Fisheries Research Unit was not set up until 1949, at Port Blair. This unit was renamed the Fisheries Development Unit in 1955 and in 1975 became the Directorate of Fisheries. A current proposal to the Indian Government is to establish a Fisheries Development Corporation.

Since World War II, a few attempts to exploit various fisheries on an industrial scale have been made, the most recent of which was a deep-sea fishing joint venture Indo/Tata-Thai, which lasted two years, 1978-1979.

The existing cold storage facilities were constructed in 1968, consisting of a 5-tonne/day capacity ice plant and 15 tonnes of cold storage.

In the early 1970s a settlement scheme was introduced to increase the size of the fishing community. So far about 125 fishermen families have settled from mainland India.

A training center was set up in 1977 to train youths in mechanized fishing. Each year, 30 candidates undertake a 9-month certificate course. Subsidies are available for purchase of small boats with outboard motors to bonafide fishermen.

In research, an Exploratory Fisheries Project began in 1971. The survey vessels are two 17-m trawlers.

The Fisheries

Annual landings have increased steadily over the past 30 years from 44 tonnes in 1950, with rapid increases in the past few years to 3,850 t in 1982 (see graph). Of 700 recorded species, there are around 200, classified into 62 groups, of economic importance.

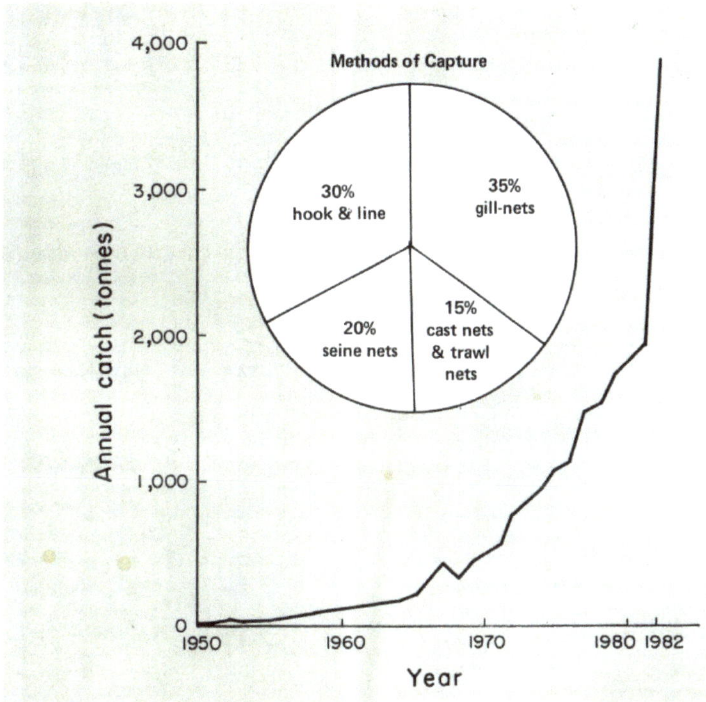

Marine fish landings in Andaman and Nicobar Islands since 1950.

The major groups, main fishing seasons and their proportion of annual landings are shown in the table.

Although molluscan shellfish are only consumed by aborigines of the Andaman Islands group, a commercial shell fishery has been in operation since 1929. There are nine demarcated zones which are publicly auctioned each year. A variety of shell handicrafts is produced, mostly for export to India.

The Island aboriginal groups still catch fish this way.

Equipment

Traditional fishing boats are simple, wooden vessels propelled by paddle, punting and sail. Three types are recognized—flat bottomed, round bottomed and Burmese dugout—ranging in length from 3 to 11 m. Crew sizes are 3-5, 4-6 and 6-16, respectively.

Major traditional gear types include drift gill nets and seine nets. The most advanced gear is an anchored seine made of cotton consisting of a codend and tapering wings. Operated from a boat, it is anchored across the current, which keeps the net open. Cast nets and hook-and-line gear are also in use.

The Island aboriginal groups still use ancient fishery technology described from the Paleolithic age in India. Boats are simple dugouts, sometimes with outriggers, rarely with sails, from 4 to 15 m long. Fishing methods range from hand-catching at night with flares, various poisons, harpoons, bow and arrow to hand-dragged scoops and tanglenets.

The author, standing beside a Burmese dugout.

Table. Fishery resources and their seasonality, Andaman and Nicobar Islands.

Resources	% of catch	Peak season
Pelagic resources		
Sardines – *Sardinella* spp., *Dussumieria* spp., and *Anodontostoma* sp.	13.9	July-December
Carangids – *Megalapsis cordyla, Caranx* spp., *Elagatis* sp., *Decapterus* sp. and *Selar* spp.	9.2	July-December
Mackerels – *Rastrelliger kanagurta* and *R. brachysoma*	9.1	March-June September-December
Anchovies – mainly *Anchoviella commersoni* and *Thrissocles mystax*	7.6	June-November
Scombrids – mainly *Scomber commersoni* and *S. guttatus;* some *Auxis thazard*	7.2	March-August
Barracuda – *Sphyraena jello*	2.9	July-September
Tuna – yellow fin *(Thunnus albacore),* bigeye *(T. obesus),* albacore *(T. alalunga)* *(Euthynnus affinis),* and skipjack *(E. pelamis)*	2.6	December-January
Demersal resources		
Perch-like fishes – especially *Lates* sp., cods (*Serranus* spp. and *Epinephelus* spp.), threadfins (*Polydactylus* spp.), emperors (*Lethrinus* spp.) and grunts (*Pomadasys* spp.)	13.4	August-November
Silver bellies or slipmouths – mainly *Leiognathus equula, L. splendens* and *Gaza* sp.	10.2	June-December
Low-value demersal fishes – croakers (scianids), goatfish (upeneoids), threadfin breams (nemipterids) and lizardfish (saurids)	5.7	–
Mullets (and minor estuarine fish), mainly *Mugil tade, M. cephalus* and *Liza* spp.	5.3	July-December
Sharks and rays – 11 shark spp.; 6 rays	2.7	–
Crustaceans – mainly the shrimp *(Metapenaeus dobsoni),* and the lobsters *Panulirus polyphagus* and *P. ornatus*	2.4	(Shrimps) November and February-March

Effort

Excluding aboriginal craft, the fishing fleet consists of about 760 traditional craft and 37 mechanized 10-m boats, using 620 gill nets, 26 boat seines, 540 cast nets, 19 shore seines and innumerable sets of hooks and lines. Most of the subsidized engine-powered vessels are now used as ferries or cargo boats. There are about 2,300 fishermen—1,700 full-time, 400 parttime and 170 occasional; nearly all are emigrants from India. Women play no roles in either fishing or marketing.

Marketing

The catch is generally sold directly by fishermen. A small proportion is bought by middlemen who sell from door to door or at the road side if there is no nearby marketplace.

Most of the catch is sold fresh; only 15% is processed–salted or sun-dried–with the usual defects of poor salting, rancidity and high sand content.

Potential

The available survey reports indicate the presence of fairly large stocks of both pelagic and demersal resources, as shown in the table. However, bottom trawling is limited due to the narrow shelf and rough bottom. Best results have been in 20 to 29 m depths, where average catch/hour was 200 kg. Longlining trials over 5 years have yielded catch rates of 9 fish per 100 hooks, and trolling, 25-40 kg/hour.

Traps have also been successful, one survey yielding 680 kg for 125 hours of effort.

Productivity data suggest that a total annual production of 470,000 tonnes could, in theory, be generated within the 600,000-km^2 territory. However, actual harvests would be a fraction of this amount, depending on the economic features of the fishing operations.

Long-term planning is needed by an outside agency in consultation with the local administration, taking into account all expected future developments. A Fisheries Development Corporation may be set up with the initial objective of on-board processing and direct export to nearby countries until shore-based facilities are developed. •

Top: Round-bottomed anchor-net craft ready to move out for fishing. ***Middle:*** Aberdeen jetty, the largest marine fish landing center of Andaman and Nicobar Islands. ***Bottom:*** Burmese dugout (side view) with snail curved keel.

ICLARM is certainly a unique organization. It is rather small, despite the passing parade of staff profiles. In fact, we have but seven professional staff in Manila at present, and another seven in field positions. The numbers haven't changed much over the last four years, which encompass ICLARM's publishing history.

Recently, we sat down to itemize our publishing record and surprised ourselves. In the first three years, ICLARM published 22 major monographs, other books and reports. This year we increased the total by 50%, another 11 titles. Some 26,000 have been distributed free or sold, while over 80,000 copies of the Newsletter have been sent out.

Through these publications and contributions to the primary literature, which together total no less than 170 contributions to date, there has been considerable impact. In this issue are featured two "impact areas." Stock assessment methodology developed at ICLARM since 1980 is being used in twelve countries so far; a report on its cost-effective use in one such country is presented (p. 3). The multidisciplinary study of a complex Philippine fishery which has already resulted in legislative change is our main feature (p. 11). Truly a model study, it should provoke further reaction in administrative circles throughout developing countries, where overfishing and ineffective resource management are the norm.

Also featured are the results of two other ICLARM projects in the fields of aquaculture economics (p. 5) and anthropology (p. 6). We thank Mr. A.M. Mustafa for a summary of the first comprehensive study of fisheries in the Andaman and Nicobar Islands. Ed.

SUBSISTENCE FISHING

KATCHAL ISLAND, CENTRAL NICOBAR

COMSUMPTION OF FRESH, DRESSED, RAW FISH AND SHELL IS A WAY OF LIFE

FISH AND FISHERIES
OF
ANDAMAN AND NICOBAR ISLANDS

by

Arif M. Mustafa

DISSERTATION

JUNE - 1981

CENTRAL INSTITUTE OF FISHERIES EDUCATION

(INDIAN COUNCIL OF AGRICULTURAL RESEARCH)

POST BOX NO. 7392, J. P. ROAD, ANDHERI [WEST],

BOMBAY
INDIA

F I S H A N D F I S H E R I E S
O F
A N D A M A N A N D N I C O B A R
I S L A N D S .

by

Arif M. Mustafa.

from :
Port Blair
Andaman Islands.

A dissertation submitted in partial fulfilment for the award of POST GRADUATE DIPLOMA in FISHERY SCIENCE. (Recognised by Government of India as equivalent to the M.Sc. degree of Indian Universities).

at

CENTRAL INSTITUTE OF FISHERIES EDUCATION
(Indian Council of Agricultural Research)
Bombay.

June - 1981.

D E D I C A T I O N

I dedicate this dissertation in honour of those:

Innocent victims who where
killed in Andaman & Nicobar
islands by the Japanese
occupation force during IInd
world war.

ACKNOWLEDGEMENTS

With pleasure, I express my sincere thanks to :-

Dr. S. N. Dwivedi, Director, Central Institute of Fisheries Education for suggesting this problem and for his constant encouragement and sincere help. I am grateful to him for permitting me to visit Andaman and Nicobar islands for collecting data contained in this work.

Prof.R. K. Gulati, C.I.F.E., my guide and mentor for providing an out line of the present work and also for his invaluable guidance throughout the present study. I thank him especially for providing me with references over Andaman Fisheries and also for going critically through bulk of this work.

Dr. M. Devaraj, C. I. F. E., for his guidance for chapters on estuarine and marine fishery section of this work and also for encouragement during the initial stages of this work. He deserves special mention for finalising the list on economically important fishes.

Prof. Y. Shrikrishna, C.I.F.E., for his guideline on the chapter dealing with indigenous crafts and also for reading through the same, critically.

Prof. P. S. Rao, C.I.F.E., for helping me with the principles of fishery regulation and economics etc.

Mr. Rajan Abraham, C.I.F.E., for critically going through the chapter on the fishery implements of the aborgines of Andaman and Nicobar Islands.

Mr. K. S. Bhatnagar, C.I.F.E., for guiding me for completing the chapter on traditional craft and gear and also for many useful suggestions.

Mr. P. M. Sherry, C.I.F.E. for his wholehearted help in preparation of the taxonomic list of commercially important fishes of Andaman and Nicobar Islands.

(ii)

Dr. S.Z. Quasim, Director, N I O, Goa, for his kind permission to use the survey done by R.V. Gavashni on Andaman sea and also for his help in providing me with many invaluable papers pertaining to this work.

Dr. K. Peethambaran Asari, Department of Zoology, Government college Port Blair, my ex-teacher and well wisher for going through the entire manuscript and also for offering very constructive criticism. I am much indebted to him for the pains he has taken in getting this work completed in time.

Dr. Talwar and Mr. H.C. Ghosh, Zoological survey of India for their sincere help in getting information on fishes and crustaceans of Andaman and Nicobar islands.

Dr. S.A.H. Abidi, Scientist, N I O, Bombay regional centre for his help in getting me expert help from crustacean and molluscan experts of the centre in solving few problems associated with those groups collected from Andaman and Nicobar islands.

The director and technical personnel of the directorate of fisheries, Andaman and Nicobar Administration for providing with some very useful information on landing data and other informations.

Dr. S.K. Dass, Officer-in-charge, Zoological survey of India regional station, Port Blair for providing the list of locally available corals, echinoderms and fishes etc. I am indebted to him for his help in obtaining many published informations about the fauna of this region.

Dr. T.N. Pandit, Officer-in-charge, Anthropological survey of India regional station for helping with the chapter on aborgines of these islands and also for permitting me to go through some unpublished informations from their library.

The wild life warden Andaman and Nicobar administration and Mr. Bhatt, in-charge of the Haddo mini zoo for their help in getting information on the protected species of this territory.

My friends, Mr. J. Chandrasekhar, K. Bahadur and GovindaRaju for their untiring efforts in getting this work completed in time.

The innumerable fishermen of this territory for helping me in getting first hand information on their profession and socio-economic conditions etc.

Mr. M. Mohan and Mr. Jamal for their timely and devoted help in getting this work typed in time.

The then director of fisheries of Andaman and Nicobar administration, Dr. S.A.H. Abidi for selecting me for undergoing this training and for his constant encouragement and help.

The Andaman and Nicobar administration for providing me stipend for this training.

All my colleagues of C I F E for their co-operation and help during my stay at C I F E.

My mother, Mrs. Hafizun Gafoor, brothersMr. M.A. Mujtaba and Yameen M. Murtaza, wife Ishrut and daughter Aurshi for their compassion and understanding during my preoccupation with my studies.

CONCLUSIONS

1. The Andaman and Nicobar group of islands of Bay of Bengal, lies at latitude 6°45'N and 13° 45' N and longitude 92° 15' E and 94° 00' E. It has a total number of 321 islands and 264 rocks and islets with a total coast line of 1,500 Km. and a total land area of 2,280 (approx.) Sq. Km.

2. The terrain of these islands is uneven with dense, evergreen low latitude tropical foests. The low lying coastal areas of these are engulfed by luxurient mangrove forests.

3. Geologically these islands are believed to have been originated as a result of submergence and tilting of the once continuous range from Arakan Yoma of Burma to the Achin head of Indonesia. The oldest sediment of these islands dates back to 100 million years.

4. Barren and Narcondum islands of the Andaman group are on the earthquake belt of South east Asia (Tectonic) belt of Tertiary era)

5. These islands experience both the monsoons. Rainfall is often very heavy(2,750 mm. to 4,550 mm.). Recognisable seasons are monsoon and dry season. Temperature fluctuation is very less (4- 8° C.) The Andaman sea has been the intensifier for many a cyclone .

6. The Andaman sea is connected with the surrounding seas by Preparis channel at North, 10° channel between Andaman and Nicobar islands and Great channel beyond Great Nicobar.

7. Low concentration of Oxygen has been found on the eastern side of Central Andamans, indicating that this region is seperated by shallow sills from North and South.

8. Andaman sea is oligo tropic, the western side (Bay of Bengal) is more productive than the eastern side (Andaman sea).

9. The tertiary production of Andaman sea has been calculated to be 18.70×10^5 tonnes C/ year Of which 25% or 4,70,000 tonnes of fish per year can safely be exploited.

10. There is no pollution in the sea around Andaman and Nicobar islands, in the strict term.

11. The first colonization of these islands was in 1789; extensive settlements however, started only after 1858. New settlements are established even now.

12. Organized fishing activity was started in 1929 in the form of shell fishery. However, the fishery developments and exploitation of marine wealth was initiated only in 1947. A research unit was established in 1949, which by gradual development has grown to a directorate of fishery in 1975.

13. Main achievements of the fishery directorate are; control of shell fishery, cold storage facility, settlement of fishermen community, training and education facility, supply of fishery requisites, mechanisation, aquaculture, extraction of sharkliver oil, formation of cooperative socities for better exploitation and marketing, provision of harbour and anchorage facility for fishing boats and construction of fish market.

14. The topography of the sea bottom does not permit bottom trawling and hence, the assessment of resources by bottom trawling of a limited area by 17.5 m. boats can not be of much use. Tuna long lining and line trawling were found to be very encouraging.

15. Out of 700 species (approx.) recorded from these islands 203 are of economic value.

16. Deep sea fishing conducted for a very short time by the Tata - Thai venture provided excellent results.

17. There are about 10 mechanised boats, 605 country crafts and 1,210 active fishermen operating in the coastal waters. There are about 550 nets operated from these non-mechanised boats. All these bring about 2,000 m.tonnes of fish valued at 6,00,000 rupees (1978-'79).

18. Of all the major divisions of Andaman and Nicobar islands viz., South, North, Middle Andamans and Nicobar, landing is more in South Andaman. But it is difficult to say that South Andaman is more productive. It may be due to more fishermen, many more crafts and gear and more landing centres located in South Andaman.

19. Estuarine and inland fishery resources are yet to develop in a big way.

20. Elasmo branch fishery is dominated by 11 species of sharks.

21. Prawn fishery is confined only to Andaman region and no special crafts and gear is used for this purpose. Annual landing is assumed to be about 70 tonnes. No survey has been done for prawn resources, although 40 species of prawns are reported.

22. Shell fishery is mostly based on Turbo, Trochus and Nautilus and is comparatively better developed. The approximate annual turn over is about 11 lakhs of rupees from 13 registered shell craft units and 7 shell fishing zones.

23. Minor fisheries are mainly lobester, crab and holothurian fisheries.

24. The sea around Andamans possess many varieties of equarium fishes of very high quality.

25. Many varieties of sea-Weeds are found in the Andaman waters. However, no proper survey is conducted regarding their quantity and quality.

26. Marine mammals, crocodiles, turtles and robber crabs of these islands are protected under wild life act, 1972

27. The main traditional fishing gears used are; cast net, drift gill net, ancher net, shore siene and hook and line.

28. The main traditional crafts are flat-bottom, round bottom and dug out canoes. None of these are mechanised.

29. Loss to wooden crafts by marine borers and foulers is maximum in these islands.

30. The aborgines and tribals viz., Great Andamanese, Oggese , Jarawas, Sentinelese, Nicobarese and Shompens of these islands still use the oldest methods of fishing. They use bows and arrows, harpoons, spears, entangling nets from natural fibres and stupifying devices.

SUGGESTIONS

First and foremost, the fisheries division of these islands requires integrated long term planning taking into consideration all the possible future developmental activities. It has been observed from the past history of the fisheries department that the chief executives always had their own ideas without really studying the multivarious necessities in depth. In most of the cases, these ideas either remained miserably incomplete or were overlooked and neglected by the successors. The change of the chief executive very often, before completing a particular project or plan and the criminal neglect shown to the same by the next incumbent and also the curious manner by which this process was repeated very often has been oneofthe main reasons for the present miserable status of the fisheries developments of these islands. It is suggested that long term planning may be drawn up meticulously by taking into consideration all the expected future developments. This planning must start from the grass root level and preferably this should be done by an outside agency of experts. It should be the duty of the chief executive of the fisheries department to mainly adhere to these broad guidelines while drawing up his own ideas for immediate local requirements.

The infrastructure facilities available at present for the development of fisheries is throughly inadequate. The idea of mechanisation was laudable, but most of the mechanised boats of the fisheries department are under constant repairs. The process of repairing is extremely slow as the fisheries department lacks repairing facility. The country canoes are made without any imagination and expertise. Similarly, the gears used, available marketing facility and also the welfare measures for the fishing community also is poor. It is therefore suggested that a boat building/repairing yard, exclusively for the department of fisheries should be constructed. A deep freezing plant with bulk storage facility and a workshop with skilled manpower for repairing and distributing country crafts of local requirements, good marketing outlets etc. should be given top priority. In addition to these, training and fishery education facilities for the local youth, community welfare measures like allotment of houses to the fishermen should also be given equal importance.

It has been proposed by the department of fisheries and few other expert committies on fisheries for setting up a modern processing plant at Campbell Bay in Great Nicobar. The establishment of such a plant shall be extremely difficult as everything should be brought from the mainland, India; moreover, this place is much under developed, without any basic amenities. The tuna survey off Great Nicobar has not been completed and their exploitable stock has not been ascertained. So, it is feared that the products processed in the proposed plant, even after the steady availability of the raw materials, may be much costlier than those available in the mainland, India. This problem will be more acute when these will have to withstand the extremely stift competition from other countries in the international market. It is therefore, suggested that instead of going for a big investment whose outcome is not sure, the fish catches, if in good quantity can be sold as raw materials to the neighbouring countries as is done by many developed and developing countries. Suitable outlet can be found at the ports of Singapore, Thailand and Taiwan which are nearer.

The present shark liver extraction centre located at Port Blair should be supplied with enough raw material. It has been observed that the present set-up is for the plant to depend upon the shark catches of fisheries jetty and Aberdeen landing centres as there is no transport facility to bring the shark catches from the remaining 18 landing centres of South Andaman; not to speak of the same from middle and North Andaman and also from Nicobars. A refrigerated storage vessel/vehicle may be obtained for the above mentioned purpose.

The lack of skilled manpower has been another problem with the department. This phenomenon exists from top to bottom. Perhaps the main reason for this may be the geographical isolation of this region and the hesitancy of skilled personnel to come over here and work. It is suggested that, while selecting personnel of all categories, consideration must be given for ability and zeal and proper training obtained from professional fisheries bodies such as C.I.F.E, C.I.F.T, C.MF.R.I, rather than pure academic qualifications. Most of the trained personnel trained at the expense of the public exchequer for working in the fisheries department were not able to prove their worth, as the department was not able to absorb them in suitable position. Many of them are working in menial position in other establishments, wasting their talents and the knowledge gained during training. It will only be in the fitness of things, if persons are trained for specific purposes to this place to be accommodated in pre conceived projects. Most of the skilled fishermen are throughly disillusioned as they do not have the basic minimum requirements like crafts, gears of their choice, proper house sites and marketing outlets. This has caused migration of skilled fishermen to other places.

Extension work is practically nil at present. The inhabited islands are widely scattered and their exists lack of communication between the fisheries head quarters at Port Blair and these farflung islands. At present, the personnel posted at these islands are mostly untrained in the basic principles of fisheries and they do not have any facility to conduct extention work. It is suggested that persons of the rank of Assistant fisheries development officers should be posted in all these islands for extension work and these persons should be provided with suitable manpower for the execution of the same. In many remote places, inland and estuarine fisheries are at its very infancy and the local population seems to be blissfully ignorant about the scientific methods of their exploitation. It should be emphasised that the knowledge gained from experience and studies and also the fruits of experimental results obtained from different sources should be supplied to the people along with encouragement and minimum basic facilities, so that these knowledge can be practically implemented at the village level. It has been found that the traditional crafts and gears used by the local fishermen are very poorly equipped and are often defective. It is a pity that these people are not guided properly for adopting better crafts and gears of scientific nature. It is felt that proper extension work by trained personale will fill this lacuna.

Knowledge on the biology of exploitable fishery resources, their seasonal availability and ecology etc., is an integral part of fishery development/exploitation. At present the only authenticated survey report (fishery resources) is from only a very small patch, supplied by the Exploratory fisheries project. Nothing is known about the crustacean resources especially, those of prawns and lobsters, eventhough the topography of these islands and its extensive mangrove fringed creeks establish their abundance. These creeks and lobsters beds should be surveyed properly to ascertain their potential. All these require a research cum-survey wing in the department. The main aims of this wing must be to study the locally available fishes from their exploitable point of view and resource survey of commercially important fisheries especially, crustacean resources. It is also essential to keep statistical records of landing, (area wise as well as group wise) to know the exact position and availability of the exploitable stock. The establishment of such a unit, either in the fisheries section or in other local bodies like the Government college where, trained, research oriented and qualified personnel is available, is strongly recommended.

The presently exploited fishery resource is much less when compared to the exploitable stock. It has been calculated that from the Andaman sea, about 4, 7000 tonnes of fish per year can safely be exploited. The present rate of the exploitation (about 1,800 tonnes in 1980) is much less. Out of the total population of two lakhs (approx.), 95% of the population are fish eaters. If, it is taken that 100 gms. of fish is the daily requirement per person, the annual landing must be upto the tune of 6,570 tonnes. This means that at present only about 28% of the requirement is met. But curiously enough a good percentage ∠ meagre 28% is exported to mainland, India as dried fish, as the market rates for

∠ of this

the same is much higher there. The introduction of purseiners for the proper exploitation of the pelagic fishery resources, introduction of Kalava traps for the perches and other traps for lobsters as well as the introduction of deep sea fishing (mainly for tunas) may be able to boost the annual landing. At the same time the 10 mechanised fishing boats owned by the department of fisheries and by private parties should be reconditioned and put to optimum use.

The much sought after algal resources remain either unexploited or underexploited. It is suggested that the Govt. of India should be approached to establish an algal research unit in the same lines with the one presently established at Mandapam, Tamil Nadu. Of all the fishery resources, only the molluscans seems to be the most exploited, probably for their established commercial value and easy mode of collection.

Aquaculture, undertaken in this territory proved to be a miserable failure mainly due to the non-availability of biology of the local culturable forms. Although we have the technology for culture, the same is impracticable without proper knowledge on biology. Hence, it is absolutely necessary that a thorough knowledge on the biology of the culturable species from this place is available. So it is impossible to make the culture techno-economically viable when worked on the basis of biology studied elsewhere;/with moreover the culture should be in accordance/ demand and and efficient extensive survey as mentioned earlier is needed. No culture should be started on hit and trial method as it is to be started under the guidance of experts.

The fisheries department should come forward for the purchase of fishes at landing centres and should take up the responsibility of marketing through cooperative societies this will successfully eradicate the middle man who ruthlessley exploit the fishermen. Planned colonisation, establishment of fishermen colony in the vicinity of the fishing harbour, adequate transportation and storage facilities for the catches etc. should be guaranteed.

The fishermen should be insured against life and equipment to ensure safety.

The establishment of proper marketing facilities is an immediate requirement. The requirement of the local market/demands must be fulfilled first as this may act as a small check over the price rise on other consumable articles because there is not much transportation cost for fish. The fishes can be carried either in refrigerated vans or through deep freezing chambers to distant places. At any case remunerative prices should be paid to the fishermen irrespective of the fluctuating market price. This shall induce a feeling of security among fishermen community and will ensure regular supply of fish. It is also suggested to create the posts of the marketing officers to supervise these activities.

...

Andaman Aborigines fishing at Sea

दूरध्वनी : ५७१४४६-८
Telephone : 571446-8

दूरलेख : फिशइंस्ट
Telegram : FISHINST
Telex : 011-6476

केन्द्रीय मत्स्य शिक्षा संस्थान
(भारतीय कृषि अनुसंधान परिषद)
पोस्ट बॉक्स ७३९२, जयप्रकाश रोड
अंधेरी (वेस्ट), बम्बई-४०० ०५८.

Central Institute of Fisheries Education
(Indian Council of Agricultural Research)
Post Box No. 7392, Jaiprakash Road,
Andheri (west), Bombay-400 058.

No. 7-19/79-81/A & N/6549 Dated: 30th June, 1981

TO WHOMSOEVER IT MAY CONCERN

CERTIFICATE

This is to certify that the chapters 13 & 14 on "Traditional fishing gears & crafts" and "Fishing implements of aborigines" respectively, of the dissertation entitle "Fish & Fisheries of Andaman & Nicobar islands submitted by Mr. Arif M. Mustafa is orignally prepared under my guidance and the technical personnel of Craft and Gear Technology Division, C.I.F.E., Bombay.

This is one of the BEST and OUT-STANDING dissertation so far submitted at C.I.F.E. in partial fulfilment for the award of P.G.D.F.Sc.

The chapter nos. 13 & 14 of the dissertation has almost filled the lacuna felt previously regarding the technological aspects of Andaman & Nicobar islands traditional/aborigines/tribals, fishing gears and crafts, and will definatly be of great use to develop and improve the various technological aspects of the crafts and gears of the area and will be an asset for those who would like to develop and improve the existing crafts & gears.

The dissertation as a whole and chapter nos. 13 & 14 in particular is a result of assiduity and sincere efforts done by the author and is a very timely contribution to the fishing technologists.

I assert, that the said dissertation is a mile-stone in the development of fisheries of Andaman & Nicobar islands.

I wish him all success and prosperity.

30/6/81

Prof. Y. Sreekrishna,
Head of the:
Division of Fisheries Technology.
C.I.F.E.,
Bombay - 61.

दूरध्वनी : ५७१४४६-८
Telephone : 5 71 4 4 6-8

दूरलेख : फिशइंस्ट 417
Telegram : FISHINST
Telex : 011-6476

केन्द्रीय मत्स्य शिक्षा संस्थान
(भारतीय कृषि अनुसंधान परिषद)
पोस्ट बॉक्स ७३९२, जयप्रकाश रोड
अंधेरी (वेस्ट), बम्बई-४०० ०५८.

Central Institute of Fisheries Education

(Indian Council of Agricultural Research)
Post Box No. 7392, Jaiprakash Road,
Andheri (west), Bombay-400 058.

No. 7-19/79-81/A &.N/6550 Dated:28th June,1981.

TO WHOMSOEVER IT MAY CONCERN

CERTIFICATE

This is to certify that Mr.Arif M.Mustafa student 19th batch, 1979-81 session of P.G.D.F.Sc course of Central Institute of Fisheries Education, Bombay, has worked under my guidance for the preparation of his dissertation entitled "Fish and Fisheries of Andaman & Nicobar islands" in partial fulfilment for the award of P.G.D.F.Sc.

The said dissertation is the BEST "Broad Spectrum" work done for the first time in C.I.F.E., which is a result of very sincere efforts done by the author over the preplasm of existing facts and figures.

I am sure that the work of Mr. Arif M.Mustafa will full-fill a long needed reference on the subject for the Fisheries Scientists/students/ administrators and planners and will be of immence value.

With the experience and expertise which I had gained in past as the Cheif executive of Andaman & Nicobar fisheries, I have no hesitation to state that this dissertation is a mile - stone in the development of Andaman & Nicobar fisheries and the Suggestions given in that deserve special consideration by the authorities.

I congratulate Mr. Arif M.Mustafa for his out standing work and wish him all success and prosperity.

28-6-81

Prof.Rajender Kumar Gulati
Division of Fisheries Biology.
Central Institute of Fisheries Education,
Bombay-61.

दूरध्वनी : ५७१४४६-८
Telephone : 571446-8

दूरलेख : फिशइंन्स्ट
Telegram : FISHINST
Telex : 011-6476

केन्द्रीय मत्स्य शिक्षा संस्थान
(भारतीय कृषि अनुसंधान परिषद)
पोस्ट बॉक्स ७३९२, जयप्रकाश रोड
अंधेरी (वेस्ट), बम्बई-४०० ०५८.

Central Institute of Fisheries Education
(Indian Council of Agricultural Research)
Post Box No. 7392, Jaiprakash Road,
Andheri (west), Bombay-400 058.

Dr.S.N.DWIVEDI,D.Sc(Paris)
DIRECTOR

7-19/79-81/AN/ 6551

Dated the 17th July, 1981.

TO WHOMSOEVER IT MAY CONCERN

This is to certify that Mr.Arif M.Mustafa student XIXth batch (1979-81) session of Post Graduate Diploma in Fishery Science Course of the Central Institute of Fisheries Education, Bombay has completed his course as a sponsored candidate of Andaman & Nicobar Islands Administration.

Mr.Arif M.Mustafa has distinguished himself as one of the industrious and sagacious student and it is only with his hard work and sincere efforts he has been able to prepare a good dissertation entitled"Fish and Fisheries of Andaman & Nicobar Islands" in partial fulfilment for the award of Post Graduate Diploma in Fishery Science which is one of the best broad spectrum dissertation.

Mr.Mustafa has collected very valuable information for the area which will be helpful for formulation of the development of fish and fisheries of the Islands.

I am confident that with the experience and knowledge Shri Mustafa has gained during the period of his stay will be valuable for fisheries development both marine or inland. Shri Mustafa is an intelligent hard working and well behaved young man.

I wish him success in life.

S.N.DWIVEDI
DIRECTOR.

Know the Man

Born on 17th day of March,1956 at Port Blair in the then Civil Hospital functioning from the Administrative block of the Cellular Jail now a National Memorial. Nurtured among educationalists, father late M.A.Gafoor a Teacher and Secretary, Bharat Scout & Guide Movement in A&N Islands, two elder sisters and a brother served as Principal, Education Department, another elder brother although an Engineer also served as HOD (Electrical) in Polytechnic. Represent second generation of unsung rebels of British era Late Pandit Ayodiya Rai Sharma alias Nazeer Mohammed the grand father who was exiled from Kothar, Shajahanpur, dumped at Netaji Subhash Chandra Bose/Ross Island on 3rd day of January 1871, imprisioned in Viper Island Jail; languished at Aberdeen Port Blair during 1891 and Late Chand Bibi the grand mother, she was the daughter of Moplah lady convict exiled from Ponnani Tehsil of Malabar.

Graduated from Punjab University, acquired Post graduation award from 'The Central Institute of Fisheries Education' / Deem University, ICAR, GoI, Bombay. Ph. D research work at Bombay and Pondicherry Universities. In service training at CMFRI; CIFA; MPEDA; NIO; NRSA; MANAGE / NFDB; SAC/DES.

Joined Department of Fisheries as Technical Assistant to Director of Fisheries, A&N islands during 1982 subsequently Superintendent of Fisheries to Assistant Fisheries Development Officer finally Assistant Director of Fisheries upto 2016.

Five years spent as Incharge, A&N Centre for Ocean Development, Deptt. of Ocean Development, New Delhi on deputation. Significant ex-post-facto assignments were Chief Liaison Officer of A&N Administration and Consultant to Indian Navy, A&N Command.

Participated as Indian counterpart in (1) The Cousteau Society, New York, (2) Russian Academy of Sciences Moscow and (3) National Geography expeditions to A&N Islands.

Formerly Member Network of Tropical Fisheries Scientists, ICLARM, Philippines. Fisheries Technologist of India. INTECH. Founding Life Member Andaman Science Association & FANS.

In recognition conferred Hon'ble Lieutenant Governor's 'Commendation' Certificate. 'Fellow' Andaman Science Association. 'Appreciation' by Indian Coast Guard.

Island Facts

Location
Long. - 92° to 94° East. : Lat - 6° to 14° North.

Highest High (Mtr.) 732

Literacy rate - 86.63%

Deepest Deep (Mtr.) 4198

Exclusive Economic Zone (Sq.Km.)	-	600000
Continental Shelf (Sq.km.)	-	35000
Coral cover area (Sq.km.)	-	20000
Coast Line (Km.)	-	1962

No. of Islands/Islets/Rocklets	-	836
Total area (Sq.km.)	-	8249
No.of Districts	-	3

No. of Inhabited Islands - 31
Biggest Inhabited Islands - Middle Andaman (Sq.Km.) - 1536
Smallest Inhabited Island - Curlew Island (Sq.Km.) - 0.03

Sea Gauge

(Port Blair harbour as base port)
(Km.)

A. Mainland India

1. Kolkatta	1255	West Bengal
2. Vishakapatnam	1200	Andhra Pradesh
3. Chennai	1190	Tamil Nadu

B. Inter Island ***Inhabited***

1. Diglipur	185	North Andaman
2. Mayabunder	159	Middle Andaman
3. Rangat	93	Middle Andaman
4. Kadamtala	93	Middle Andaman
5. Long Island	85	Middle Andaman
6. Baratang Is.	65	Middle Andaman
7. Swaraj Dweep Is.	39	South Andaman
8. Shaheed Dweep Is.	37	South Andaman
9. Little Andaman	122	South Andaman
10.Car Nicobar	278	Nicobar Disrict
11.Katchal	422	Nicobar District
12. Nancowry	435	Nicobar District
13. Great Nicobar	544	Nicobar District

Un-inhabited

1. Narcondum (Extinct volcano)	259	North Andaman
2. Barren Island (Active volcano)	139	Middle Andaman
3. North Passage Is. (Merk Bay)	80	Middle Andaman

Land Gauge

(Aberdeen bus terminus as base station)
(Km.)

A. Around Port Blair

1. Chath am Is. (Saw Mill)	05
2. Haddo Wharf (Mainland Is. Port)	05
3. Sippighat (Horticulture Farm)	14
4. Humphrygunj (Balidan Bedi)	20
5. Chidiyatapu (Beach, dive site)	28
6. Wandoor (Marine Park)	30
7. Bamboo flat Jetty (Mount Harriet)	46

B. Andaman Trunk Road

1. Jirkatang	64
2. Middle Strait	100
3. Kadamtala	125
4. Rangat	175
5. Billiground	200
6. Mayabunder	240
7. Diglipur	310

Air Distance
(Km)

Chennai	-	1330
Kolkata	-	1303

Road Length
(Km)

Black topped	-	2165
Andaman Trunk Road	-	333

CARRYING CAPACITY

Experts postulates that these islands could sustainably hold :

1.6 lakh persons (@ 3 / Ha.) under moderate Agri. technology or

*2.5 lakh persons (@ 5 / Ha.) under advance Agri. technology or

*8.5 lakh persons when Agriculture & Fisheries potential are put to optimum use

Population that could be here at a point in time, but not for an extended period.

Ref. **Gazetteer GoI-2005**

www.ingramcontent.com/pod-product-compliance
Ingram Content Group UK Ltd.
Pitfield, Milton Keynes, MK11 3LW, UK
UKHW060118300726
14090UKWH00002B/253